Ergebnisse der Mathematik und ihrer Grenzgebiete

3. Folge · Band 32

A Series of Modern Surveys in Mathematics

Springer
Berlin
Heidelberg
New York
Barcelona
Hong Kong
London
Milan
Paris
Singapore
Tokyo

János Kollár

Rational Curves on Algebraic Varieties

Springer

János Kollár
Department of Mathematics
Fine Hall, Washington Road
Princeton University
Princeton, NJ 08544-1000, USA
e-mail: kollar@math.princeton.edu

The Library of Congress has cataloged the original printing as follows:

Library of Congress Cataloging-in-Publication Data

Kollár, János.
Rational curves on algebraic varieties / János Kollár.
p. cm. -- (Ergebnisse der Mathematik und ihrer Grenzgebiete ; 3. Folge, Bd. 32)
Includes bibliographical references (p. -) and index.
ISBN 3-540-60168-6 (hardcover : alk. paper)
1. Curves, Algebraic. 2. Algebraic varieties. I. Title.
II. Series.
QA565.K74 1996
516.3'5--dc20 96-42735
CIP

Corrected Second Printing 1999

Mathematics Subject Classification (1991):
14-02, 14C05, 14E30, 14J26, 14J45, 14M20, 14C40, 14E35, 14H10, 14J10

ISSN 0071-1136
ISBN 3-540-60168-6 Springer-Verlag Berlin Heidelberg New York

Typesetting: Data conversion by Springer-Verlag
SPIN 10930984 44/3111 - 5 4 3 2 1 - Printed on acid-free paper

Preface

The aim of this book is to provide an introduction to the structure theory of higher dimensional varieties by studying the geometry of curves, especially rational curves, on varieties. According to the accepted terminology, "higher" means dimensions 3 and up. Investigating the geometry of curves lying on a variety is the classical approach to higher dimensional geometry, and it also seems to be the most accessible.

The first part of the book develops a variety of methods useful in many different situations. The second part illustrates the utility of these techniques in various settings.

The main application is in the study of varieties with lots of rational curves on them. These are rather special among all varieties, but they may be the ones that appear most frequently. Also, they possess very rich internal geometry, which makes their study especially difficult and rewarding. Many of the deepest questions are so elementary that they can easily be explained in the first lecture of an introductory algebraic geometry course, yet their solutions seem still completely out of reach. It is hard to believe, that with all the machinery of algebraic geometry at hand, many basic questions about cubic hypersurfaces are still completely open.

I hope that this book provides a systematic introduction to this field of study.

Acknowledgement. I had the opportunity to discuss various parts of this book with A. Bertram, H. Clemens, K. Corlette, A. Corti, W. Fulton, M. Green, T. de Jong, Y. Kawamata, S. Kovács, W. Lang, N. Lauritzen, Yu. Manin, D. Miličić, Y. Miyaoka, S. Mori, R. Narasimhan, G. Prasad, P. Roberts, E. Szabó and D. Toledo. Their comments have been very valuable.

A preliminary version of the complete manuscript was in limited circulation since the spring of 1994. I have received numerous comments and corrections from J. Cheah, I. Ciocan-Fontanine, Y. Lee, K. Oguiso, M. Schneider and C. Voisin. S. Kleiman and K. Matsuki sent me very detailed suggestions; these resulted in substantial revisons in Sect. VI.2 and I.2–3. I am very grateful for their attention and help.

Partial financial support was provided by the NSF under grant numbers DMS-8707320 and DMS-8946082.

The second printing corrects several mistakes of the first printing. The mathematically significant improvements are in (I.1.7.2), (I.7.4.1) and (IV.4.15.2). I thank O. Debarre, M. Knebusch and N. Tziolas for calling these to my attention. The pagination and the numbering of the theorems is unchanged.

Several of the open problems have been solved in the last 5 years. These are indicated by a footnote.

Salt Lake City, February 1999 — János Kollár

Contents

Introduction

The aim of this monograph is to give a comprehensive introduction to some of the recent progress in higher dimensional algebraic geometry. One of the most profound developments in the last two decades consists of the evolution of a structure theory for higher dimensional varieties. Its roots can be traced back to the classical theory of minimal models of surfaces and to the works of Fano in the thirties.

The works of [Iskovskikh77,78] on Fano threefolds, [Reid80] on canonical models and [Mori82] on the minimal model program provided the major impetus for the recent resurgence of this field.

Since 1982 the theory has developed in two distinct directions. One is the cohomological approach of Kawamata, Reid and Shokurov whose main achievement is the proof of the Cone Theorem and the Contraction Theorem for mildly singular varieties. This direction is written up in the first Utah Summer Seminar notes [CKM88]. A technically complete discussion is given in [KaMaMa87].

The other direction is harder to classify. It concentrates on trying to find a more geometric approach to the structure of varieties by studying the geometry of curves, especially rational curves, on them. This approach relies on classical geometric ideas and strives to understand the intrinsic geometry of varieties. The most significant single achievement of this direction is the proof of the existence of flips in dimension three [Mori88]. Unfortunately, the proof is rather long and complicated, and it is not typical of the ideas and techniques in this field of study.

Therefore I present the circle of ideas which originate around the theme "rational curves on algebraic varieties", without touching on the theory of minimal models. The minimal model program, while not directly considered, always lurks in the background as motivation for the results and source of the techniques.

The core of these techniques is the theory of deformation of curves on varieties. This in turn requires the understanding of universal families of curves on varieties and of morphisms of curves to varieties. The necessary foundational results about Hilbert and Chow schemes have not been treated in the literature in the generality that is needed for our purposes. Therefore Chap. I contains a detailed introduction to the theory of Hilbert schemes and Chow varieties in the relative setting.

Technically this is the most demanding part of the book. Later chapters use only the basic definitions and theorems, but very few of the techniques are employed. Thus it may be easier to start with Chap. II, and refer back to Chap. I only as necessary.

The general results about Hilbert schemes and Chow varieties are considered for curves in Chap. II. There are several simplifications which greatly enhance their usefulness. The Cone of Curves, one of the fundamental objects of higher dimensional geometry, is introduced in (II.4). The previous results are employed in (II.5) to develop the bend-and-break method. The basic principle of the method asserts that if a curve in a variety has lots of deformations then there are rational curves around. In applications it is important to get detailed information about the resulting rational curves; this accounts for the technicalities involved.

The geometric methods allow us to view several aspects of the theory of surfaces in a unified way. The arguments are rather short and clear; they are presented in Chap. III. Hopefully this can serve as an introduction to the higher dimensional cases where the technical details become rather formidable.

The most important part of this book is Chap. IV. Its aim is to present a theory of uniruled varieties, i.e. those varieties which are covered by rational curves. Our circle of ideas applies most successfully to these problems. For this class of varieties, Mori's program does not produce a minimal model and is not expected to give a definitive structure theory. The techniques of this book are birational in nature. Thus we get a good understanding of this class without reference to minimal models.

The results of the previous chapters are applied to the study of Fano varieties in Chap. V. The theory of Fano varieties developed considerably in the last five years, but many good open problems remain.

The last chapter covers two topics that did not fit anywhere else but are needed in order to get a complete picture.

The book contains a large number of exercises. Their role is to give further applications and examples. They are an integral part of the book.

References to the literature are given by [full name+year], except for papers with at least 3 authors where the names are abbreviated. An internal reference, say (I.1.13.2) should be decoded as: Chap. I, Sect. 1, Part 13, Subpart 2. The chapter number is omitted for references within the same chapter.

There are two important recent approaches to the study of rational curves on varieties which are not covered in this book. The first is the study of rational curves on varieties with trivial canonical class. For such varieties one expects relatively few, but frequently infinitely many, rational curves. The study of this situation is intimately connected with the general theory of mirror symmetry. The second is the development of a quantum cohomology theory of varieties with some very deep applications to enumerative questions involving rational curves.

Both of these directions are intensely developing, and their final form is not clear yet. Also, I am not an expert in either field. Thus it seemed safest not to attempt a treatment of them.

Prerequisites

It is assumed that the reader of this book is familiar with basic algebraic geometry. The main reference that I use is [Hartshorne77]. Results contained there are frequently applied without explicit mention. Chapter I is rather technical, and a working knowledge of [Hartshorne77] is indispensable. There are also many references to [Matsumura80,86]. The other chapters are more geometric in nature, but probably nothing is very elementary. There are occasional references to other topics, but I tried to keep the exposition as self-contained as possible.

Terminology

The terminology follows [Hartshorne77] for algebraic geometry and [Matsumura80] for commutative algebra. Some other notions, which are in general use in higher dimensional algebraic geometry, are defined below.

1. A line bundle L on an irreducible scheme X is called *big* if $h^0(X, L^n) > \text{const} \cdot n^{\dim X}$ for $n \gg 1$. Equivalently, the global sections of L^n define a birational map of X for $n \gg 1$.

2. A *morphism* is everywhere defined; it is denoted by a solid arrow $\to$. A *map* is defined only on a dense set; it is denoted by a broken arrow $\dashrightarrow$.

3. Let $f : X \dashrightarrow Y$ be a map, $Z \subset X$ a closed subscheme such that f is defined over a dense set $Z^0 \subset Z$. The closure of $f(Z^0) \subset Y$ is denoted by $f_*(Z)$. It is called the *birational transform* of Z. If $g : Y \dashrightarrow X$ is birational, then we use $g_*^{-1}(Z)$ instead of $(g^{-1})_*(Z)$. The birational transform is also frequently called the proper or strict transform in the literature.

4. Let P be a property of points of a scheme. We say that a *general point* of X satisfies P, if there is an open and dense set $X^0 \subset X$ such that P holds for every point of X^0. We say that a *very general point* of X satisfies P, if there are countably many open and dense sets $X_i \subset X$ such that P holds for every point of $\cap X_i$. I frequently use the mathematically less precise form of this terminology:

   If $x \in X$ is a (very) general point, then P holds.

5. Discrete valuation ring is frequently abbreviated as DVR.

Assumptions

We always work over a fixed scheme, which is usually denoted by S. We assume that S is locally Noetherian and all schemes are essentially of finite type over the base scheme (that is, their local rings are localizations of schemes of finite type over S). In some cases this assumption can be weakened, but I have not made any systematic effort to discover the natural setting for the results. Starting with I.3 we frequently require that the normalization of a scheme is finite over the scheme. To be on the safe side, from I.3 on I assume that all schemes are essentially of finite type over an excellent base scheme. In practice the base schemes we need are spectra of fields or $\operatorname{Spec} \mathbb{Z}$, but the above generality is frequently convenient.

Chapter I. Hilbert Schemes and Chow Varieties

Let X be an algebraic variety, say for simplicity $X = \mathbb{P}^n$. Studying the geometry of X frequently involves understanding some of the subvarieties of X. Already at the very beginning of these studies it becomes clear that subvarieties come in families. The simplest case arises in studying hypersurfaces in $\mathbb{P}^n$, when we discover that smooth hypersurfaces of the same degree are very much alike – for instance, they are diffeomorphic to each other. Similarly, it is usually not difficult to understand families of hypersurfaces on an arbitrary variety.

The question becomes more subtle if we look at higher codimension subvarieties. The first case that attracted serious attention is the study of curves in $\mathbb{P}^3$. In two remarkable papers [Cayley1860,1862] the following approach is developed:

Let $G(2,4)$ denote the Grassmann variety of lines in $\mathbb{P}^3$. Given a curve $D \subset \mathbb{P}^3$ let $Ca(D) \subset G(2,4)$ be the subset of lines that intersect D. Cayley proved that $Ca(D) \subset G(2,4)$ is a divisor. Furthermore, it is possible to give an internal characterization of those divisors which are of the form $Ca(D)$ for some curve D. Thus the question of understanding curves in $\mathbb{P}^3$ is reduced to a question of understanding certain divisors in $G(2,4)$. In concrete questions this approach is not always very useful, but it gives a convenient way to think about all curves in $\mathbb{P}^3$ together.

Unfortunately this process leads to parameter spaces which are not compact. For instance, smooth hypersurfaces of a given degree in $\mathbb{P}^n$ are parametrized by an affine variety (except for hypersurfaces of degree 1). For various purposes it is important to understand what happens in a family of hypersurfaces as we approach the boundary of the parameter space. In case of hypersurfaces it is quite natural to allow singular hypersurfaces. This way we obtain a nice compact variety whose points correspond to possibly singular hypersurfaces. The same approach works for hypersurfaces in an arbitrary (normal) variety.

It is less clear what to do in the other cases, and different approaches are possible. If we are looking at curves in $\mathbb{P}^3$, we can just look at the corresponding Cayley form $Ca(D) \subset G(2,4)$. We already know how to compactify the space of all smooth hypersurfaces in $G(2,4)$. Take the closure of the space of all Cayley forms. This gives a compactification of the space parametrizing all curves, but it is not a priori clear that there is some reasonable geometric meaning attached to the boundary points. Fortunately the boundary points do have a clear geometric interpretation and this way we obtain a nice compactification of the space parametrizing all curves in $\mathbb{P}^3$.

This construction has been viewed as part of basic algebraic geometry until about the late 1950's. It is discussed in detail in [Hodge-Pedoe52; Samuel55]. It is not clear to me why the name changed from Cayley forms to Chow forms and Chow varieties. Chow proved that the above outlined procedure indeed gives a compactification with a geometric interpretation [Chow-v.d.Waerden37], but the basic construction has not changed much since it was invented by Cayley. The drawback of this approach is that it seems very hard to get information about these parameter spaces. For instance, even computing their dimension was possible only by other methods and only in a few cases.

The major breakthrough came in the deformation theory of [Kodaira-Spencer58]. In the original setup the theory was restricted to families of smooth subvarieties of a smooth variety, but in a large number of cases it was used successfully to compute the dimension of the spaces parametrizing them.

With Grothendieck's overhaul of algebraic geometry, the old methods of Cayley were also discarded. A new way of compactifying was introduced in [Grothendieck62] and baptized the Hilbert scheme. This seems the natural setup for the deformation theory of [Kodaira-Spencer58]. The powerful modern methods worked well and it was possible to get some information about Hilbert schemes. The original methods of infinitesimal computations yielded a dimension estimate only in the "unobstructed case". An observation of Mumford extended these to the general case.

Deformation theory is a difficult subject to write about. The existing literature can be divided into two large groups.

There are books that consider a very special case and work it out in detail in a way that is not hard to understand. The best examples I have in mind is [Artin76] on the deformation of singularities, [Kodaira86] on the deformation theory of manifolds and [Mumford66] on the Hilbert scheme of curves on an algebraic surface. The tacit claim of these authors is that if someone really understands these special cases, it is not too hard to work out the more general ones. In my experience this is true, but needs substantial hard work.

The other approach is to try to develop deformation theory in the most general setting, and prove the theorems that settle the subject once and for all. The best examples here are [Illusie71] and [Bingener87]. Unfortunately it seems that general deformation theory needs a rather formidable arsenal of machinery, which makes these treatises difficult to read.

When I gave a series of lectures on the subject in 1990 at the University of Utah, my aim was to achieve the happy middle ground. With this in mind I chose to give a complete treatment of the construction of the Hilbert scheme and related infinitesimal computations. This fills a gap in the literature which has been getting more and more annoying since many recent articles use the

Hilbert scheme in a rather general setting (see e.g. [Mori79; KoMiMo92a,b,c]). Since then I realized that I am no closer to the happy middle ground than my predecessors. It would have been desirable to treat a slightly more general case: the Quot-schemes of Grothendieck and also the theory of deformations with restrictions on some subschemes. Also, the treatment ended up more complicated than I envisioned.

Thus in Sect. 1–2 there is a detailed introduction to the theory of Hilbert schemes. I tried to keep the machinery as concrete as possible. As for the above-mentioned more general problems, "anyone who really understands the Hilbert scheme, should have no problem with the more general cases".

The first main result is the existence of the Hilbert schemes, given in Sect. 1. The section is not long, but the proof uses two results from [Mumford66]. I could not improve his presentation and it seemed pointless to copy his proofs. The Hom-schemes appear as special cases of the Hilbert schemes.

The infinitesimal computations are discussed in Sect. 2. This is rather self-contained, and follows the presentation of [Artin76]. The main results are (2.15) and (2.17).

Chow varieties are discussed in Sect. 3. In recent years Chow varieties have made a comeback. It is easier to relate them to down-to-earth geometric questions, and this is useful in many applications (see e.g. [Kollár91a; Catanese92]).

The definition of Chow varieties is not completely settled yet, and the one I follow is somewhat new. Set-theoretically there is no problem, but a given set may have different scheme structures (like a cuspidal curve and its normalization). In characteristic zero it is possible to study Chow varieties with some additional scheme structure [Barlet75], but in positive characteristic this may be difficult [Nagata55]. I chose to get rid of this problem by always looking at the coarsest scheme structure possible. In my view this is more in keeping with the basically set-theoretic nature of the Chow varieties.

In characteristic zero the resulting notion is very well-behaved. One can define the notion of Chow functor (3.18) and it turns out that the Chow variety represents this functor (3.21). In particular there is a universal family of cycles over the Chow variety. These matters are presented in Sect. 3.

The additional difficulties arising in positive characteristic are treated in Sect. 4. One of the new results is (4.5), which asserts that the field of definition of the Chow form of a cycle is the same as the intersection of all fields of definition of that cycle. This implies that the Chow variety coarsely represents the Chow functor in general (4.13).

My methods are projective in nature, thus they do not apply for more general schemes or analytic spaces. I list the known results about these other settings in Sect. 5, mostly without even a hint of proof.

Chow varieties and Hilbert schemes agree for "nice subvarieties". For smooth subvarieties this has been well-known, but it is true in much greater generality. Some of these results are discussed in Sect. 6.

Chow varieties and Hilbert schemes are not in competition with each other. The way I like to look at them is that they both compactify the same space of "nice subvarieties". Depending on the question at hand, one compactification may be better than the other. Sometimes they both need to be studied together. There are other methods of compactifying the space of "nice subvarieties" that I could not get into (see e.g [Sacks-Uhlenbeck81, Kontsevich-Manin94]). Several recent results emphasize the need to understand the relationship between several of the geometrically meaningful compactifications. See for instance [BDW96; Dolgachev-Hu98; Thaddeus94].

Finally, Sect. 7 contains auxiliary results from commutative algebra for which I could not find a convenient reference in [Matsumura80,86].

I.1 Construction of the Hilbert Scheme

The aim of this section is to construct the Hilbert scheme $\operatorname{Hilb}(Y/S)$ of a scheme Y/S projective over S. We always assume that S is locally Noetherian. The proof is due to [Grothendieck62] with simplifications by [Mumford66]. The main idea is rather simple and goes back over a hundred years. A subscheme $Z \subset \mathbb{P}^n_k$ is determined by its equations. This gives an injection of sets

$$\{\text{subschemes of } \mathbb{P}^n_k\} \hookrightarrow \{\text{subvector spaces of } k[x_0, \dots, x_n]\}.$$

The right hand side is an infinite dimensional Grassmannian, thus a linear object. The two main technical points that need to be addressed are the following:

(i) replace the infinite dimensional Grassmannian by a finite dimensional one. This is accomplished by (1.5).

(ii) show that the image of the above injection is an algebraic subvariety of the Grassmannian. Here the main step is (1.6).

The rest is a rather straightforward, though lengthy, exercise in flatness and base change.

As an application we construct a scheme $\operatorname{Hom}(X, Y)$ which parametrizes morphisms between two schemes. A morphism can be identified with its graph, thus it is natural to construct $\operatorname{Hom}(X, Y)$ as a subscheme of $\operatorname{Hilb}(X \times Y)$. This is accomplished in (1.10).

1.1 Definition. Let S be a scheme and F a contravariant functor

$$F : \{\text{schemes over } S\} \to \{\text{sets}\}.$$

A scheme $X(F)/S$ and an element $U(F) \in F(X(F))$ *represents* F if for every S-scheme Z

$\mathrm{Hom}_S(Z, X(F)) \to F(Z/S)$, given by $g \mapsto g^*U(F)$, is an isomorphism.

$U(F)$ is frequently called the *universal element* or *universal family* over $X(F)$. It is clear that the pair $(X(F), U(F))$ is unique up to isomorphism.

1.2 Definition – Proposition. Let $g : Y \to Z$ be a projective morphism and $\mathcal{O}(1)$ a g-ample line bundle. Assume that Z is connected. Let F be a sheaf on Y which is flat over Z.

(1.2.1) $g_*F(n)$ is locally free for $n \gg 0$. This is proved in [Mumford66, Lecture 7, Corollary 3]; this is also proved in [Hartshorne77, III.9.9] since the argument for (i) $\Rightarrow$ (ii) does not use the assumption that A is integral. There is a polynomial $P(n)$ such that rank $g_*F(n) = P(n)$ for $n \gg 0$. P is called the *Hilbert polynomial* of F.

(1.2.2) Let $V \subset Y$ be a closed subscheme, flat over Z. The Hilbert polynomial of $\mathcal{O}_V$ is called the *Hilbert polynomial* of V.

(1.2.3) If $p : Z' \to Z$ is any morphism and F' is the pull back of F to $Y \times_Z Z'$, then the Hilbert polynomial of F' agrees with the Hilbert polynomial of F. In particular the Hilbert polynomial of F can be computed on any fiber of g.

(1.2.4) $P(n)$ has the property that it takes integer values at integers. Any such polynomial can be written as

$$P(n) = \sum_{i=0}^{\deg P} a_i \binom{n}{i}.$$

(1.2.5) Assume that

$$0 \to F(-1) \to F \to G \to 0$$

is exact and G is also flat over Z. Let P_F (resp. P_G) be the Hilbert polynomials of F (resp. G). If

$$P_F(n) = \sum_{i=0}^{k} a_i \binom{n}{i} \quad \text{then} \quad P_G(n) = \sum_{i=0}^{k-1} a_{i+1} \binom{n}{i}.$$

1.3 Definition. (1.3.1) Let X/S be a scheme. The *Hilbert functor*

$$\mathit{Hilb}(X/S) : \{\text{schemes over } S\} \to \{\text{sets}\}$$

is defined as follows:

$$\mathit{Hilb}(X/S)(Z) = \left\{\begin{array}{l}\text{Subschemes}\\ V \subset X \times_S Z\\ \text{which are proper}\\ \text{and flat over } Z\end{array}\right\} = \left\{\begin{array}{l}\text{Quotient sheaves } F \text{ of the}\\ \text{structure sheaf of } X \times_S Z\\ \text{which are flat and have}\\ \text{proper support over } Z\end{array}\right\}$$

(1.3.2) Let X/S be a scheme, $\mathcal{O}(1)$ a relatively ample line bundle and P a polynomial. The functor

$$Hilb_P(X/S) : \{\text{schemes over } S\} \to \{\text{sets}\}$$

is defined as follows:

$$Hilb_P(X/S)(Z) = \left\{ \begin{matrix} \text{Subschemes } V \subset X \times_S Z \\ \text{which are proper and flat} \\ \text{over } Z \text{ and have Hilbert} \\ \text{polynomial } P \end{matrix} \right\}$$

If Z is connected, then

$$Hilb(X/S)(Z) = \cup_P Hilb_P(X/S)(Z).$$

and the union of the right hand side is disjoint.

The following is the main result of this section.

1.4 Theorem. *Let X/S be a projective scheme, $\mathcal{O}(1)$ a relatively ample line bundle and P a polynomial. The functor $Hilb_P(X/S)$ is represented by a morphism*

$$\operatorname{Univ}_P(X/S) \xrightarrow{u} \operatorname{Hilb}_P(X/S) \quad \textit{where} \quad \operatorname{Univ}_P(X/S) \subset X \times_S \operatorname{Hilb}_P(X/S).$$

$\operatorname{Hilb}_P(X/S)$ *is projective over S.*

1.4.1 Exercises. (1.4.1.1) Show that $\operatorname{Hilb}_1(X/S) = X/S$ (here 1 denotes the constant 1 polynomial).

(1.4.1.2) Let C be a smooth curve over a field k. Show that

$$\operatorname{Hilb}_m(C) = S^m C, \quad \text{where} \quad S^m C := \underbrace{C \times \ldots \times C}_{m\text{-times}} / \Sigma_m,$$

and Σ_m is the symmetric group on m letters acting by permuting the factors.

(1.4.1.3) Let X be a smooth surface over a field k. Is it true that $\operatorname{Hilb}_m(X) = S^m X$?

(1.4.1.4) Let E be a vector bundle of rank $m+1$ over S and

$$P_d(n) = \binom{m+n}{m} - \binom{m+n-d}{m}.$$

Show that

$$\operatorname{Hilb}_{P_d}(\mathbb{P}(E)/S) = \mathbb{P}((S^d E)^*).$$

(1.4.1.5) Let $Z \to S$ be a morphism. Show that

$$\operatorname{Hilb}(X \times_S Z/Z) \cong \operatorname{Hilb}(X/S) \times_S Z.$$

Before starting the proof, we need to recall two general results. There is nothing that I could add to the proofs presented in [Mumford66, Lectures 14 and 8]; also the arguments given there are fairly self contained.

1.5 Theorem. *For every polynomial P, there is an integer $N(P)$ with the following property:*

If $\mathbb{P}$ is a projective space over a field k and $I \subset \mathcal{O}_{\mathbb{P}}$ a subsheaf with Hilbert polynomial P, then for every $n \geq N(P)$,

(1.5.1) $h^i(\mathbb{P}, I(n)) = 0$ *for* $i \geq 1$;

(1.5.2) $I(n)$ *is generated by global sections;*

(1.5.3) $H^0(\mathbb{P}, I(n)) \otimes H^0(\mathbb{P}, \mathcal{O}(1)) \to H^0(\mathbb{P}, I(n+1))$ *is surjective.* □

1.6 Theorem. *Let X/S be a projective scheme and $\mathcal{O}(1)$ relatively ample. Let F be a coherent sheaf on X. For every polynomial P there is a locally closed subscheme $i_P : S_P \hookrightarrow S$ with the following property:*

Given any morphism $p : Z \to S$ the pull back p^*F on $Z \times_S X$ is flat over Z with Hilbert polynomial P iff p can be factored as

$$p : Z \to S_P \xrightarrow{i_P} S. \qquad \square$$

1.6.1 Comment. The proof of (1.5) is tricky but uses little machinery. (1.6) is deep, the hardest part of the proof of (1.4).

Before starting the proof of (1.4) we discuss a special case which is needed during the arguments: the construction of Grassmannians. The proof given here is not the simplest (see e.g. [Mumford66, Lecture 5]) but it shows the machinery of the general case in an elementary setup.

1.7 Grassmannians

1.7.1 Definition. Let S be a scheme, E a vector bundle on S and r a natural number. The functor

$$Grass(r, E) : \{\text{schemes over } S\} \to \{\text{sets}\}$$

is defined as follows:

$$Grass(r, E)(Z) = \left\{ \begin{matrix} \text{Subvector bundles} \\ \text{of rank } r \text{ of } E \times_S Z \end{matrix} \right\}$$

1.7.2 Theorem. *Let S be a scheme, E a vector bundle on S and r a natural number. Then $Grass(r, E)$ is represented by a scheme and a subvector bundle*

$$U \subset E \times_S \mathrm{Grass}(r, E).$$

Proof. Set $Y := \wedge^r E$ with projection $p : Y \to S$ and let $\mathcal{O}_Y \to p^*(\wedge^r E)$ be the universal section. The wedge product map

$$E \otimes \wedge^r E \to \wedge^{r+1} E \quad \text{induces} \quad M : p^*E \cong p^*E \otimes \mathcal{O}_Y \to p^*(\wedge^{r+1} E),$$

which is locally given by a matrix $M = (m_{ij})$ whose entries are functions on Y. Let $Y_r \subset Y$ be the subscheme defined by all $r \times r$-subdeterminants of M. By (1.7.3) the corank of M is everywhere at most r. Thus $m|Y_r$ has constant corank r and so its kernel $K := \ker(M|Y_r)$ is locally free of rank r.

We obtain that if $f : Z \to Y$ is a morphism, then $\ker(f^*M)$ is locally free of rank r iff f factors as $f : Z \to Y_r \to Y$.

1.7.3 Lemma. *Let E_Z be a vector bundle on Z and $F_Z \subset E_Z$ a rank r subbundle. Then F_Z and $\wedge^r F_Z$ are each others annihilators under the wedge product map*

$$w : E_Z \otimes \wedge^r E_Z \to \wedge^{r+1} E_Z.$$

Proof. This is a local question, so we may assume that E_Z is trivial and F_Z is a direct summand. (1.7.3) is reduced to an easy explicit computation. □

Let $\mathbb{P}(Y) := \mathbb{P}((\wedge^r E)^*)$ with projection q to S. It is clear that Y_r descends to a subscheme $S_r \subset \mathbb{P}(Y)$ and K descends to a subsheaf $K \subset q^*E|S_r$.

1.7.4 Claim. *The pair (S_r, K) represents $Grass(r, E)$.*

Proof. Let $F \subset E \times_S Z$ be a subbundle of rank r. We obtain a subline bundle $\wedge^r F \subset \wedge^r E_Z$ which corresponds to a morphism $\sigma_F : Z \to \mathbb{P}(Y)$. The pull back of w gives

$$E_Z \otimes \wedge^r F \to \wedge^{r+1} E_Z,$$

whose kernel is precisely $F = \sigma^* K$ by (1.7.3). Thus, by definition, $\sigma_F : Z \to \mathbb{P}(Y)$ factors through S_r.

Conversely, assume that we have a morphism $\rho : Z \to S_r$. This gives a subvector bundle $\rho^* K \subset E_Z$ of rank r. By (1.7.3), $\wedge^r(\rho^* K) \subset \wedge^r E$ is the unique subline bundle which annihilates $\rho^* K$, thus $\rho = \sigma_{\rho^* K}$. □

1.7.5 Remarks. It is not hard to prove that $S_r \subset \mathbb{P}(Y)$ is closed, hence $\mathrm{Grass}(r, E)$ is projective over S.

The present construction does not show that $\mathrm{Grass}(r, E)$ is given by a rather simple set of quadratic equations (see e.g. [Mumford66, Lecture 5]).

1.8 Proof of (1.4)

Step 0. Idea of the Proof. Assume for simplicity that S is a point and X is a projective space $\mathbb{P}$. Let $Y \subset \mathbb{P}$ be a closed subscheme with ideal sheaf I_Y. Choose N such that $I_Y(N)$ is generated by global sections. Then the subvectorspace

$$H^0(\mathbb{P}, I_Y(N)) \subset H^0(\mathbb{P}, \mathcal{O}_{\mathbb{P}}(N))$$

determines Y since

$$I_Y(N) = \mathrm{im}[\mathcal{O}_{\mathbb{P}} \times H^0(\mathbb{P}, I_Y(N)) \to \mathcal{O}_{\mathbb{P}} \times H^0(\mathbb{P}, \mathcal{O}_{\mathbb{P}}(N)) \to \mathcal{O}_{\mathbb{P}}(N)].$$

The choice of N depends on Y but (1.5) says that if we fix the Hilbert polynomial of I_Y to be $Q(n)$, then we obtain an N which works for all such Y. This way we obtain an injective map

$$\left\{\begin{matrix}\text{Subschemes of } \mathbb{P} \text{ with} \\ \text{Hilbert polynomial } P\end{matrix}\right\} \to \mathrm{Grass}(Q(N), H^0(\mathbb{P}, \mathcal{O}_{\mathbb{P}}(N))).$$

(1.6) can be used to ensure that the image is an algebraic subset of the Grassmannian. Since we want to keep track of the scheme structures as well, a more careful version of the above argument is required in general.

Step 1 (A Simpler Case). Since X/S is projective there is a closed immersion $X/S \hookrightarrow \mathbb{P}/S$ where $\mathbb{P} = \mathrm{Proj}_S\, E$ for some vector bundle E of constant rank $m+1$ over S. This induces an injection of functors

$$\mathit{Hilb}(X/S) \hookrightarrow \mathit{Hilb}(\mathbb{P}/S).$$

Next we construct $\mathrm{Hilb}(\mathbb{P}/S)$ and later $\mathrm{Hilb}(X/S)$ as a subscheme of $\mathrm{Hilb}(\mathbb{P}/S)$.

Step 2 (Construction of $\mathit{Hilb}(\mathbb{P}/S) \to \mathit{Grass}$). For any $p : Z \to S$ we have the base-change diagram

$$\text{(1.8.1)} \qquad \begin{array}{ccc} \mathbb{P} \times_S Z & \longrightarrow & \mathbb{P} \\ {\scriptstyle p^*f}\downarrow & & \downarrow{\scriptstyle f} \\ Z & \xrightarrow{p} & S. \end{array}$$

Let $Y \subset \mathbb{P} \times_S Z$ be a closed subscheme, flat over Z with Hilbert polynomial P. Let k be a field and $\mathrm{Spec}\, k \to Z$ a morphism. Pulling everything back to $\mathrm{Spec}\, k$, we obtain $Y_k \subset \mathbb{P}_k$. Let I_k be the ideal sheaf of Y_k. Then

$$\chi(I_k(n)) = \chi(\mathbb{P}_k, \mathcal{O}(n)) - \chi(Y_k, \mathcal{O}(n)) = \binom{m+n}{m} - P(n) := Q(n).$$

In particular, Q depends only on P and m. Fix N as in (1.5), depending only on P and m. From the sequence

$$0 \to I_Y \to \mathcal{O}_{\mathbb{P}} \to \mathcal{O}_Y \to 0,$$

we obtain the long exact sequence

$$0 \to (p^*f)_* I_Y(N) \to (p^*f)_* \mathcal{O}_{\mathbb{P}}(N) \to (p^*f)_* \mathcal{O}_Y(N) \to R^1(p^*f)_* I_Y(N).$$

Since I_Y is flat and $H^i(\mathbb{P}_k, I_k(N)) = 0$ for $i \geq 1$ for every closed fiber, we conclude that

$$R^1(p^*f)_* I_Y(N) = 0.$$

Also, from the above sequence we see that

$$h^0(Y_k, \mathcal{O}(N)) = P(N) \quad \text{and} \quad h^i(Y_k, \mathcal{O}(N)) = 0 \quad \text{for} \quad i \geq 1.$$

Therefore $(p^*f)_*\mathcal{O}_Y(N)$ is locally free of rank $P(N)$. Thus

$$(p^*f)_* I_Y(N) \hookrightarrow (p^*f)_*\mathcal{O}_{\mathbb{P}}(N) = p^*S^N E$$

is a subvectorbundle of rank $Q(N)$. This gives a map

$$Hilb(\mathbb{P}/S) \to Grass(Q(N), S^N E).$$

Step 3 (Construction of Hilb($\mathbb{P}/S$)). Consider the diagram

$$\begin{array}{ccc} \mathrm{Grass}(Q(N), S^N E) \times_S \mathbb{P} & \xrightarrow{\pi_2} & \mathbb{P} \\ {\scriptstyle \pi_1}\downarrow & & \downarrow{\scriptstyle f} \\ \mathrm{Grass}(Q(N), S^N E) & \xrightarrow{g} & S. \end{array}$$

Let $U_{Gr} \hookrightarrow g^*S^N E$ be the universal subbundle. Let F be the cokernel of the composition

$$\pi_1^* U_{Gr} \to \pi_1^* g^* S^N E = \pi_1^* g^* f_* \mathcal{O}_{\mathbb{P}}(N) = \pi_2^* f^* f_* \mathcal{O}_{\mathbb{P}}(N) \to \pi_2^* \mathcal{O}_{\mathbb{P}}(N).$$

Applying (1.6) to the sheaf $F(-N)$ over

$$\mathrm{Grass}(Q(N), S^N E) \times_S \mathbb{P} \xrightarrow{\pi_1} \mathrm{Grass}(Q(N), S^N E),$$

there is a largest subscheme $i : G_P \hookrightarrow \mathrm{Grass}(Q(N), S^N E)$ such that $i^*F(-N)$ as a sheaf on

$$G_P \times_{\mathrm{Grass}(Q(N), S^N E)} \mathrm{Grass}(Q(N), S^N E) \times_S \mathbb{P} \cong G_P \times_S \mathbb{P}$$

is flat over G_P with Hilbert polynomial P. $i^*F(-N)$ is a quotient sheaf of the structure sheaf of $G_P \times_S \mathbb{P}$, thus it corresponds to a subscheme

$$U \subset G_P \times_S \mathbb{P}$$

which is flat over G_P with Hilbert polynomial P. Thus

$$U \in Hilb_P(\mathbb{P}/S)(G_P).$$

1.8.2 Claim. *The pair (G_P, U) represents $Hilb_P(\mathbb{P}/S)$.*

Proof. Let $Y \subset \mathbb{P} \times_S Z$ be a closed subscheme, flat over Z with Hilbert polynomial P. In Step 2, we obtained a rank $Q(N)$ subvectorbundle

$$(p^*f)_* I_Y(N) \hookrightarrow (p^*f)_* \mathcal{O}_{\mathbb{P}}(N) = p^* S^N E.$$

This corresponds to a morphism $\bar{p} : Z \to \mathrm{Grass}(Q(N), S^N E)$ such that

$$\bar{p}^*[U \to g^* S^N E] \cong [(p^*f)_* I_Y(N) \to p^* S^N E].$$

This implies that $\bar{p}^* F \cong \mathcal{O}_Y(N)$, thus $\bar{p}^* F(-N)$ is flat over Z with Hilbert polynomial P. By the definition of G_P, this means that $\bar{p}$ can be factored as

$$\bar{p} : Z \to G_P \xrightarrow{i} \mathrm{Grass}(Q(N), S^N E). \qquad \square$$

Step 4 (Construction of $\mathrm{Hilb}(X/S)$). As was mentioned earlier $Hilb(X/S)$ is a subfunctor of $Hilb(\mathbb{P}/S)$ therefore one should be able to represent it by a subscheme of $\mathrm{Hilb}(\mathbb{P}/S)$. In order to find this subscheme let $U \subset \mathbb{P} \times_S \mathrm{Hilb}_P(\mathbb{P}/S)$ be the universal subscheme. Let

$$V = U \cap (X \times_S \mathrm{Hilb}(\mathbb{P}/S)) \subset \mathbb{P} \times_S \mathrm{Hilb}_P(\mathbb{P}/S)$$

be the scheme theoretic intersection. By (1.6), there is a subscheme $i : H_P \hookrightarrow \mathrm{Hilb}_P(\mathbb{P}/S)$ such that for any $p : Z \to \mathrm{Hilb}_P(\mathbb{P}/S)$, the pull back $V \times_{\mathrm{Hilb}_P(\mathbb{P}/S)} Z \subset X \times_S Z$ is flat over Z with Hilbert polynomial P iff p can be factored through H_P.

1.8.3 Claim. *$i^*V \to H_P$ represents $Hilb_P(X/S)$.*

Proof. By construction, $i^*V \in Hilb(X/S)(H_P)$. Conversely, given any $p : Z \to S$ and a subscheme

$$W \subset X \times_S Z \subset \mathbb{P} \times_S Z$$

which is proper and flat over Z, there is a lifting $\bar{p} : Z \to \mathrm{Hilb}(\mathbb{P}/S)$ such that $\bar{p}^* U = W$. Therefore

$$W = \bar{p}^* U = \bar{p}^*(U \cap (X \times_S \mathrm{Hilb}(\mathbb{P}/S)))$$

is flat over Z with Hilbert polynomial P. Hence, by the definition of H_P, we can factor $\bar{p}$ as

$$\bar{p} : Z \to H_P \xrightarrow{i} \mathrm{Hilb}(\mathbb{P}/S). \qquad \square$$

Step 5 ($\mathrm{Hilb}_P(X/S)$ is Proper over S). By construction, $\mathrm{Hilb}_P(X/S)$ is a locally closed subscheme of a Grassmannian, therefore it is of finite type and separated. Thus we need to show that it satisfies the valuative criterion of properness. Let T be the spectrum of a DVR with generic point T_g. Given a commutative diagram

$$\begin{array}{ccc} T_g & \xrightarrow{\overline{p\circ i}} & \mathrm{Hilb}_P(X/S) \\ {\scriptstyle i}\downarrow & & \downarrow{\scriptstyle h} \\ T & \xrightarrow{p} & S, \end{array}$$

we have to find a lifting $\bar{p}: T \to \mathrm{Hilb}_P(X/S)$. $\overline{p \circ i}$ corresponds to a subsheaf I_g of the structure sheaf of $T_g \times_S X$ and we would like to extend it to a subsheaf I of the structure sheaf of $T \times_S X$ such that the quotient is flat over T. The desired extension is given by the definition

$$\Gamma(U, I) = \ker[\Gamma(U, \mathcal{O}_U) \to \Gamma(U \cap (T_g \times_S X), \mathcal{O}_{U\cap(T_g\times_S X)}/I_g)],$$

where $U \subset T \times_S X$ is any open set. This defines an ideal sheaf I of $\mathcal{O}_{T\times_S X}$ such that $\mathcal{O}_{T\times_S X}/I$ has no embedded points supported over the closed point of T. Therefore $\mathcal{O}_{T\times_S X}/I$ is flat over T. □

Most representability results in algebraic geometry use the existence of the Hilbert scheme as their starting point. The following is one of the simplest examples which is used frequently.

1.9 Definition. Let X/S and Y/S be schemes. $Hom(X, Y)$ is the functor

$$Hom_S(X, Y)(T) = \{T\text{-morphisms: } X \times_S T \to Y \times_S T\}.$$

The graph of a morphism identifies the functor $Hom_S(X, Y)$ with a subfunctor of the functor of subschemes of $X \times_S Y$. One can take this as a definition of $Hom_S(X, Y)$.

1.10 Theorem. *Let X/S and Y/S be projective schemes over S. Assume that X is flat over S. Then $Hom_S(X, Y)$ is represented by an open subscheme*

$$\mathrm{Hom}_S(X, Y) \subset \mathrm{Hilb}(X \times_S Y/S).$$

Proof. Let $f: X \times_S T \to Y \times_S T$ be a morphism. Its graph

$$\Gamma: X \times_S T \xrightarrow{(id,f)} X \times_S T \times_T Y \times_S T \cong X \times_S Y \times_S T$$

is a closed immersion, thus $\Gamma(X \times_S T) \cong X \times_S T$ is flat over T. This gives a map $Hom_S(X, Y) \to Hilb_S(X \times_S Y/S)$.

We have a commutative diagram

$$\begin{array}{ccc} \mathrm{Univ} & \xrightarrow{\pi} & X \times_S \mathrm{Hilb} \\ {\scriptstyle u}\downarrow & & \downarrow{\scriptstyle p} \\ \mathrm{Hilb} & = & \mathrm{Hilb}, \end{array}$$

where Hilb $= \mathrm{Hilb}_S(X \times_S Y/S)$ and Univ is the universal family. π is the restriction of the natural projection $X \times_S Y \times_S \mathrm{Hilb} \to X \times_S \mathrm{Hilb}$ and p is the second projection.

Assume that $\pi_z : u^{-1}(z) \to p^{-1}(z)$ is an isomorphism for some $z \in \mathrm{Hilb}$; that is, $u^{-1}(z)$ is graph. We need to show that π is an isomorphism over an open neighborhood of z.

1.10.1 Lemma. *Let $0 \in T$ be the spectrum of a local ring. Let U/T be flat and proper and V/T arbitrary. Let $p : U/T \to V/T$ be a morphism. If $p_0 : U_0 \to V_0$ is a closed immersion (resp. an isomorphism), then p is a closed immersion (resp. an isomorphism).*

Proof. By assumption, p is finite. We can replace V by the scheme theoretic image of p, thus we may assume that p_0 is an isomorphism. $p_*\mathcal{O}_U$ is a finite $\mathcal{O}_V$ module which is flat by assumption. Apply (7.4.1.2) to the map $\mathcal{O}_V \to p_*\mathcal{O}_U$ to conclude that p is an isomorphism. □

Applying (1.10.1) to the morphism π concludes the proof of (1.10). □

1.10.2 Exercise. Define the notion of a *group scheme.* Show that if X/S is a projective scheme, flat over S, then $\mathit{Aut}_S(X) = \mathit{Hom}_S(X, X)$ is represented by a scheme $\mathrm{Aut}_S(X)$.

Let $\mathcal{O}(1)$ be ample on X/S, $p : X \times_S Z \to X$ the projection and define

$$\mathit{Aut}^{\tau}_S(X)(Z) := \{g \in \mathit{Aut}_S(X)(Z) | g^*p^*\mathcal{O}_X(1) \equiv p^*\mathcal{O}_X(1)\}.$$

($\equiv$ denotes numerical equivalence (II.4.2)). Show that if S is the spectrum of a field or X/S has geometrically integral fibers, then $\mathit{Aut}^{\tau}_S(X)$ is represented by an open and closed subgroup scheme of finite type $\mathrm{Aut}^{\tau}_S(X) \subset \mathrm{Aut}_S(X)$.

1.11 Definition – Proposition. Let $f : X \to S$ be flat and $D \subset X$ an *effective Cartier divisor*; i.e. for every $x \in X$ there is an $f_x \in \mathcal{O}_{x,X}$ which is not a zero divisor such that $D = \operatorname{Spec} \mathcal{O}_{x,X}/(f_x)$ in a neighborhood of x. D is called a *relative Cartier divisor* if the following two equivalent conditions are satisfied:

(1.11.1) D is flat over S;
(1.11.2) For every x, f_x is not a zero divisor in $\mathcal{O}_{x,X} \otimes_{\mathcal{O}_S} k(f(x))$.

Proof. This follows at once from (7.4.1.1) applied to the multiplication $f_x : \mathcal{O}_{x,X} \to \mathcal{O}_{x,X}$. □

1.12 Definition. Let X/S be flat. The functor

$$\mathit{CDiv}(X/S) : \{\text{schemes over } S\} \to \{\text{sets}\}$$

is defined as follows:

$$CDiv(X/S)(Z) = \left\{ \begin{matrix} \text{Relative effective Cartier} \\ \text{divisors of } V \subset X \times_S Z \end{matrix} \right\}.$$

1.13 Theorem. *Let X be a scheme, flat and projective over S.*

(1.13.1) *$CDiv(X/S)$ is representable by an open subscheme* $\mathrm{CDiv}(X/S) \subset \mathrm{Hilb}(X/S)$.

(1.13.2) *If X/S is smooth, then* $\mathrm{CDiv}(X/S) \subset \mathrm{Hilb}(X/S)$ *is universally closed.*

Proof. In order to see the first part, we need to show that if $g : Y = X \times_S Z \to Z$ is flat and $U \subset Y$ is flat over Z, then the set of those points $z \in Z$ such that $U_z \subset Y_z$ is a Cartier divisor is open. Let $I \subset \mathcal{O}_Y$ be the ideal sheaf of U and $y \in f^{-1}(z)$. Since U is flat over Z, $I_z = I \otimes_{\mathcal{O}_Z} k(z)$. Let $f_{z,y}$ be a local generator of I_z at y. Let f_y be a local section of I at z such that $f_{z,y} = f_y \otimes_{\mathcal{O}_Z} 1$. By the Nakayama lemma f_y is a local generator of I at y thus U is a relative Cartier divisor at y.

By the valuative criterion of properness, (1.13.2) is implied by the following special case:

Let T be the spectrum of a DVR and X/T smooth. Let $U \subset X$ be flat over T. Assume that $U_{gen} \subset X_{gen}$ is a Cartier divisor. Then U is a relative Cartier divisor.

To see this, note that X is smooth, U has codimension one and has no embedded points. Thus $U \subset X$ is a Cartier divisor. For $x \in X_0$, let f_x be its defining equation. Since $U_0 \subset X_0$ has codimension one, f_x is not a zero divisor in $\mathcal{O}_{X_0}$. Thus, by (1.11.2), U is a relative Cartier divisor. □

1.13.3 Complement. If X is a geometrically normal scheme over a field, then $\mathrm{CDiv}(X/S)$ is closed in $\mathrm{Hilb}(X/S)$. Unfortunately I do not know any simple proof of this. See e.g. [Grothendieck62, 236-11].

1.13.4 Exercise. Let $Q/T \subset \mathbb{P}^3_T$ be a family of smooth quadrics degenerating to a quadric cone. Follow a line on Q_t to show that $\mathrm{CDiv}(Q/T)$ is not proper. Note also the change in the selfintersection of the line.

1.14 Exercises. (1.14.1) Let $f : X \to S$ be flat and projective with integral fibers such that $H^1(X_s, \mathcal{O}_{X_s}) = 0$ for every $s \in S$. Assume that S is reduced and connected.

(1.14.1.1) Show that if M is a line bundle on X such that $M_s \cong \mathcal{O}_{X_s}$ for some $s \in S$, then $M \cong f^*(f_*M)$.

(1.14.1.2) Let L be a line bundle on X such that $h^0(X_s, L_s)$ is independent of $s \in S$. Show that f_*L is locally free and $\mathrm{Proj}_S(f_*L)$ is a connected component of $\mathrm{CDiv}(X/S)$ and also of $\mathrm{Hilb}(X/S)$.

(1.14.2) Let X be smooth, irreducible and projective of dimension n over a field k. Let

$$\mathrm{Hilb}_{\dim \le m}(X) := \cup_{P : \deg P \le m} \mathrm{Hilb}_P(X)$$

be the part of the Hilbert scheme parametrizing subschemes of dimension at most m. Let $[X] \in \operatorname{Hilb}(X)$ be the point corresponding to X. Prove that

$$\operatorname{Hilb}(X) - [X] = \operatorname{CDiv}(X) \times \operatorname{Hilb}_{\dim \leq n-2}(X).$$

Finally we relate our construction to the classical notion of field of definition.

1.15 Definition. Let X be a scheme over a field k. Let $k \subset K$ be a field extension and I_K an ideal sheaf on $X \times_{\operatorname{Spec} k} \operatorname{Spec} K$. A field $k \subset F \subset K$ is called a *field of definition of* I_K if there is an ideal sheaf I_F on $X \times_{\operatorname{Spec} k} \operatorname{Spec} F$ such that $I_F \otimes_F K = I_K$.

It is easy to see that in this case $I_F = \mathcal{O}_{X_F} \cap I_K$. In particular, I_F is unique if it exists.

Let $Z \subset X_K$ be a closed subscheme. A field of definition of its ideal sheaf I_K is also called a *field of definition of* Z. Also, if $X_k \subset Y_k$ is a closed embedding then $Z \subset X_K$ is defined over F iff $Z \subset Y_K$ is defined over F.

1.16 Proposition (Notation as above). *Assume in addition that X is projective. Corresponding to Z there is a morphism* $\operatorname{Spec} K \to \operatorname{Hilb}(X_k)$. *Let* $[Z] \in \operatorname{Hilb}(X_k)$ *be the image and F the residue field of* $\operatorname{Hilb}(X_k)$ *at* $[Z]$. *Then F is the unique smallest field of definition of Z.*

Proof. By construction, there is an ideal sheaf I_F of $\mathcal{O}_{X_F}$ such that $I_K = I_F \otimes_F K$. Conversely, if E is a field and I_E exists, then this gives a morphism $\operatorname{Spec} E \to \operatorname{Hilb}(X_k)$. Its image is $[Z]$ hence $F \subset E$. □

1.17 Corollary (Notation as above). *Let* $G \subset \operatorname{Aut}(K/k)$ *be a subgroup. Assume that $I_K^g = I_K$ for every $g \in G$. Then I_K is defined over K^G.*

Proof. The ideal sheaf I_K corresponds to a morphism $i : F \to K$. Since $I_K^g = I_K$, we must have $g \circ i = i$. Therefore $\operatorname{im} i \subset K^G$. □

1.18 Exercise. Let k be a field and $\mathbb{P} \times \cdots \times \mathbb{P}$ a product of projective spaces over k. Let $K \supset k$ be a field extension and $H \subset (\mathbb{P} \times \cdots \times \mathbb{P})_K$ a hypersurface. H is defined by a multihomogeneous polynomial h which is unique up to a multiplicative constant.

Show that if $\{b_\lambda : \lambda \in \Lambda\}$ are all the nonzero coefficients of h in a fixed coordinate system, then

$$F = k\left(\frac{b_\lambda}{b_\mu} : \lambda, \mu \in \Lambda\right)$$

is the smallest field of definition of H.

1.19 Exercise. We outline Weil's construction of the smallest field of definition [Weil62, I.7 Lemma 2].

Let R be a k-algebra and $I_K \subset R \otimes_k K$ an ideal. Let $M_\lambda : \lambda \in \Lambda$ be a k-basis of R. Assume that the index set Λ is well ordered by an ordering $<$. Let

$$\Lambda' := \{\lambda \in \Lambda | M_\lambda \in I_K + \sum_{\mu<\lambda} K \cdot M_\mu\} \quad \text{and} \quad \Lambda'' = \Lambda - \Lambda'.$$

Claim 1. For every $\lambda \in \Lambda'$, there are unique $i_\lambda \in I_K$ and $c_{\lambda,\mu} \in K$ for every $\mu < \lambda; \mu \in \Lambda''$ such that

$$M_\lambda = i_\lambda + \sum_{\mu<\lambda;\mu\in\Lambda''} c_{\lambda,\mu} M_\mu.$$

Claim 2. The set

$$\left\{ M_\lambda - \sum_{\mu<\lambda;\mu\in\Lambda''} c_{\lambda,\mu} M_\mu | \lambda \in \Lambda' \right\}$$

generates I_K. In particular, $F = k(\{c_{\lambda,\mu}\})$ is a field of definition of I_K.

Claim 3. Assume that I_K is definable over a field L. Then the construction using I_L gives the same Λ'. Now conclude that $F = k(\{c_{\lambda,\mu}\})$ is the unique smallest field of definition of I_K.

Claim 4. For an arbitrary scheme X, let $U_i = \operatorname{Spec} R_i$ be an affine cover. Then the unique smallest field of definition of I_K is the composite of the smallest fields of definition of $I_K \cap R_i$.

I.2 Infinitesimal Study of the Hilbert Scheme

The aim of this section is to study infinitesimal properties of $\mathrm{Hilb}(Y/S)$. The main ideas are again due to [Grothendieck62]; my presentation follows [Artin76]. First we study flat families where the central fiber is an affine subscheme. This part is rather straightforward commutative algebra. The main problem is to set up reasonably coherent notation. This is done in (2.1–4). The globalization of the local deformations is not very difficult (2.5–7). The main applications are (2.8, 2.10) which identify the tangent spaces to $\mathrm{Hilb}(Y)$ and give lower bounds for its dimension in terms of the "obstruction group" (2.6).

The general methods give an obstruction group which is too large in most applications. Therefore it is crucial to get more precise information about the obstructions. The main result is (2.14) which gives much better estimates in certain situations. Everything is brought together in (2.15), which is the main theorem of the section.

As in Sect. 1, this gives the corresponding result about $\mathrm{Hom}(X, Y)$ rather easily (2.16–17).

In order to get a heuristic explanation of the method, assume for simplicity that S is the spectrum of an algebraically closed field k. Let $Z \subset Y$ be a closed subscheme corresponding to a point $[Z] \in \mathrm{Hilb}(Y)$. Thus Z corresponds to a morphism $\phi_0 : \mathrm{Spec}\, k \to \mathrm{Hilb}(Y)$. Extensions of ϕ_0 to $\phi_1 : \mathrm{Spec}\, k[\epsilon]/(\epsilon^2) \to \mathrm{Hilb}(Y)$ correspond to tangent vectors of $\mathrm{Hilb}(Y)$ at $[Z]$. More generally, as Grothendieck pointed out, studying the completion of the local ring of $[Z] \in \mathrm{Hilb}(Y)$ is equivalent to studying flat families of subschemes of Y over Artin local rings where the central fiber is Z.

Let B be an Artin local ring with residue field k and maximal ideal m. We would like to understand all extensions of ϕ_0 to morphisms $\phi_B : \mathrm{Spec}\, B \to \mathrm{Hilb}(Y)$. We do this in several steps as follows. Choose a sequence of ideals $m = I_0 \supset I_1 \supset \ldots \supset I_r = 0$, such that $mI_i \subset I_{i+1}$. We study the extensions of morphisms $\phi_i : \mathrm{Spec}\, B/I_i \to \mathrm{Hilb}(Y)$ to $\phi_{i+1} : \mathrm{Spec}\, B/I_{i+1} \to \mathrm{Hilb}(Y)$.

The advantage of this case is, that the extension problem translates to a linear algebra question, which can be made very explicit in many examples. Without the assumption $mI_i \subset I_{i+1}$, we get a nonlinear problem, which is very hard to study.

We start with the affine version of the problem. Fix a local ring (m_B, B) and an ideal $J < B$ such that $m_B J = 0$. We are mostly interested in the case when B is Artin, but it is useful to study the situation for arbitrary B. Let $\mathrm{Spec}\, T^B \subset Y^B := Y \times_k \mathrm{Spec}\, B$ be an affine open set. Assume that we already have an extension of Z to a subscheme $Z^A \subset Y^A$, where $A = B/J$. We want to study the corresponding extension problem of the ideal sheaf $I^A \subset T^A$ to an ideal sheaf $I^B \subset T^B$.

Once the local problem is dealt with in a sufficiently functorial way, everything will glue together to give a solution of the global problem.

In the general case we have to make two assumptions which are automatic when $S = \mathrm{Spec}\, k$. First, we need to assume that Y/S is flat. This is essential for the computations to work. Any projective Y/S can be embedded into a projective space bundle over S, so the general case can be reduced to the flat (even smooth) one in principle.

Also, we need to assume that T^B really arises as a deformation of something over the closed point. The main case we want to avoid is something like $B = k[x]$ and $T^B = k[x, x^{-1}]$, where the central fiber is empty. Thus we assume that the m_B-adic topology on T^B is separated (this always holds if B is Artin, so there is no need to worry what this means; cf. (7.4.1)).

2.1 Notation. Let A and B be Noetherian local rings with maximal ideals m_A resp. m_B and with the same residue field k. Assume that B is an extension of A by the ideal J

$$0 \to J \to B \to A \to 0$$

such that $m_B J = 0$.

Let T be a k-algebra and $I \subset T$ an ideal with quotient ring $Q = T/I$. Let

$$R \xrightarrow{r} G \xrightarrow{g} I \to 0 \tag{2.1.1}$$

be a presentation of I by free T-modules. Thus G is a complete set of generators of I and R is a complete set of relations.

Let T^B be a Noetherian ring flat over B such that $T^B \otimes_B k \cong T$ and the m_B-adic topology on T^B is separated. Let $T^B \otimes_B A \cong T^A$. Extensions of the above ideals, modules or morphisms to T^A or T^B-ideals, modules or morphisms are denoted by a corresponding upper index.

Let $I^A \subset T^A$ be an ideal such that $Q^A = T^A/I^A$ is flat over A.

2.1.2 Lemma (Notation as above). *Let*

$$(F_\cdot) \qquad \cdots \xrightarrow{d_2} F_2 \xrightarrow{d_1} F_1 \xrightarrow{g} I \longrightarrow 0$$

be a free resolution of I. Then there is a free resolution

$$(F_\cdot^A) \qquad \cdots \xrightarrow{d_2^A} F_2^A \xrightarrow{d_1^A} F_1^A \xrightarrow{g^A} I^A \longrightarrow 0$$

such that

$$(F_\cdot^A) \otimes k \cong F_\cdot .$$

Proof. Let $\{x_i\}$ be a basis of F_1 and $g_i = g(x_i)$. Q^A is flat over A, thus $I^A \otimes k \cong I$. Therefore the generators $g_i \in I$ can be lifted to $g_i^A \in I^A$ and by (7.4.1), $\{g_i^A\}$ generates I^A. Let F_1^A be the free module over T^A generated by the symbols x_i^A and set $g^A(x_i^A) = g_i^A$.

The kernel of g^A is again flat over A, hence F_2^A can be constructed similarly. And so on. □

Let G^B and R^B be free T^B-modules such that $G^B \otimes_{T^B} T^A \cong G^A$ and $R^B \otimes_{T^B} T^A \cong R^A$. By (2.1.2), we can lift (2.1.1) to an exact sequence of T^A-modules

$$R^A \xrightarrow{r^A} G^A \xrightarrow{g^A} I^A \longrightarrow 0.$$

Let $g^B : G^B \to T^B$ be any morphism lifting g^A and $r^B : R^B \to G^B$ any morphism lifting r^A. We obtain a commutative diagram:

$$\begin{array}{ccccccccl} R^B & \xrightarrow{r^B} & G^B & \xrightarrow{g^B} & I^B & \longrightarrow & 0 & & \text{(not a complex!)} \\ \downarrow & & \downarrow & & \downarrow & & & & \\ R^A & \xrightarrow{r^A} & G^A & \xrightarrow{g^A} & I^A & \longrightarrow & 0 & & \text{(exact row)} \end{array} \tag{2.1.3}$$

where the top row is not necessarily a complex and $I^B := \operatorname{im} g^B \subset T^B$. I^B is an ideal which reduces to $I^A \operatorname{mod} JT^B$. Let $Q^B = T^B/I^B$.

2.2 Lemma (Notation as above).

(2.2.1) *To a diagram as above one can associate a morphism* $e(g^B) : R \to Q \otimes J$ *which is independent of the choice of* r^B.

(2.2.2) *The module*

$$E(g^B) = \operatorname{coker}[R \xrightarrow{e(g^B)+r} Q \otimes J + G]$$

can be written as an extension

(2.2.2.1) $$0 \to Q \otimes J \to E(g^B) \to I \to 0;$$

hence we obtain an element $[E(g^B)] \in \operatorname{Ext}^1_T(I, Q \otimes J)$.

(2.2.3) *Any different lifting* $\bar{g}^B : G^B \to T^B$ *can be written as* $\bar{g}^B = g^B + \phi^B$ *where* $\phi^B \in \operatorname{Hom}(G^B, T \otimes J)$. *This gives a natural extension isomorphism*

$$E(\phi^B) : E(g^B) \to E(\bar{g}^B).$$

In particular, $[E(g^B)] \in \operatorname{Ext}^1_T(I, Q \otimes J)$ *is independent of the choice of* g^B *and depends only on the data over* A. *It is denoted by* $E^B(Q^A)$ *and called the obstruction.*

(2.2.4) *The following are equivalent:*

(2.2.4.1) Q^B *is flat over* B,
(2.2.4.2) $e(g^B) = 0$,
(2.2.4.3) $g^B \circ r^B = 0$ *for a suitable choice of* r^B.

(2.2.5) *For a suitable choice of* g^B *the quotient* Q^B *is flat over* B *iff* $[E^B(Q^A)] = 0$. *If this is the case, we have (noncanonical) affine linear isomorphisms*

$$\left\{\begin{matrix}\textit{Set of quotients } Q^B \\ \textit{which are flat over } B\end{matrix}\right\} \cong \{\textit{Splittings of } (2.2.2.1)\} \cong \operatorname{Hom}_T(I, Q \otimes J).$$

(2.2.6) *If*

$$R_1^A \xrightarrow{r_1^A} G_1^A \xrightarrow{g_1^A} I^A \to 0$$

is a different presentation of I^A, *then we obtain a natural extension isomorphism* $E(g^B) \cong E(g_1^B)$. *Thus* $[E^B(Q^A)]$ *depends only on* Q^A. *Also if* $[E^B(Q^A)] = 0$, *then the principal homogeneous space structure on the set of quotients* Q^B *which are flat over* B *depends also on* Q^A *only.*

Proof. $g^B \circ r^B : R^B \to T^B$ is the zero map modulo J, hence its image is contained in $T^B J \cong T \otimes J$. Therefore the kernel contains $m_B R^B$ and we obtain a morphism

$$e(g^B) : R \xrightarrow{g^B \circ r^B} T \otimes J \xrightarrow{q} Q \otimes J.$$

Any other lifting of r^A can be written as $\bar{r}^B = r^B + \psi^B$ where $\psi^B \in \operatorname{Hom}(R^B, G \otimes J)$. Then

$$g^B \circ \bar{r}^B = g^B \circ r^B + g^B \circ \psi^B = g^B \circ r^B + (g^k \otimes id_J) \circ \psi^B.$$

Since $\mathrm{im}((g^k \otimes id_J) \circ \psi^B) \subset I \otimes J$, we conclude that

$$q \circ g^B \circ \bar{r}^B = q \circ g^B \circ r^B.$$

This shows (2.2.1).

The exact sequence

$$R + Q \otimes J \xrightarrow{(r+e(g^B), id_{Q\otimes J})} G + Q \otimes J \xrightarrow{(g,0)} I \to 0$$

gives an exact sequence

$$Q \otimes J \xrightarrow{i} E(g^B) \to I \to 0.$$

i is injective iff $\ker e(g^B) \supset \ker r$. By (2.1.2) a^k can be lifted to $a^A \in \ker r^A$. Let a^B be a lifting of a^A to R^B. $r^A(a^A) = 0$, thus $r^B(a^B) \in JG^B$, and

$$g^B \circ r^B(a^B) \in JI^B = I \otimes J.$$

Thus $e(g^B)(a^k) = 0$, which completes (2.2.2).

Any other choice $\bar{g}^B$ can be written as $\bar{g}^B = g^B + \phi^B$. Left multiplication with the matrix

$$\begin{pmatrix} id_{Q\otimes J} & -\phi^B \\ 0 & id_G \end{pmatrix}$$

on $Q \otimes J + G$ gives the required isomorphism $E(\phi^B)$. This shows (2.2.3).

By [Matsumura80, 20.C.3], Q^B is flat over B iff $J \otimes T/J \otimes I \cong J \otimes Q^B$ and $JT^B/(I^B \cap JT^B) \cong JQ^B$ are naturally isomorphic. This condition is equivalent to $I^B \cap T^B J = I \otimes J$. Let x_i be a set of generators of G^B and $g^B(\sum b_i x_i) \in I^B \cap T^B J$. The reduction of $\sum b_i x_i$ modulo J is contained in the image of R^A. Since $R^B \to R^A$ is surjective, we can find an $r' \in R^B$ such that $r^B(r') = \sum b_i' x_i$ and $b_i - b_i' \in J$ for every i. Thus

$$g^B(\sum b_i x_i) = \sum (b_i - b_i') g^B(x_i) + (g^B \circ r^B)(r').$$

Here the first component is in JI^B. Thus we see that Q^B is flat over B iff $\mathrm{im}(g^B \circ r^B) \subset I \otimes J$, hence (2.2.4.1) and (2.2.4.2) are equivalent.

Assume now that $\mathrm{im}(g^B \circ r^B) \subset I \otimes J$. Since $G \otimes J \to I \otimes J$ is surjective, one can get a lifting

$$g^B \circ r^B : R^B \xrightarrow{\rho^B} G \otimes J \xrightarrow{g\otimes id_J} I \otimes J.$$

Setting $\bar{r}^B = r^B - \rho^B$, we see that $g^B \circ \bar{r}^B = 0$. This shows that (2.2.4.2) implies (2.2.4.3), while the converse is clear.

If Q^B is flat, then, by (2.2.4), the extension $E(g^B)$ splits. Assume conversely that $E(g^B)$ splits. This means that there is a $\phi \in \mathrm{Hom}(G, Q \otimes J)$

such that $e(g^B) = r^*(\phi)$. Since G^B is free, we can lift ϕ to a morphism $\phi^B \in \mathrm{Hom}(G^B, T \otimes J)$. Now $\bar{g}^B = g^B - \phi^B$ satisfies $\mathrm{im}(\bar{g}^B \circ r^B) \subset I \otimes J$.

If g^B gives a flat extension, then $\bar{g}^B = g^B + \phi^B$ is also flat iff $\mathrm{im}(\phi^B \circ r^B) \subset I \otimes J$. This holds iff $\phi : G \to T \otimes J$ descends to a morphism $I \to Q \otimes J$. This constructs the map whose existence is asserted in (2.2.5). Somewhat lengthy but straightforward diagram chasing shows that it is in fact affine linear. We leave this as an exercise.

(2.2.6) can be established as follows. There are connecting maps between the two presentations that make the following diagram commutative:

$$\begin{array}{ccccccc} R^A & \xrightarrow{r^A} & G^A & \xrightarrow{g^A} & I^A & \longrightarrow & 0 \\ \downarrow{\scriptstyle \pi_R^A} & & \downarrow{\scriptstyle \pi_G^A} & & \downarrow{\scriptstyle \cong} & & \\ R_1^A & \xrightarrow{r_1^A} & G_1^A & \xrightarrow{g_1^A} & I^A & \longrightarrow & 0. \end{array}$$

Lifting all morphisms to B, we obtain a noncommutative diagram

$$\begin{array}{ccccc} R^B & \xrightarrow{r^B} & G^B & \xrightarrow{g^B} & T^B \\ \downarrow{\scriptstyle \pi_R^B} & & \downarrow{\scriptstyle \pi_G^B} & & \downarrow{\scriptstyle \cong} \\ R_1^B & \xrightarrow{r_1^B} & G_1^B & \xrightarrow{g_1^B} & T^B, \end{array}$$

which is commutative modulo J. As a consequence we obtain natural maps that make the following diagram commutative

$$\begin{array}{ccc} R & \xrightarrow{e(g^B)+r} & Q \otimes J + G \\ \downarrow{\scriptstyle \pi_R^k} & & {\scriptstyle id_{Q\otimes J}} \downarrow{\scriptstyle \pi_G^k} \\ R_1 & \xrightarrow{e(g_1^B)+r_1} & Q \otimes J + G_1. \end{array}$$

This induces a map between the extensions

$$\begin{array}{ccccccccc} 0 & \longrightarrow & Q \otimes J & \longrightarrow & E(\bar{g}^B) & \longrightarrow & I & \longrightarrow & 0 \\ & & \downarrow{\scriptstyle \cong} & & \downarrow{\scriptstyle \pi_E^k} & & \downarrow{\scriptstyle \cong} & & \\ 0 & \longrightarrow & Q \otimes J & \longrightarrow & E(g_1^B) & \longrightarrow & I & \longrightarrow & 0, \end{array}$$

which is necessarily an isomorphism. Thus

$$\pi_E^k \circ E(\phi^B) : E(g^B) \to E(g_1^B)$$

is the required isomorphism. The remaing parts of (2.2.6) are again left as an exercise. □

2.2.7 Special Case. The above computations can be made very explicit in the special case when $A = k$ and $B = k + J$. Then $T^B = T^B \otimes_B k + T^B \otimes_B J = T + T \otimes_k J$ (this is not a B-module decomposition). In this case, there is a trivial extension $I^B_{triv} = I + I \otimes_k J$. Any other extension can be obtained as follows:

Note first that any flat extension I^B of I contains $I \otimes_k J$. Therefore it is sufficient to describe $I^B/(I \otimes_k J) \subset T + (T/I) \otimes_k J$. For $\phi \in \operatorname{Hom}(I, T/I \otimes_k J)$, let

$$I^B_\phi/(I \otimes_k J) = \operatorname{im}[I \xrightarrow{(id_I,\phi)} T + T/I \otimes_k J].$$

This gives a natural isomorphism

$$\left\{\begin{matrix}\text{Set of quotients } Q^B \\ \text{which are flat over } B\end{matrix}\right\} \cong \operatorname{Hom}_T(I, Q \otimes J).$$

There is a natural morphism

$$T \xrightarrow{(id_T,0)} T + T \otimes_k J \to (T + T/I \otimes_k J)/\operatorname{im}(id_I, \phi) = Q^B_\phi$$

which sits in the following commutative diagram:

$$(2.2.8) \qquad \begin{array}{ccccccccc} 0 & \longrightarrow & I & \longrightarrow & T & \longrightarrow & Q & \longrightarrow & 0 \\ & & \phi\downarrow & & \downarrow & & \downarrow\cong & & \\ 0 & \longrightarrow & T/I \otimes_k J & \longrightarrow & Q^B_\phi & \longrightarrow & Q & \longrightarrow & 0. \end{array}$$

2.3 Notation. We also need to compare the obstructions when we change the setup in two different ways.

(2.3.1) Let $\tilde{T}^B$ be a flat T^B-algebra. Tensoring by $\tilde{T}^B$ we obtain $\tilde{T}, \tilde{I}, \tilde{Q}$ and so on. The constructions of (2.1–2) commute with tensoring by a flat T^B-algebra. In particular, if

$$0 \to Q \otimes J \xrightarrow{i} E \to I \to 0$$

is an extension, then

$$0 \to (Q \otimes_T \tilde{T}) \otimes_k J \xrightarrow{i} E \otimes_T \tilde{T} \to I \otimes_T \tilde{T} \to 0$$

is also exact. The corresponding extension is denoted by $[E^B(\tilde{Q}^A)]$.

There is an isomorphism

$$\operatorname{Ext}^1_{\tilde{T}}(\tilde{I}, \tilde{Q} \otimes J) \cong \operatorname{Ext}^1_T(I, Q \otimes J) \otimes_T \tilde{T}.$$

Thus for $[E^B(Q^A)] \in \operatorname{Ext}^1_T(I, Q \otimes J)$ we obtain $[E^B(Q^A)] \otimes_T \{1\} \in \operatorname{Ext}^1_{\tilde{T}}(\tilde{I}, \tilde{Q} \otimes J)$.

(2.3.2) In the second variant, T is unchanged. Instead, consider a homomorphism between ring extensions

$$\begin{array}{ccccccccc} 0 & \longrightarrow & J & \longrightarrow & B & \longrightarrow & A & \longrightarrow & 0 \\ & & h\downarrow & & \downarrow & & \downarrow & & \\ 0 & \longrightarrow & J' & \longrightarrow & B' & \longrightarrow & A' & \longrightarrow & 0, \end{array}$$

where the rings have the same residue field k and J, J' are k-vectorspaces. Let $Q^{A'} = Q^A \otimes_A A'$. The aim is to compare the two obstructions $[E^B(Q^A)]$ and $[E^{B'}(Q^{A'})]$.

Since J is free over k, we have natural isomorphisms

$$\mathrm{Ext}^1_T(I, Q \otimes J) \cong \mathrm{Ext}^1_T(I, Q \otimes k) \otimes J \cong \mathrm{Ext}^1_T(I, Q) \otimes J,$$

and similarly for J'. Therefore we have a natural k-linear map

$$V : \mathrm{Hom}(J, J') \to \mathrm{Hom}(\mathrm{Ext}^1_T(I, Q \otimes J), \mathrm{Ext}^1_T(I, Q \otimes J')).$$

On the level of extensions the map V can be constructed as follows. Let

$$0 \to Q \otimes J \xrightarrow{i} E \to I \to 0$$

be an extension. If E' is defined by the sequence

$$Q \otimes J \xrightarrow{(id_Q \otimes h, i)} Q \otimes J' + E \to E' \to 0$$

then E' fits into an exact sequence

$$0 \to Q \otimes J' \to E' \to I \to 0,$$

and $V(h)([E]) = [E']$.

More generally, one can consider the behaviour of obstructions when both of the above changes happen, and we even allow a residue field extension. We do not need this case.

The constructions of (2.1–2) commute with these changes in the expected way. This is clear in the first case, the second case needs a little work.

2.4.1 Proposition (Notation as above).

(2.4.1.1) $[E^B(\tilde{Q}^A)] = [E^B(Q^A)] \otimes_T \{1\}$.

(2.4.1.2) *The base change map*

$$\left\{\begin{array}{l}\textit{Set of quotients } Q^B\\ \textit{which are flat over } B\end{array}\right\} \to \left\{\begin{array}{l}\textit{Set of quotients } \tilde{Q}^B\\ \textit{which are flat over } B\end{array}\right\}$$

is affine linear. □

2.4.2 Proposition (Notation as above).

(2.4.2.1) $[E^{B'}(Q^{A'})] = V(h)([E^B(Q^A)])$.

(2.4.2.2) *For any Q^A, the map*

$$\mathrm{Hom}(J, J') \to \mathrm{Ext}^1_T(I, Q) \otimes J' : \quad h \mapsto [E^{B'}(Q^{A'})]$$

is k-linear.

(2.4.2.3) *The restriction map*

$$\left\{\begin{matrix}\textit{Set of quotients } Q^B \\ \textit{which are flat over } B\end{matrix}\right\} \to \left\{\begin{matrix}\textit{Set of quotients } Q^{B'} \\ \textit{which are flat over } B'\end{matrix}\right\}$$

is affine linear.

Proof. Given $g^B : G^B \to T^B$ we obtain

$$g^{B'} := g^B \otimes_B B' : G^{B'} = G^B \otimes_B B' \to T^B \otimes_B B' = T^{B'}.$$

This gives a commutative diagram

$$\begin{array}{ccc} R & \xrightarrow{e(g^B)+r} & Q \otimes J + G \\ \downarrow \cong & & id_Q \otimes h \downarrow id_G \\ R & \xrightarrow{e(g^{B'})+r} & Q \otimes J' + G \end{array}$$

This induces a map between the extensions

$$\text{(2.4.2.4)} \qquad \begin{array}{ccccccccc} 0 & \longrightarrow & Q \otimes J & \longrightarrow & E(g^B) & \longrightarrow & I & \longrightarrow & 0 \\ & & id_Q \otimes h \downarrow & & \downarrow & & \downarrow \cong & & \\ 0 & \longrightarrow & Q \otimes J' & \longrightarrow & E(g^{B'}) & \longrightarrow & I & \longrightarrow & 0. \end{array}$$

This is exactly the map induced by $V(h)$ between the two Ext-groups.

(2.4.2.2) is a special case of (2.4.2.1).

The restriction map is clearly affine linear on the set of splittings of the sequences in (2.2.2.1). This implies (2.4.2.3) by (2.2.5). □

2.5 Proposition. *Let $Y = Y^k$ be a scheme over a field k and $Z \subset Y$ a closed subscheme defined over k with ideal sheaf $I = I_Z$. Let Y^B be a scheme flat over B such that $Y^B \times_{\mathrm{Spec}\ B} \mathrm{Spec}\ k \cong Y$. Let $Y^A = Y^B \times_{\mathrm{Spec}\ B} \mathrm{Spec}\ A$ and $Z^A \subset Y^A$ a closed subscheme, flat over A extending Z.*

(2.5.1) *The obstruction $[E^B(Z^A)]$ to extend Z^A to a subscheme $Z^B \subset Y^B$ flat over B lies in $\mathrm{Ext}^1_Y(I_Z, \mathcal{O}_Z \otimes J)$.*

(2.5.2) *The set of extensions is either empty or they form a principal homogeneous space over $\mathrm{Hom}_Y(I_Z, \mathcal{O}_Z \otimes J)$.*

Proof. Take an affine cover $\{U_i^B\}$ of Y^B and let $U_{ij}^B = U_i^B \cap U_j^B$. Over each U_i^A, we can pick a presentation

$$R_i^A \xrightarrow{r_i^A} G_i^A \xrightarrow{g_i^A} I_i^A \to 0 \quad \text{and liftings} \quad g_i^B : G_i^B \to \mathcal{O}_{U_i^B}.$$

By (2.2) this gives extensions $E(g_i^B)$ over U_i and (2.4.1) gives isomorphisms between the restrictions

$$E(g_i^B)|U_{ij} \xrightarrow{\cong} E(g_j^B)|U_{ij}.$$

These glue together to an extension $E(\{g_i^B\})$ of I by $\mathcal{O}_Z \otimes J$.

Assume that the obstruction is zero, i.e. the above extension splits. Then first of all every $E(g_i^B)$ splits, thus we can choose $\{g_i^B\}$ in such a way that each I_i^B is flat over B. Over U_{ij}, we get two possible extensions. By (2.4.1) and (2.2.5) this gives an element $\phi_{ij} \in \operatorname{Hom}(I_{ij}, \mathcal{O}_{Z_{ij}} \otimes J)$. $\{\phi_{ij}\}$ satisfies the cocycle condition and it gives an element

$$\{\phi_{ij}\} \in H^1(Y, \mathcal{H}om_Y(I, \mathcal{O}_Z \otimes J)) \hookrightarrow \operatorname{Ext}^1_Y(I, \mathcal{O}_Z \otimes J).$$

This is the obstruction class associated to $\{g_i^B\}$, thus zero by assumption. Hence there are $\phi_i \in \operatorname{Hom}_Y(I_i, \mathcal{O}_{Z_i} \otimes J)$ such that $\phi_{ij} = \phi_i - \phi_j$. Lift every ϕ_i to a morphism $\phi_i^B \in \operatorname{Hom}_{Y^B}(G_i^B, \mathcal{O}_{Z_i} \otimes J)$. By construction, the new maps $\bar{g}_i^B = g_i^B - \phi_i^B$ give ideal sheaves I_i^B which are isomorphic over U_{ij}^B. Thus they glue together to a global extension I^B.

Conversely, assume that I^B exists. Then we obtain split extensions $E(g_i^B)$ and the isomorphisms $E(g_i^B)|U_{ij} \cong E(g_j^B)|U_{ij}$ respects the splittings. Thus $E^B(Z^A) = 0$.

If $\{g_i^B\}$ gives a flat extension, then $\bar{g}_i^B = g_i^B - \phi_i^B$ is also a flat extension if $\phi_i^B|U_{ij}^B = \phi_j^B|U_{ij}^B$. Thus $\{\phi_i^B\}$ gives an element in $\operatorname{Hom}_Y(I, \mathcal{O}_Z \otimes J)$. □

2.5.3 Special Case. As in (2.2.7), the above computations can be made very explicit in the special case when $A = k$, $B = k+J$ and $Y^B = Y \times_{\operatorname{Spec} k} \operatorname{Spec} B$. There is a trivial extension $Z^B_{triv} = Z \times_{\operatorname{Spec} k} \operatorname{Spec} B$. As in (2.2.7), there is a natural isomorphism

$$\left\{\begin{matrix}\text{Set of subschemes } Z^B \\ \text{which are flat over } B.\end{matrix}\right\} \cong \operatorname{Hom}_Y(I_Z, \mathcal{O}_Z \otimes J).$$

For any $\phi \in \operatorname{Hom}_Y(I_Z, \mathcal{O}_Z \otimes J)$ there is a natural morphism $u^* : u^*\mathcal{O}_Y \to \mathcal{O}_{Z^B_\phi}$ which sits in the following commutative diagram:

$$\begin{array}{ccccccccc} 0 & \longrightarrow & u^*I_Z & \longrightarrow & u^*\mathcal{O}_Y & \longrightarrow & \mathcal{O}_Z & \longrightarrow & 0 \\ & & \phi\downarrow & & \downarrow u^*V & & \downarrow\cong & & \\ 0 & \longrightarrow & \mathcal{O}_Z \otimes_k J & \longrightarrow & \mathcal{O}_{Z^B_\phi} & \longrightarrow & \mathcal{O}_Z & \longrightarrow & 0. \end{array} \tag{2.5.4}$$

It turns out that in many important cases, we are able to show that the above obstruction space is too large. The following definition identifies the smallest subvectorspace which carries all the obstructions.

2.6 Definition. Let Y^S/S be a scheme, flat over S. Let $s \in S$ be a point with residue field k and Y^k the fiber of Y^S over s. Let $Z \subset Y^k$ be a closed subscheme. The *obstruction space* $\operatorname{Obs}(Z) \subset \operatorname{Ext}^1_{Y^k}(I_Z, \mathcal{O}_Z)$ is the smallest k-sub vector space such that

$$[E^B(Z^A)] \in \mathrm{Obs}(Z) \otimes J$$

holds for every choice of A, B and Z^A where A and B are local S-algebras with residue field k as in (2.1), $Y^B = Y \times_S \operatorname{Spec} B$ and $Z^A \subset Y^A$ is any subscheme, flat over Z such that $Z^A \times \operatorname{Spec} k = Z$.

We say that the deformations of $Z \subset Y^S$ are *unobstructed* if $\mathrm{Obs}(Z) = 0$.

2.7 Proposition (Notation as in (2.3) and (2.6)).

(2.7.1)

$$[E^{B'}(Z^{A'})] = V(h)([E^B(Z^A)]).$$

In particular, the mapping

$$\mathrm{Hom}(J, J') \to \mathrm{Ext}^1_Y(I_Z, \mathcal{O}_Z) \otimes J' \quad \textit{given by} \quad h \mapsto [E^{B'}(Z^{A'})]$$

is k-linear.

(2.7.2) *Given B and Z^A, there is a unique maximal quotient $q : J \to J^m$ such that $[E^{B^m}(Z^A)] = 0$, where $B^m = B/\ker q$.*

(2.7.3) $\dim_k \ker q \le \dim_k \mathrm{Obs}(Z)$.

Proof. Let $E(\{g_i^B\})$ be the extension corresponding to the liftings $\{g_i^B\}$. The map $h : J \to J'$ gives a new extension

$$\begin{array}{ccccccccc} 0 & \longrightarrow & \mathcal{O}_Z \otimes J & \xrightarrow{i} & E(\{g_i^B\}) & \longrightarrow & I_Z & \longrightarrow & 0 \\ & & \big\downarrow{\scriptstyle id_Z \otimes h} & & \big\downarrow & & \big\downarrow{\scriptstyle \cong} & & \\ 0 & \longrightarrow & Q \otimes J' & \longrightarrow & E' & \longrightarrow & I_Z & \longrightarrow & 0 \end{array} \qquad (2.7.4)$$

as in (2.3).

As in (2.4.2) we obtain the liftings $\{g_i^{B'}\}$ and hence an extension $E(\{g_i^{B'}\})$. We need to show that $E' \cong E(\{g_i^{B'}\})$. (2.4.2.1) shows that $E'|U_i \cong E(g_i^{B'})$ and (2.4.2.3) implies that for every i, j the isomorphisms over U_i and U_j are compatible over U_{ij}. This shows (2.7.1).

Set $K = \ker[\mathrm{Hom}(J, k) \to \mathrm{Ext}^1_S(I_Z, \mathcal{O}_Z)]$ and let $q_i : J \to k_i \cong k$ be a basis of K. Let

$$J^m := \sum k_i \quad \text{and} \quad q := \sum q_i : J \to J^m.$$

Since

$$\mathrm{Ext}^1_S(I_Z, \mathcal{O}_Z) \otimes J^m \cong \sum \mathrm{Ext}^1_S(I_Z, \mathcal{O}_Z) \otimes k_i,$$

we conclude that $[E^{B^m}(Z^{A'})] = 0$. Conversely, if $J \to J'$ is a quotient such that $[E^{B'}(Z^A)] = 0$ and $p : J \to J' \to k$ is any quotient, then also $[E^{B^p}(Z^A)] = 0$ where $B^p = B/\ker p$. This implies that every quotient of J' is a quotient of J^m, hence J' itself is a quotient of J^m.

Finally, by construction,

$$\dim \ker q = \dim J - \dim K \le \dim \mathrm{Obs}(Z).$$

This implies (2.7.3). □

As an application, first consider the Hilbert scheme of a scheme over a field.

2.8 Theorem. *Let Y^k be a projective scheme over a field k and $Z \subset Y^k$ a closed subscheme. Then:*

(2.8.1) *The Zariski tangent space of* $\operatorname{Hilb}(Y^k)$ *at* $[Z]$ *is naturally isomorphic to*

$$\operatorname{Hom}_{Y^k}(I_Z, \mathcal{O}_Z) = \operatorname{Hom}_Z(I_Z/I_Z^2, \mathcal{O}_Z).$$

(2.8.2) *Let (Q, m_Q) be the local ring of* $\operatorname{Hilb}(Y^k)$ *at $[Z]$. Then Q can be written as the quotient of a regular local k-algebra R of dimension $d =$* $\dim_k \operatorname{Hom}_{Y^k}(I_Z, \mathcal{O}_Z)$.

(2.8.3) *The kernel $K = \ker[R \to Q]$ is generated by* $\dim \operatorname{Obs}(Z)$ *elements.*

(2.8.4) *The dimension of every irreducible component of* $\operatorname{Hilb}(Y^k)$ *at $[Z]$ is at least*

$$\dim \operatorname{Hom}_{Y^k}(I_Z, \mathcal{O}_Z) - \dim \operatorname{Obs}(Z).$$

Proof. The Hilbert scheme of a projective scheme exists by (1.4). The Zariski tangent space at $[Z]$ is the set of flat extensions of Z to a closed subscheme of $Y^k \times_{\operatorname{Spec} k} \operatorname{Spec} k[\epsilon]$. Choosing the trivial extension $Z \times_{\operatorname{Spec} k} \operatorname{Spec} k[\epsilon]$ as the origin, by (2.5.2), this set is

$$\operatorname{Hom}_{Y^k}(I_Z, \mathcal{O}_Z) = \operatorname{Hom}_Z(I_Z/I_Z^2, \mathcal{O}_Z).$$

Therefore the embedding dimension of Q is d and Q is the quotient of a regular local k-algebra R of dimension d.

(2.8.3–4) are special cases of (2.10). □

2.9 Notation. Let Y^S/S be a scheme, flat and projective over S. Let $s \in S$ be a point with residue field k and $(\mathcal{O}_{s,S}, m_S)$ the local ring of S at s. Let Y^k be the fiber over s and $Z \in Y^k$ a closed subscheme. Let I_Z denote the ideal sheaf of Z in Y^k and $\operatorname{Obs}(Z)$ the obstruction space for deformations of Z in Y^S/S. Let (Q, m_Q) be the local ring of $\operatorname{Hilb}(Y^S/S)$ at $[Z]$.

2.10 Theorem (Notation as above). *Assume that Y^S is flat over S.*

(2.10.1) *The $\mathcal{O}_{s,S}$-algebra Q can be written as the quotient of a local $\mathcal{O}_{s,S}$-algebra R, where*

$$\operatorname{Spec} R \to S$$

is smooth of relative dimension $d = \dim_k \operatorname{Hom}_{Y^k}(I_Z, \mathcal{O}_Z)$.

(2.10.2) *The kernel $K = \ker[R \to Q]$ is generated by* $\dim_k \operatorname{Obs}(Z)$ *elements.*

(2.10.3) *Assume that S is equidimensional at s. Then the dimension of every irreducible component of* $\operatorname{Hilb}(Y^S/S)$ *at $[Z]$ is at least*

$$\dim_k \operatorname{Hom}_{Y^k}(I_Z, \mathcal{O}_Z) + \dim_s S - \dim_k \operatorname{Obs}(Z).$$

(2.10.4) *If*

$$\dim_{[Z]} \operatorname{Hilb}(Y^k) = \dim_k \operatorname{Hom}_{Y^k}(I_Z, \mathcal{O}_Z) - \dim_k \operatorname{Obs}(Z),$$

then $\operatorname{Hilb}(Y^S/S) \to S$ *is a local complete intersection morphism at* $[Z]$. *In particular it is flat at* $[Z]$.

Proof. By (1.4), the Hilbert scheme of Y^S/S exists. Its tangent space is not easy to compute, however, the relative tangent space over S is given by (2.8.1) as:

$$\left(m_Q/(m_S Q, m_Q^2)\right)^* \cong \operatorname{Hom}_{Y^k}(I_Z, \mathcal{O}_Z).$$

Therefore Q is the quotient of a local $\mathcal{O}_{s,S}$-algebra R which is smooth over S of relative dimension $\dim_k \operatorname{Hom}_{Y^k}(I_Z, \mathcal{O}_Z)$. This is (2.10.1).

Since Q and R have the same relative tangent space over S, we conclude that $K \subset (m_S R, m_R^2)$.

Let $B = R/(m_R K)$ and m_B the corresponding maximal ideal. The kernel J of $B \to Q$ is isomorphic to $K/(m_R K)$.

By (2.7.2) there is a largest quotient $q: J \to J^m$ such that $[E^{B^m}(Z^Q)] = 0$, and by (2.7.3),

$$\text{(2.10.5)} \qquad \begin{aligned} \dim J^m &\geq \dim J - \dim \operatorname{Obs}(Z) \\ &= \dim K/(m_R K) - \dim \operatorname{Obs}(Z). \end{aligned}$$

By the definition of J^m, there is a subscheme $Z^{B^m} \subset Y^S \times_S B^m$ which is flat over B^m. This gives a morphism $\operatorname{Spec} B^m \to \operatorname{Hilb}(Y/S)$, thus we obtain a map $s: Q \to B^m$. Since Hilb represents the Hilbert functor, s provides a splitting of the natural quotient map $B^m \to Q$. Apply (2.10.6) with $B^m = C$ to conclude that $J^m = 0$.

Therefore, by (2.10.5), we obtain that

$$\text{number of generators of } K = \dim K/(m_R K) \leq \dim \operatorname{Obs}(Z).$$

Thus $\operatorname{Hilb}(Y^S/S)$ is locally defined by $(\dim \operatorname{Obs}(Z))$-equations in $\operatorname{Spec} R$, and so (2.10.3) follows from (2.10.2).

Finally let $e = \dim \operatorname{Obs}(Z)$ and $g_1, \dots, g_e$ generators of K. Let g_i' be the image of g_i in $R \otimes_S k$. By assumption

$$\dim_{[Z]} \operatorname{Hilb}(Y^k) = \dim(R \otimes_S k)/(g_1', \dots, g_e') = \dim R \otimes_S k - e.$$

Thus $g_1', \dots, g_e'$ is a regular sequence. Applying (7.4.1.1) to the multiplication map $g_1 : R \to R$, we conclude that $R/(g_1)$ is flat over S. By the same argument applied to $g_2 : R/(g_1) \to R/(g_1)$, we see that $R/(g_1, g_2)$ is also flat over S, and so on. Finally, we obtain that $Q = R/(g_1, \dots, g_e)$ is flat over S. □

2.10.6 Lemma. *Let* $\mathcal{O}$ *be a local ring with maximal ideal* m_S *and* C *a local* $\mathcal{O}$*-algebra with maximal ideal* m_C. *Let* $J \subset (m_S C, m_C^2)$ *be an ideal and* q :

$C \to C/J$ the quotient. Assume that there is an $\mathcal{O}$-algebra homomorphism $s : C/J \to C$ such that the composite $C/J \to C \to C/J$ is an isomorphism. Then $J = 0$.

Proof. We can take the quotient of C by $m_C J$ while keeping the other assumptions valid. Thus we may also assume that $m_C J = 0$. Since $J \subset (m_S C, m_C^2)$, any $j \in J$ can be written as

$$j = a_0 c_0 + \sum_{i>0} a_i c_i \quad \text{where } a_0 \in m_S, c_0 \in C \text{ and } a_i, c_i \in m_C.$$

$T := s \circ q : C \to C$ is an $\mathcal{O}$-algebra homomorphism, thus $0 = T(j) = a_0 T(c_0) + \sum_{i>0} T(a_i)T(c_i)$. $c - T(c) \in J$ for any $c \in C$ and $m_C J = 0$. Thus $a_0 T(c_0) = a_0 c_0$ and $T(a_i)T(c_i) = a_i c_i$ for every i. Therefore $0 = T(j) = j$, hence $J = 0$. □

In view of the above results, it is of considerable interest to find the group $\operatorname{Obs}(Z)$ or at least to estimate its dimension from above.

2.11 Definition. Let Y^S/S be a scheme, flat over S. Let $s \in S$ be a point with residue field k and $Y = Y^k$ the fiber of Y^S over s. Let $Z \subset Y^k$ be a closed subscheme.

(2.11.1) We say that $Z \subset Y$ is *locally unobstructed* if there is an affine cover $\{U_i\}$ of Y^S such that there are no obstructions for deforming $Z \cap U_i \subset U_i$ for every i.

(2.11.2) We say that $Z \subset Y$ is *generically unobstructed* if Z has no embedded points and for every generic point z_{gen} of Z the deformations of $\operatorname{Spec} \mathcal{O}_{z_{gen}, Z}$ in $\operatorname{Spec} \mathcal{O}_{z_{gen}, Y^S}$ are unobstructed.

2.12 Lemma (Notation as above). *$Z \subset Y$ is locally unobstructed if any of the following three conditions is satisfied:*

(2.12.1) *$Z \subset Y$ is a local complete intersection.*

(2.12.2) *Y is smooth and Z is a CM subscheme of codimension two.*

(2.12.3) *Y is a smooth threefold and Z is purely one dimensional without embedded points (e.g. Z is reduced).*

2.13 Lemma (Notation as above). *$Z \subset Y$ is generically unobstructed if any of the following three conditions is satisfied:*

(2.13.1) *Z has no embedded points and Z is a local complete intersection in Y at every generic point of Z.*

(2.13.2) *Z is reduced and no irreducible component of Z is contained in* $\operatorname{Sing} Y$.

(2.13.3) *Z has no embedded points, no irreducible component of Z is contained in* $\operatorname{Sing} Y$ *and every irreducible component of Z has codimension at most two.*

Proof. (2.12.3) is a special case of (2.12.2). (2.13.1) follows from (2.12.1) and (2.13.2) is a special case of (2.13.1). (2.13.3) follows from (2.12.2). Thus we need to prove (2.12.1–2). The question is local so we may assume that Y^B is affine.

If $Z \subset Y$ is a complete intersection given by $I = I_Z = (g_1 = \cdots = g_n = 0)$, then consider the Koszul resolution (cf. [Matsumura80, 18.D])

$$\wedge^2(\mathcal{O}_Y^n) \xrightarrow{\wedge^2(g_1,\ldots,g_n)} \mathcal{O}_Y^n \xrightarrow{(g_1,\ldots,g_n)} I \to 0.$$

By (2.1.2) we can lift the equations to g_i^A such that $Z^A \subset Y^A$ is given by $(g_1^A = \cdots = g_k^A = 0)$ and, then I^A is resolved by

$$\wedge^2(\mathcal{O}_{Y^A}^n) \xrightarrow{\wedge^2(g_1^A,\ldots,g_n^A)} \mathcal{O}_{Y^A}^n \xrightarrow{(g_1^A,\ldots,g_n^A)} I^A \to 0.$$

If g_i^B are arbitrary liftings of g_i^A, then

$$\wedge^2(\mathcal{O}_{Y^B}^n) \xrightarrow{\wedge^2(g_1^B,\ldots,g_n^B)} \mathcal{O}_{Y^B}^n \xrightarrow{(g_1^B,\ldots,g_n^B)} I^A \to 0$$

is a complex, hence by (2.2) $[E^B(Z^A)] = 0$.

The proof of (2.12.2) is similar and we do not use it much. See [Artin76, §5] for details. □

2.14 Proposition (Assumptions as in (2.11)).

(2.14.1) $\mathrm{Obs}(Z) \subset \mathrm{Ext}^1_Z(I_Z/I_Z^2, \mathcal{O}_Z)$ *if* $Z \subset Y$ *is generically unobstructed.*

(2.14.2) $\mathrm{Obs}(Z) \subset H^1(Z, Hom_Z(I_Z/I_Z^2, \mathcal{O}_Z))$ *if* $Z \subset Y$ *is locally unobstructed.*

Proof. The precise meaning of the first assertion is the following. Consider the natural maps:

$$\text{(2.14.3)} \qquad \mathrm{Ext}^1_Z(I_Z/I_Z^2, \mathcal{O}_Z) \xrightarrow{j} \mathrm{Ext}^1_Y(I_Z/I_Z^2, \mathcal{O}_Z) \xrightarrow{ext} \mathrm{Ext}^1_Y(I_Z, \mathcal{O}_Z).$$

j takes an extension over $\mathcal{O}_Z$, then looks at it as an $\mathcal{O}_Y$-sheaf and ext comes from the map $I_Z \to I_Z/I_Z^2$. Then (2.14.1) asserts that $\mathrm{Obs}(Z) \subset \mathrm{im}(ext \circ j)$. One can see that $ext \circ j$ is an injection, but this is not important for us.

Assume that an extension

$$\text{(2.14.4)} \qquad 0 \to \mathcal{O}_Z \otimes J \xrightarrow{i} E \to I_Z \to 0$$

arises as an actual obstruction $E = E^B(Z^A)$. We would like to prove that this sequence comes from an extension over Z. The natural candidate to replace E is the $\mathcal{O}_Z$-sheaf $E/I_Z E$. The key step in the proof is:

2.14.5 Claim. $i(\mathcal{O}_Z \otimes J) \cap I_Z E = 0$.

Proof. We can view $i(\mathcal{O}_Z \otimes J) \cap I_Z E$ as a subsheaf of $\mathcal{O}_Z \otimes J$. Since $Z \subset Y$ is generically unobstructed, the obstruction is zero over the generic points of Z, therefore (2.14.4) is split at every generic point of Z. In particular, $i(\mathcal{O}_Z \otimes J) \cap I_Z E$ is the zero sheaf at every generic point of Z. Since Z has no embedded points, $i(\mathcal{O}_Z \otimes J) \cap I_Z E \subset \mathcal{O}_Z \otimes J$ is the zero sheaf everywhere. □

Thus we obtain an extension over Z

(2.14.6) $$0 \to \mathcal{O}_Z \otimes J \to E/I_Z E \to I_Z/I_Z^2 \to 0,$$

whose image under $ext \circ j$ is the sequence (2.14.4).

To see the second part, note first that if (2.14.4) is locally split, then the above argument constructs (2.14.6) without using the assumption that Z has no embedded points. In order to get an even smaller obstruction group, look at the exact sequence:

$$H^1(Z, Hom(I_Z/I_Z^2, \mathcal{O}_Z \otimes J)) \hookrightarrow \operatorname{Ext}^1_Z(I_Z/I_Z^2, \mathcal{O}_Z \otimes J) \xrightarrow{rest} H^0(Z, Ext^1_Z(I_Z/I_Z^2, \mathcal{O}_Z \otimes J)).$$

(This arises as the beginning of the local-global spectral sequence for Ext but it is easy to construct it by hand. We essentially did it during the proof of (2.5).)

The map *rest* is the restriction map from the global to the local obstructions. If the local obstructions vanish for an extension $E = E^B(Z^A)$, then $rest(E^B(Z^A)) = 0$, thus the obstruction lies in $H^1(Z, Hom(I_Z/I_Z^2, \mathcal{O}_Z \otimes J))$. □

The next result is a combination of (2.10) and (2.14). We change notation slightly and let Y denote what was formerly Y^S.

2.15 Theorem. *Let Y/S be a scheme, flat and projective over S and $s \in S$ a point. Let F be a field and p : $\operatorname{Spec} F \to s \in S$ a morphism. Let $Y^F = Y \times_S \operatorname{Spec} F$ be the fiber over* Spec *F and $Z \in Y^F$ a closed subscheme. Let I_Z denote the ideal sheaf of Z in Y^F. Let $\bar{p} : \operatorname{Spec} F \to \operatorname{Hilb}(X/S)$ be the corresponding morphism and set $[Z] = \operatorname{im} \bar{p}$. Assume that S is equidimensional at s. Then*

(2.15.1) *The dimension of every irreducible component of* Hilb(Y/S) *at $[Z]$ is at least*

$$\dim_F \operatorname{Hom}_{Y^F}(I_Z, \mathcal{O}_Z) - \dim_F \operatorname{Ext}^1_{Y^F}(I_Z, \mathcal{O}_Z) + \dim_s S.$$

(2.15.2) *If $Z \subset Y^F$ is generically unobstructed, then the dimension of every irreducible component of* Hilb(Y/S) *at $[Z]$ is at least*

$$\dim_F \operatorname{Hom}_Z(I_Z/I_Z^2, \mathcal{O}_Z) - \dim_F \operatorname{Ext}^1_Z(I_Z/I_Z^2, \mathcal{O}_Z) + \dim_s S.$$

(2.15.3) *If $Z \subset Y^F$ is locally unobstructed, then the dimension of every irreducible component of* Hilb(Y/S) *at $[Z]$ is at least*

$$\dim_F \operatorname{Hom}_Z(I_Z/I_Z^2, \mathcal{O}_Z) - \dim_F H^1(Z, Hom(I_Z/I_Z^2, \mathcal{O}_Z)) + \dim_s S.$$

(2.15.4) *If in any of the above cases* $\dim_{[Z]} \operatorname{Hilb}(Y^F)$ *is the smallest possible, then* $\operatorname{Hilb}(Y/S) \to S$ *is a local complete intersection morphism at* $[Z]$. *In particular, it is flat at* $[Z]$.

Proof. If $p : \operatorname{Spec} F \to s$ is an isomorphism, then this is a combination of (2.10) and (2.14). The general case is reduced to this one.

By (7.4.2) there is a faithfully flat morphism $q : S^F \to S$ such that $q^{-1}(s) = \operatorname{Spec} F$. As we remarked, (2.15) is valid for $\operatorname{Hilb}(Y \times_S S^F/S^F)$. Since

$$\operatorname{Hilb}(Y \times_S S^F/S^F) \cong \operatorname{Hilb}(Y/S) \times_S S^F,$$

this implies the result for Y/S. □

2.15.5 Exercises. (2.15.5.1) Let $Z \subset \mathbb{P}^n$ be a subvariety which is the complete intersection of hypersurfaces. Describe $\operatorname{Hilb}(\mathbb{P}^n)$ in a neighborhood of $[Z]$.

(2.15.5.2) Let F be a rank two vector bundle on a smooth curve, $F \to L$ a quotient line bundle and $C \subset \mathbb{P}(F)$ the corresponding section. Describe $\operatorname{Hilb}(\mathbb{P}(F))$ near $[C]$. Find some examples where $\operatorname{Hilb}(\mathbb{P}(F))$ is not a reduced scheme at $[C]$.

(2.15.5.3) Let $p \in X$ be a closed point of a scheme. Describe $\operatorname{Hilb}(X)$ near $[p]$. What is the corresponding obstruction group? When is $[p]$ locally unobstructed?

2.16 Theorem. *Let X and Y be projective schemes over a field k. Assume that X is without embedded points. Let $f : X \to Y$ be a morphism. Assume that Y has no embedded points contained in $f(X)$ and the image of every irreducible component of X intersects the smooth locus of Y. Then*

(2.16.1) *The tangent space of* $\operatorname{Hom}(X, Y)$ *at* $[f]$ *is naturally isomorphic to*

$$\operatorname{Hom}_X(f^*\Omega^1_Y, \mathcal{O}_X).$$

(2.16.2) *The dimension of every irreducible component of* $\operatorname{Hom}(X, Y)$ *at* $[f]$ *is at least*

$$\dim \operatorname{Hom}_X(f^*\Omega^1_Y, \mathcal{O}_X) - \dim \operatorname{Ext}^1_X(f^*\Omega^1_Y, \mathcal{O}_X).$$

Proof. Let $Z \subset X \times Y$ be the graph of f. By definition, $\operatorname{Hom}(X, Y)$ at $[f]$ and the Hilbert scheme of $X \times Y$ at $[Z]$ are locally isomorphic. The exact sequence

$$I_Z/I_Z^2 \to \Omega^1_{X \times Y} \otimes \mathcal{O}_Z \to \Omega^1_Z \to 0$$

splits by [Matsumura80, 26.I] since

$$\mathcal{O}_Z \cong \mathcal{O}_X \xrightarrow{(id_X, 1)} \mathcal{O}_X \otimes \mathcal{O}_Y \cong \mathcal{O}_{X \times Y}$$

provides a splitting of $\mathcal{O}_{X \times Y} \to \mathcal{O}_Z$.

Thus we have a commutative diagram

$$\begin{array}{ccccccccc} 0 & \longrightarrow & I_Z/I_Z^2 & \longrightarrow & \Omega^1_{X\times Y}\otimes\mathcal{O}_Z & \longrightarrow & \Omega^1_Z & \longrightarrow & 0 \\ & & \cong\downarrow & & \downarrow\cong & & \downarrow\cong & & \\ 0 & \longrightarrow & f^*\Omega^1_Y & \xrightarrow{(df,-id)} & \Omega^1_X + f^*\Omega^1_Y & \xrightarrow{(id,df)} & \Omega^1_X & \longrightarrow & 0 \end{array}$$

This provides a canonical isomorphism

$$I_Z/I_Z^2 \cong f^*\Omega^1_Y. \tag{2.16.3}$$

The assumptions imply that $Z \subset X \times Y$ is generically unobstructed, hence (2.16) follows from (2.8). □

2.16.4 Exercise. Let X be a projective scheme without embedded points over a field k. Show that the tangent space to $\operatorname{Aut}(X)$ at the identity is $\operatorname{Hom}(\Omega^1_X, \mathcal{O}_X)$.

By a theorem of Cartier, in characteristic zero, every group scheme is smooth (see e.g. [Mumford66, Chapt. 25]). Thus, in that case,

$$\dim \operatorname{Aut}(X) = \dim \operatorname{Hom}_k(\Omega^1_X, \mathcal{O}_X).$$

2.16.5 Exercise. Let X be a reduced variety with a closed point $x \in X$ and G an algebraic group acting on X. Show that there is a natural homomorphism $G \to \operatorname{Aut}(X)$. Using (2.16.4) we obtain natural maps

$$\phi_x : T_1G \to T_1 \operatorname{Aut}(X) \cong \operatorname{Hom}(\Omega^1_X, \mathcal{O}_X) \to T_xX.$$

$x \in X$ defines a G-equivariant morphism $m_x : G \to X$. Show that ϕ_x is the differential of m_x at $1 \in G$.

The relative setting, which is a straightforward consequence of (2.15) and (2.16), can be formulated as follows:

2.17 Theorem. *Let X/S and Y/S be schemes flat and projective over S. Let F be a field and $p : \operatorname{Spec} F \to S$ a morphism. Let X^F and Y^F be the fibers over* $\operatorname{Spec} F$ *and $f : X^F \to Y^F$ a morphism. Assume that X^F has no embedded points and that the image of every irreducible component of X^F intersects the smooth locus of Y^F.*

(2.17.1) *Assume that S is equidimensional at s. Then, the dimension of every irreducible component of* $\operatorname{Hom}_S(X,Y)$ *at $[f]$ is at least*

$$\dim_F \operatorname{Hom}(f^*\Omega^1_{Y^F}, \mathcal{O}_{X^F}) - \dim_F \operatorname{Ext}^1(f^*\Omega^1_{Y^F}, \mathcal{O}_{X^F}) + \dim_s S.$$

(2.17.2) *If*

$$\begin{aligned}\dim_{[f]} \operatorname{Hom}(X^F, Y^F) = {} & \dim_F \operatorname{Hom}(f^*\Omega^1_{Y^F}, \mathcal{O}_{X^F}) \\ & - \dim_F \operatorname{Ext}^1(f^*\Omega^1_{Y^F}, \mathcal{O}_{X^F}),\end{aligned}$$

then $\mathrm{Hom}_S(X,Y) \to S$ *is a local complete intersection morphism at* $[f]$. *In particular it is flat at* $[f]$. □

Finally, we consider the infinitesimal properties of the second projection morphism

$$\mathrm{Univ}(X/S) \hookrightarrow \mathrm{Hilb}(X/S) \times_S X \to X.$$

Since we understand only the relative differentials of $\mathrm{Hilb}(X/S) \to S$, we consider only the case when $S = \operatorname{Spec} k$ for a field k.

2.18 Notation. Let X be a scheme over a field k and $g : U \to H$ a flat family of closed subschemes of X. By definition, $U \subset H \times X$. Let $u : U \to X$ be the second projection, $z \in H$ be a point and $m \subset \mathcal{O}_H$ the ideal sheaf of $z \in H$. Let $Z \subset X$ be the subscheme corresponding to z, $I_Z \subset \mathcal{O}_X$ the ideal sheaf of $Z \subset X$ and $J_z \subset \mathcal{O}_U$ the ideal sheaf of $g^{-1}(z) \cong Z \subset U$.

By (2.5.3), the extensions of Z to $\mathrm{Spec}(\mathcal{O}_H/m^2) = \mathrm{Spec}(k(z) + m/m^2)$ are parametrized by $\mathrm{Hom}(I_Z/I_Z^2, \mathcal{O}_Z \otimes_{k(z)} m/m^2)$. Let $\phi_{z,H} \in \mathrm{Hom}(I_Z/I_Z^2, \mathcal{O}_Z \otimes_{k(z)} m/m^2)$ be the morphism corresponding to $f : U \to H$ at z.

We have a commutative diagram

$$\begin{array}{ccccccccc} 0 & \longrightarrow & u^*I_Z & \longrightarrow & u^*\mathcal{O}_X & \longrightarrow & \mathcal{O}_Z & \longrightarrow & 0 \\ & & \downarrow{\scriptstyle \Phi} & & \downarrow{\scriptstyle u^*} & & \downarrow{\scriptstyle \cong} & & \\ 0 & \longrightarrow & J_z & \longrightarrow & \mathcal{O}_U & \longrightarrow & \mathcal{O}_Z & \longrightarrow & 0. \end{array}$$

$\Phi : u^*I_Z \to J_z$ gives a map

$$\phi : I_Z/I_Z^2 \cong u^*(I_Z/I_Z^2) \to J_z/J_z^2 \cong \mathcal{O}_Z \otimes_{k(z)} m/m^2.$$

By (2.5.4), $\phi = \phi_{z,H}$.

Passing to differentials, we obtain the commutative diagram:

$$(2.18.1)\qquad \begin{array}{ccccccc} & & I_Z/I_Z^2 & \longrightarrow & u^*\Omega_X^1 \otimes \mathcal{O}_Z & \longrightarrow & \Omega_Z^1 & \longrightarrow & 0 \\ & & \downarrow{\scriptstyle \phi} & & \downarrow{\scriptstyle du} & & \downarrow{\scriptstyle \cong} & & \\ J_z/J_z^2 & = & \mathcal{O}_Z \otimes_{k(z)} m/m^2 & \longrightarrow & \Omega_U^1 \otimes \mathcal{O}_Z & \longrightarrow & \Omega_Z^1 & \longrightarrow & 0. \end{array}$$

Similar computations can be done for a family of morphisms. Let $F : X \times H \to Y$ be a morphism. Via the graph, we interpret it as a family of subschemes of $X \times Y$ parametrized by H. For $z \in H$, let $f : X \to Y$ be the corresponding morphism and $Z \subset X \times Y$ the graph of f. In this setting, the diagram (2.18.1) becomes:

$$(2.18.2)\qquad \begin{array}{ccccccccc} 0 & \longrightarrow & f^*\Omega_Y^1 & \longrightarrow & u^*\Omega_{X\times Y}^1 \otimes \mathcal{O}_Z & \longrightarrow & \Omega_Z^1 & \longrightarrow & 0 \\ & & \downarrow{\scriptstyle \phi} & & \downarrow{\scriptstyle du} & & \downarrow{\scriptstyle \cong} & & \\ 0 & \longrightarrow & \mathcal{O}_X \otimes_{k(z)} m/m^2 & \longrightarrow & \Omega_{X\times H}^1 \otimes \mathcal{O}_X & \longrightarrow & \Omega_X^1 & \longrightarrow & 0. \end{array}$$

The exactness of the top row was established during the proof of (2.16). The exactness of the bottom row follows from

$$\Omega_{X\times H}^1 \cong \pi_X^*\Omega_X^1 + \pi_H^*\Omega_H^1 \quad \text{and} \quad \mathcal{O}_X \otimes_{k(z)} m/m^2 \cong \mathcal{O}_X \otimes \pi_H^*\Omega_H^1.$$

F is represented by a morphism $H \to \mathrm{Hom}(X,Y)$. Taking the differential at z gives

$$\phi_{z,H} : (m/m^2)^* \to \mathrm{Hom}_X(f^*\Omega^1_Y, \mathcal{O}_X),$$

which can be considered as

$$\phi_{z,H} \in \mathrm{Hom}_X(f^*\Omega^1_Y, \mathcal{O}_X \otimes_{k(z)} m/m^2).$$

By the above results again, $\phi = \phi_{z,H}$.

2.19 Proposition (Notation as above). *Let $x \in Z \cong X$ be a smooth point on Z and X. The following diagram is commutative and has exact rows:*

$$\begin{array}{ccccccccc} 0 & \longrightarrow & I_Z/I_Z^2 \otimes_{k(z)} k(x) & \longrightarrow & u^*\Omega^1_X \otimes_{k(z)} k(x) & \longrightarrow & \Omega^1_Z \otimes_{k(z)} k(x) & \longrightarrow & 0 \\ & & \downarrow{\scriptstyle \phi\otimes_{k(z)}k(x)} & & \downarrow{\scriptstyle du(x)} & & \downarrow{\scriptstyle \cong} & & \\ 0 & \longrightarrow & m/m^2 \otimes_{k(z)} k(x) & \longrightarrow & \Omega^1_U \otimes_{k(z)} k(x) & \longrightarrow & \Omega^1_X \otimes_{k(z)} k(x) & \longrightarrow & 0. \end{array}$$

Proof. The above diagram is obtained from (2.18.1) by tensoring with $k(x)$. The top row is exact since X and Z are smooth at x. It remains to show that $p : m/m^2 \otimes_{k(z)} k(x) \to \Omega^1_U \otimes_{k(z)} k(x)$ is injective.

Let $W = U \times_H \mathrm{Spec}(k(z) + m/m^2)$. Then $\Omega^1_U \otimes_{k(z)} \mathcal{O}_Z \cong \Omega^1_W \otimes_{k(z)} \mathcal{O}_Z$. The sequence

$$0 \to m/m^2 \otimes_{k(z)} \mathcal{O}_X \to \mathcal{O}_W \to \mathcal{O}_Z \to 0$$

is split-exact in a neighborhood of x. From this sequence, it is easy to compute Ω^1_W and we obtain that p is injective in a neighborhood of x. □

Tensoring (2.18.2) with $k(x)$ for some $x \in X$ gives:

2.20 Proposition (Notation as above). *Assume that X is smooth at x. Then the map*

$$dF(x,z) : \Omega^1_Y \otimes_{k(F(x,z))} k(x,z) \to \Omega^1_{X\times H} \otimes k(x,z)$$

is injective if

$$\phi \otimes_{k(z)} k(x) : f^*\Omega^1_Y \otimes_{k(z)} k(x) \to m/m^2 \otimes_{k(z)} k(x)$$

is injective. □

I.3 Chow Varieties

The aim of this section is to construct Chow varieties. We start with the basic definitions involving algebraic cycles; some of these are also contained in [Fulton84]. Our aim is to study families of cycles over arbitrary schemes. The first step is to study cycles over spectra of fields. This case has been carefully worked out in the classical literature (see e.g. [Hodge-Pedoe52; Samuel55]). Also, it turns out that all the difficult aspects appear already at the level of fields. These matters are discussed in (3.3–9).

The definition of a "well defined family of algebraic cycles" is given in (3.10–11). If $f : X \to Y$ is a proper morphism all of whose fibers have dimension d for some d and Y is normal, then $f : X \to Y$ is automatically a well defined family of algebraic cycles by (3.17). If Y is not normal, the definition becomes somewhat more complicated. The basic properties of this notion are investigated in (3.12–19). One of the most important results is (3.19), which enables us to pull back families of cycles under certain mild conditions. These conditions are always satisfied in characteristic zero, but not in positive characteristic.

This enables us to define a Chow functor for schemes X/S in characteristic zero. The main theorem of the section is (3.21) which proves that the Chow functor is represented by a scheme $\mathrm{Chow}(X/S)$.

The proof has several steps. As in the case of the Hilbert scheme, we first consider the case of $\mathrm{Chow}(\mathbb{P}/S)$ where $\mathbb{P}$ is a projective space bundle over S. The heart of the proof is to understand the construction over the spectrum of a field k. If $Z \subset \mathbb{P}^n_k$ is a d-dimensional subvariety, then it is easy to see that the set of all $(d+1)$-tuples of hyperplanes

$$\{(H_0, \ldots, H_d) \quad \text{such that} \quad Z \cap H_0 \cap \ldots \cap H_d \neq \emptyset\}$$

is a divisor in $(\mathbb{P}^\vee)^{d+1}$ where $\mathbb{P}^\vee$ denotes the space of hyperplanes in $\mathbb{P}^n_k$. This construction provides a map

$$Ch : \{d\text{-dimensional cycles}\} \to \{\text{Cartier divisors in } (\mathbb{P}^\vee)^{d+1}\}.$$

(3.23.3) proves that this is a natural transformation of functors.

The inverse of Ch is constructed in (3.24), at least over fields. There are no problems in characteristic zero, but in positive characteristic the inverse does not always exist. Going to arbitrary base schemes is again rather easy. The remaining parts of the proof (3.25–27) are standard.

In (3.25), we first construct a scheme denoted by $\mathrm{Chow}'(\mathbb{P}/S)$ and $\mathrm{Chow}(\mathbb{P}/S)$ is the semi normalization of $\mathrm{Chow}'(\mathbb{P}/S)$. In characteristic zero, one can use $\mathrm{Chow}'(\mathbb{P}/S)$ to get a definition of the Chow variety which carries more scheme theoretic information than my definition [Barlet75].

(3.28) is a lengthy exercise which gives an explicit upper bound for the number of irreducible components of $\mathrm{Chow}(\mathbb{P}^n)$.

3.1 Definition. (3.1.1) Let X be a scheme of finite type over S. A *d-dimensional algebraic cycle on X/S* is a formal linear combination $\sum a_i[V_i]$, where the V_i are irreducible, reduced and closed subschemes of dimension d whose support is over a 0-dimensional subscheme of S, the a_i are integers and the sum is finite. A cycle is called *nonnegative* or *effective* if all the a_i are nonnegative. The set of d-dimensional cycles of a scheme X forms a group under addition which is denoted by $Z_d(X)$.

(3.1.2) Let $f : X \to Y$ be a morphism of schemes and $\sum a_i[V_i]$ an algebraic cycle on X. Let $W_i \subset Y$ be the closure of $f(V_i)$ with reduced scheme structure. Let $\deg(V_i/W_i)$ be the degree of $V_i \to W_i$ if the morphism is generically finite and 0 otherwise. Then

$$f_*(\sum a_i[V_i]) := \sum a_i \deg(V_i/W_i)[W_i]$$

is an algebraic cycle on Y. $f_* : Z_d(X) \to Z_d(Y)$ is called the *push forward* of cycles.

(3.1.3) Let V be a scheme of dimension d over an Artin ring, v_i the generic points of V and V_i the corresponding irreducible components of $\operatorname{red} V$. Let m_i be the length of the local ring $\mathcal{O}_{v_i,V}$ (this ring is Artin since v_i is a generic point). Then

$$[V] := \sum_{i:\dim V_i = d} m_i[V_i]$$

is called the *fundamental cycle* of V. If $V \subset X$ is a subscheme then we can view $[V]$ as an effective d-cycle on X.

(3.1.4) Let $f : X \to Y$ be a flat morphism of relative dimension n and $\sum a_i[V_i]$ an algebraic cycle on Y. Then

$$f^*(\sum a_i[V_i]) := \sum a_i[f^{-1}(V_i)]$$

is called the *flat pull back* of cycles. f^* increases dimension by n. If f is surjective, then $f^* : Z_d(Y) \to Z_{n+d}(X)$ is injective.

(3.1.5) Let $Z = \sum a_i[V_i]$ be an algebraic cycle of dimension d on a proper scheme Y and D a Cartier divisor or a line bundle on Y. Then

$$(Z \cdot D^d) := \sum a_i(V_i \cdot \underbrace{D \cdot \ldots \cdot D}_{d\text{-times}})$$

is called the *degree of Z* with respect to D. (See (VI.2) for a definition of the intersection product.) It is denoted by $\deg_D Z$.

(3.1.6) Let X be a scheme over a field k and $K \supset k$ a field extension. The scheme $X \times_{\operatorname{Spec} k} \operatorname{Spec} K$ is denoted by X_K. This notation is employed frequently for other objects as well under field extensions. There is a natural morphism $f : X_K \to X$ which is flat of relative dimension zero. If Z is a cycle on X, then f^*Z is denoted by Z_K. This defines an injection $Z_d(X) \to Z_d(X_K)$. It should be emphasized that here we go from k-schemes to K-schemes.

(3.1.7) Let Z be a d-cycle on X_K. We say that Z is *defined over a subfield* $k \subset F \subset K$ if there is a cycle $Z_F \in Z_d(X_F)$ such that $Z = (Z_F)_K$. Z_F is unique if it exists.

3.1.8 Lemma. *Push forwards and flat pull backs commute with base field extensions.*

Proof. For flat pull backs, this follows from the general fact that fiber products commute with base field extensions.

In order to consider push forwards, we use the notation of (3.1.2). By linearity, it is sufficient to consider the case when $f : V_k \to W_k$ is a morphism between irreducible and reduced schemes over k. The assertion is trivial if f is not generically finite. Let (R_j, m_j) be the local rings of the generic points of W_K and (R_{ij}, m_{ij}) the local rings of the generic points of V_K. We choose indices such that f_K maps $\operatorname{Spec} R_{ij}$ to $\operatorname{Spec} R_j$.

Let $k(V)$ (resp. $k(W)$) be the field of rational functions of V (resp. W) over k. Then $R_j \otimes_{k(W)} k(V) \cong \sum_i R_{ij}$, and by (7.1.1.4),

$$[k(V) : k(W)] \cdot l(R_j) = \sum_i [R_{ij}/m_{ij} : R_j/m_j] \cdot l(R_{ij}).$$

This exactly means that $(f_K)_*[V_K] = \deg(V/W)[W_K]$. □

The following result shows that both f^* and f_* commute with specialization.

3.2 Proposition. *Let X, Y be pure dimensional schemes, T the spectrum of a DVR and consider a commutative diagram*

$$\begin{array}{ccc} X & \xrightarrow{f} & Y \\ {\scriptstyle p}\downarrow & & \downarrow{\scriptstyle q} \\ T & = & T. \end{array}$$

(3.2.1) *If q and f are flat, then p is flat and $p^{-1}(0) = f^{-1}(q^{-1}(0))$. In particular*

$$(q \circ f)^*[0] = f^*(q^*[0]).$$

(3.2.2) *Assume that p is flat and f is proper. Let X_i be the irreducible components of X and $W_i := f(X_i)$ with reduced structure, $a_i = \deg(X_i/W_i)$. Then*

$$f_*[p^{-1}(0)] = \sum a_i[(q|W_i)^{-1}(0)].$$

Proof. (3.2.1) is obvious. In order to see (3.2.2), we may localize at a generic point of $q^{-1}(0) \cap f(X)$, thus we may assume that $\cup W_i$ is local and has relative dimension zero over T. We may replace Y with a suitable scheme whose support is $\cup W_i$. If $\dim X > \dim Y$, then every fiber of f has positive dimension, thus $a_i = 0$ for every i and $f_*[p^{-1}(0)] = 0$. Hence we may assume that f is finite.

Let Z be the spectrum of a DVR with structure morphism $r : Z \to T$ such that there is a finite morphism $h : Y \to Z$. Write $b_i = \deg(W_i/Z)$. Then (3.2.2) follows from the two equalities

$$h_*(f_*[p^{-1}(0)]) = \sum a_i b_i [r^{-1}(0)] = \sum a_i h_*[(q|W_i)^{-1}(0)].$$

These two equalities are just (3.2.2) applied to $h \circ f$ resp. $h|W_i$. Thus it is sufficient to prove (3.2.2) in case Y is the spectrum of a DVR with closed point $y \in Y$.

Let $f^{-1}(y) = \{x_1, \ldots, x_k\}$. Let $\mathcal{O}_{x_i}$ be the local ring of $x_i \in X$. Let t be a local parameter on T. The length of a module over $\mathcal{O}_Y$ (resp. $\mathcal{O}_{x_i}$) is denoted by l_y (resp. l_{x_i}). By definition, the coefficient of $[y]$ in $f_*[p^{-1}(0)]$ is

$$\sum_i l_{x_i}(\mathcal{O}_{x_i}/t\mathcal{O}_{x_i})[k(x_i) : k(y)] = l_y(\mathcal{O}_X/t\mathcal{O}_X),$$

and the coefficient of $[y]$ in $[q^{-1}(0)]$ is $l_y(\mathcal{O}_Y/t\mathcal{O}_Y)$.

$\mathcal{O}_X$ is finite over $\mathcal{O}_Y$ and $\mathcal{O}_Y$ is integral, thus there is a free submodule $M \subset \mathcal{O}_X$ such that $\mathcal{O}_X/M$ has finite length. Thus

$$\begin{aligned} l_y(\mathcal{O}_X/t\mathcal{O}_X) &= l_y(\mathcal{O}_X/tM) - l_y(t\mathcal{O}_X/tM) = l_y(\mathcal{O}_X/tM) - l_y(\mathcal{O}_X/M) \\ &= l_y(M/tM) = l_y(\mathcal{O}_Y/t\mathcal{O}_Y)\,\mathrm{rank}_{\mathcal{O}_Y} M = \deg f \cdot l_y(\mathcal{O}_Y/t\mathcal{O}_Y). \end{aligned}$$

□

As we saw in (1.16) and (1.19), a subscheme of a scheme always has a unique smallest field of definition. This is not quite true for cycles, but, as the next few results show, the only problem is with certain cycles in positive characteristic.

3.3 Proposition. *Let X be a scheme over a field k of characteristic p (set $p = 1$ for characteristic zero). Let $K \supset k$ be an algebraically closed field. Let $V_i \subset X_K$ be reduced, irreducible and distinct subvarieties and $Z = \sum a_i[V_i]$ a cycle over K. For every r let $I_r = \{i|a_i = r\}$ and $p^{\nu(r)}$ the largest p-power dividing r. Then $\sum a_i[V_i]$ is defined over a field $F \supset k$ iff $p^{\nu(r)} \sum_{i \in I_r}[V_i]$ is defined over F for every r.*

Proof. By definition,

$$Z = \sum_r r\Big(\sum_{i \in I_r}[V_i]\Big).$$

Therefore, if every $p^{\nu(r)} \sum_{i \in I_r}[V_i]$ is defined over F, then Z is also defined over F.

Conversely, assume that Z is defined over a field $F \subset K$. Write $Z_F = \sum b_j[P_j]$ where $P_j \subset X_F$ are irreducible and reduced over F. By (7.1.1),

$$(P_j)_K = p^{c_j}\sum_i [P_j^i] \quad \text{for some } c_j \geq 0,$$

where the P_j^i are irreducible and reduced over K. Let $d_j = p^{c_j}b_j$. We obtain that

$$p^{\nu(r)}\sum_{i\in I_r}[V_i] = \sum_{j:d_j=r} p^{\nu(b_j)}(P_j)_K,$$

hence $p^{\nu(r)}\sum_{i\in I_r}[V_i]$ is defined over F for every r. □

3.4 Proposition. *Let $V_i \subset X$ be distinct, irreducible, reduced subvarieties of dimension d and $\sum a_i[V_i] \in Z_d(X)$ a cycle. There is a unique subscheme $W \subset X$ such that every associated prime of W has dimension d and $[W] = \sum a_i[V_i]$ if any of the following conditions is satisfied:*

(3.4.1) $a_i = 1$ *for every* i;

(3.4.2) X *is nonsingular in codimension one and has pure dimension* $d+1$.

Proof. In the first case, $W = \cup V_i$ is the only choice. In the second case W is the d-dimensional part of $\prod I_{V_i}^{a_i}$ where I_{V_i} is the ideal sheaf of V_i. □

3.5 Proposition (Notation as in (3.3)). *There is a unique smallest field of definition of Z if any of the following conditions is satisfied:*

(3.5.1) $\operatorname{char} k = 0$;

(3.5.2) $(p, a_i) = 1$ *for every* i;

(3.5.3) X *is regular in codimension one and Z has pure codimension one in X.*

Proof. Let $Z = \sum a_i[V_i]$ be a cycle over K. Assume that there is a unique pure dimensional subscheme $W_K \subset X_K$ without embedded points such that $[W] = Z$. Then the smallest field of definition of W_K (cf. (1.16), (1.19)) is the smallest field of definition of Z.

Thus (3.4.2) implies (3.5.3). The first two claims follow from (3.3) and (3.4.1). □

3.6 Corollary. *Let $Z = \sum a_i[V_i]$ be a cycle defined over a field K and $G \subset \operatorname{Aut}(K/k)$ a subgroup. If $g(Z) = Z$ for every $g \in G$, then Z is defined over K^G.*

Proof. The assumptions imply that $\cup_{i\in I_r} V_i$ is G-invariant, hence defined over K^G by (1.17). Thus Z is defined over K^G. □

3.7 Proposition. *Let X_k be a scheme, L and K two field extensions of k. Let $Z_L \in Z_d(X_L)$ and $Z_K \in Z_d(X_K)$ be two cycles. The following two statements are equivalent*

(3.7.1) *For every field F and for every embedding $i_K : K \to F$ and $i_L : L \to F$,*

$$Z_L \times_{\operatorname{Spec} L} \operatorname{Spec} F = Z_K \times_{\operatorname{Spec} K} \operatorname{Spec} F.$$

(3.7.2) *There is a purely inseparable field extension $k' \supset k$ and (necessarily unique) embeddings $k' \to L^{p^{-\infty}}$ and $k' \to K^{p^{-\infty}}$ such that $Z_{L^{p^{-\infty}}}$ and $Z_{K^{p^{-\infty}}}$ are both defined over k' by the same $Z_{k'} \in Z_d(X_{k'})$.*

Proof. Clearly, (3.7.2) implies (3.7.1).

Conversely, choose F to be algebraically closed and let $g \in \operatorname{Aut}(F/k)$. By assumption, $g(Z_L)_F = Z_K$ for every g. Hence by (3.6) Z_L is defined over $k^{p^{-\infty}}$. The same holds for Z_K which shows the result. □

3.8 Definition. If any of the above two conditions are satisfied, we say that Z_L and Z_K are *essentially the same cycles.* This relationship is denoted by

$$Z_L \overset{ess}{=\!=} Z_K.$$

If Z_λ is a family of cycles and any two are essentially the same, we say that Z_λ is *essentially independent* of λ.

3.9 Definition. Let $g_i : U_i \to W$ be a proper morphism of schemes over W. Assume that W is reduced and U_i is irreducible. By generic flatness (which is a special case of (1.6)) there is an open subset $W_i \subset g_i(U_i) \subset W$ such that g_i is flat of relative dimension d over W_i. Let T be the spectrum of a DVR and $h : T \to W$ a morphism such that $h(T_g) \in W_i$ and $h(T_0) = w \in W$. Let $h^*U_i = U_i \times_h T$ and $J \subset \mathcal{O}_{h^*U_i}$ the ideal of those sections whose support is contained in the special fiber of $h^*U_i \to T$. Let

$$(U_i)'_T := \operatorname{Spec} \mathcal{O}_{h^*U_i}/J.$$

By construction $(U_i)'_T \to T$ is flat. Therefore the central fiber Z_0 of $(U_i)'_T \to T$ has pure dimension d. Let $[Z_0]$ be the fundamental cycle of the central fiber of $(U_i)'_T \to T$ and set

$$\lim_{h \to w} (U_i/W) := [Z_0] \in Z_d(g_i^{-1}(w) \times_w T_0).$$

$\lim_{h\to w}(U_i/W)$ is called the *cycle theoretic fiber* of g_i at w along h.

It is not completely clear how to define the notion of a "family of algebraic cycles". The following definition is the right one if W is defined over a field of characteristic zero and it is normal. In positive characteristic, there are several possible definitions but there does not seem to be any with all the desired properties. We discuss the problems in section four. For reduced varieties over $\mathbb{C}$, [Barlet75] gave a definition which is more refined than (3.10–11). It would be interesting to find a definition which allows infinitesimal computations. Unfortunately, it is not clear that this is possible.

3.10 Definition. A *well defined family of d-dimensional proper algebraic cycles over W* is a pair $(g : U \to W)$ satisfying the following properties:

(3.10.1) There is a reduced scheme $\operatorname{Supp} U$ with irreducible components U_i such that $U = \sum m_i[U_i]$ is an algebraic cycle. (We tacitly assume most of the time that $m_i \neq 0$ for every i. It is, however, very convenient to allow $U = \emptyset$.)

(3.10.2) W is a reduced scheme and $g : \operatorname{Supp} U \to W$ is a proper morphism.

(3.10.3) Let $g_i := g|U_i$. Then every g_i maps onto an irreducible component of W and every fiber of g_i is either empty or has dimension d. In particular there is a dense open subset $W_0 \subset W$ such that every g_i is flat over W_0.

(3.10.4) For every $w \in W$ there is a cycle

$$g^{[-1]}(w) \in Z_d(g^{-1}(w) \times_w \operatorname{Spec} k(w)^{p^{-\infty}}), \quad (\text{where } p = \operatorname{char} k(w))$$

with the following property:

For every $h : T \to W$ such that $h(T_0) = w$ and $h(T_g) \in W_0$ we have

$$g^{[-1]}(w) \stackrel{ess}{=\!=} \sum m_i \lim_{h \to w} (U_i/W).$$

$g^{[-1]}(w)$ is called the *cycle theoretic fiber* of $(g : U \to W)$ at w.

3.10.5 Comment. Conditions (3.10.1–3) are geometrically clear but (3.10.4) seems rather complicated. It is, however, a very weak condition whose main purpose is to guarantee that if two components of W meet at a point $w \in W$, then the two limit cycles that one gets from the two components agree. We prove in (3.17) that if W is normal, then the first three conditions imply (3.10.4).

3.10.6 Naive Version of the Definition. The following is a simpleminded version of the definition of "families of algebraic cycles".

Let X and W be varieties. A family of algebraic cycles in X pararametrized by W is a function

$$G : \{\text{points of } W\} \to \{\text{algebraic cycles of } X\}$$

which has the following properties:

(i) $G(w)$ depends "algebraically" on w, and

(ii) intersection numbers with Cartier divisors on X are constant in the family.

The first requirement could reasonably be interpreted to mean:

(i′) there is a variety $U \subset W \times X$ and a morphism $g : U \to W$ such that $\operatorname{Supp} G(w) = \operatorname{Supp} g^{-1}(w) \subset X$.

One can extend the second condition to:

(ii′) intersection numbers with Cartier divisors on U are constant in the family, and this holds after any base change $W' \to W$.

Even for families of 1-cycles, this definition does not specify how to distribute multiplicities between numerically equivalent cycles. (3.10) is a way of putting the ideas behind this naive version into precise form.

3.11 Definition. Let X be a scheme over S. A *well defined family of proper algebraic cycles of X/S over W/S* is a pair $(g : U/S \to W/S)$ satisfying the following properties:

(3.11.1) $\operatorname{Supp} U$ is a closed subscheme of $X \times_S W$ and g is the natural projection morphism.

(3.11.2) $(g : U \to W)$ is a well defined family of d-dimensional proper algebraic cycles over W for some d.

It is clear from the definition that $(g : U/S \to W/S)$ is determined by its fibers over the generic points of W.

It would be more natural to assume that $g^{[-1]}(w) \in Z_d(g^{-1}(w))$. Indeed, only this stronger restriction ensures that the definition has good functorial properties. On the other hand, there are examples where the stronger assumption is not satisfied in nice situations. We come back to the problem of changing the condition (3.10.4) in the next section.

3.12 Proposition. *Let $(g : U \to V)$ be a well defined family of proper algebraic cycles of X and D a Cartier divisor on X. Assume that V is connected. Then $\deg_D g^{[-1]}(v)$ is independent of $v \in V$.*

Proof. Let v_g be a generic point of V and v_0 a specialization. Let $h : T \to V$ be a morphism from the spectrum of a DVR such that $h(T_g) = v_g$ and $h(T_0) = v_0$. Let $(U_i)'_T$ be as in (3.9). Then every $p_i : (U_i)'_T \to T$ is flat over T hence

$$\deg_D p_i^{-1}(T_0) = \deg_D p_i^{-1}(T_g).$$

Summing over i gives that

$$\deg_D g^{[-1]}(v_0) = \deg_D g^{[-1]}(v_g). \qquad \square$$

The following result shows that the above definitions do not depend much on the scheme structure of W.

3.13 Proposition. *Let X/S be a scheme proper over S. Let W/S be a reduced scheme and W^{wn}/S its weak normalization (cf. (7.2)). There is a one-to-one correspondence*

$$\left\{\begin{matrix}\textit{Well defined families of}\\ \textit{algebraic cycles of } X \textit{ over } W\end{matrix}\right\} \leftrightarrow \left\{\begin{matrix}\textit{Well defined families of}\\ \textit{algebraic cycles of } X \textit{ over } W^{wn}\end{matrix}\right\}.$$

Proof. Let $(g^{wn} : U^{wn} \to W^{wn})$ be a well defined family of algebraic cycles of X. Let $U_i \subset X \times_S W$ be the image of U_i^{wn} under the natural map $U_i^{wn} \to X \times_S W$. Conversely, if $(g : U \to W)$ is a well defined family of algebraic cycles of X, then set $U_i^{wn} = \operatorname{red}(U_i \times_W W^{wn})$. We set in both cases $m_i^{wn} = m_i$. It is clear that this gives the required correspondence. □

The following useful technical result is a "descent theorem" for well defined families of algebraic cycles.

3.14 Lemma. *Assume that $(g : U \to W)$ satisfies the conditions* (3.10.1–3). *Let $\tilde{U}$ be a scheme over W such that* $\operatorname{Supp} U = \operatorname{red} \tilde{U}$. *Let $p : W' \to W$ be a proper, surjective morphism and $U_j' \subset \tilde{U} \times_W W'$ subschemes, flat over W' for every j. We have commutative diagrams*

$$(3.14.1) \qquad \begin{array}{ccc} U_j' & \longrightarrow & \tilde{U} \\ {\scriptstyle g_j'}\downarrow & & \downarrow{\scriptstyle \tilde{g}} \\ W' & \xrightarrow{\ p\ } & W. \end{array}$$

Assume that there are numbers m_j' such that:

(3.14.2) *There is a dense open set $W_0 \subset W$ such that g is flat over W_0 and for every $w \in W_0$ and every $w' \in p^{-1}(w)$*

$$\sum_j m_j'[g_j'^{-1}(w')] \overset{ess}{=\!=} \sum_i m_i[g_i^{-1}(w)].$$

(3.14.3) *For every $w \in W$ and every $w_1', w_2' \in p^{-1}(w)$*

$$\sum_j m_j'[g_j'^{-1}(w_1')] \overset{ess}{=\!=} \sum_j m_j'[g_j'^{-1}(w_2')].$$

Then $(g : U \to W)$ also satisfies the condition (3.10.4).

Conversely, if $(g : U \to W)$ is a well defined family of proper algebraic cycles, then there are diagrams (3.14.1) *satisfying the conditions* (3.14.2–3).

Proof. We use the notation of (3.9). Given $h : T \to W$ we can lift it to $h' : T' \to W'$ where T' is another DVR dominating T. Let $w' = h'(T_0')$. Then $U_j' \times_{h'} T'$ and $(U_i)_{T'}'$ are all subschemes of $(h')^*\tilde{U}$, flat over T'. Let $V_k \subset \operatorname{red} h'^*\tilde{U}$ be those irreducible components that dominate T'. We can write

$$\sum_j m_j'[U_j' \times_{h'} T'] = \sum_k c_k V_k \quad \text{and} \quad \sum_i m_i[(U_i)_{T'}'] = \sum_k d_k V_k,$$

for some integers c_k and d_k. By (3.14.2), the above two cycles agree over the generic point of T', hence $c_k = d_k$ for every k. Therefore

$$\left(\lim_{h\to w}(U/W)\right)_{k(w')} = \sum_j m'_j[U'_j \times_{h'} T'_0] = \sum_j m'_j[g'^{-1}_j(w')]$$

is essentially independent of $w' \in p^{-1}(w)$ by (3.14.3).

The converse is established by the following:

3.14.4 Construction. Assume that $(g : U \to W)$ satisfies the conditions (3.10.1–3). Let $W_0 \subset W$ be a dense open set over which g_i is flat for every i. Define $g_i^{-1} : W_0 \to \mathrm{Hilb}(\operatorname{Supp} U/W) \cup W$ (disjoint union) by

$$g_i^{-1}(w) = \begin{cases} g_i^{-1}(w) \in \mathrm{Hilb}(\operatorname{Supp} U/W) & \text{if } g_i^{-1}(w) \neq \emptyset \\ w \in W & \text{if } g_i^{-1}(w) = \emptyset. \end{cases}$$

Thus we get a section

$$\begin{aligned} s : W_0 &\to (\mathrm{Hilb}(\operatorname{Supp} U/W) \cup W) \times_W \dots \times_W (\mathrm{Hilb}(\operatorname{Supp} U/W) \cup W) \\ w &\mapsto (g_1^{-1}(w), \dots, g_r^{-1}(w)). \end{aligned}$$

Let W' be the closure of $s(W_0)$, $g'_i : U'_i \to W'$ the universal families and $g' : U' \to W'$ their union. (In this construction the U'_j are in one-to-one correspondence with the irreducible components of $\operatorname{Supp} U$, but in general this is not necessary.)

By construction (3.14.2) is satisfied and the argument at the beginning of the proof shows that $(g : U \to W)$ is a well defined family of algebraic cycles iff the condition (3.14.3) is satisfied. □

3.14.5 Complement. *Let X/S be a scheme and $(f : U \to W)$ a well defined family of algebraic cycles on X. Then the data* (3.14.1) *and the condition* (3.14.2) *uniquely determine $(f : U \to W)$.*

Proof. (3.14.2) determines the generic fibers of f in terms of the data (3.14.1) and by (3.11) the generic fibers determine $(f : U \to W)$. □

3.15 Corollary. *Assume that $g : U \to W$ is proper and flat of relative dimension d and W is reduced. Let $\sum m_i[U_i]$ be the fundamental cycle of U. Then $(g : [U] \to W)$ is a well defined family of algebraic cycles over W.*

Proof. Apply (3.14) with $\tilde{U} = U, W' = W, U'_1 = U, m'_1 = 1$. □

3.16 Corollary. *Let $(g : U \to W)$ be a well defined family of algebraic cycles over W. Let T be the spectrum of a DVR and $h : T \to W$ a morphism. Then*

$$\operatorname{Supp} g^{-1}(w) = \operatorname{Supp} g^{[-1]}(w) = \cup_i \operatorname{Supp} \lim_{h\to w}(U_i/W).$$

Proof. Let $p : W' \to W$ be as in (3.14.4). Given $h : T \to W$, h can be lifted to $h' : T' \to W'$ where T' is another DVR dominating T. Let $w' = h'(T'_0)$. Then

$$\cup \operatorname{Supp} \lim_{h \to w} (U_i/W) = \cup \operatorname{Supp} \lim_{h' \to w'} (U'_i/W') =$$
$$= \operatorname{Supp} g'^{-1}(w') = \operatorname{Supp} g^{[-1]}(w).$$

In order to see the first equality, let $k = \dim_w W$ and let $h_1, \dots, h_{k-1}$ be general elements of the ideal of $w \in W$. Then $h_1 = \dots = h_{k-1} = 0$ defines a curve $w \in C \subset W$. $U_C = (g^* h_1 = \dots = g^* h_{k-1} = 0)$ is a subset of U of pure dimension $d + 1$ which contains $\operatorname{Supp} g^{-1}(w)$. Thus every component of $\operatorname{Supp} g^{-1}(w)$ can be obtained as the specialization of a generic fiber of $U_C \to C$. Since the h_i are general, every generic point of C is contained in W_0. Thus by definition, the specialization of a generic fiber of $U_C \to C$ is independent of the generic fiber chosen. Thus $\operatorname{Supp} g^{-1}(w) \subset \operatorname{Supp} g^{[-1]}(w)$. □

3.17 Theorem. *Assume that* $(g : U \to W)$ *satisfies the conditions* (3.10.1–3) *and* W *is normal. Then* $(g : U \to W)$ *is a well defined family of algebraic cycles over* W.

Proof. We construct $g' : U' \to W'$ as in (3.14.4). We want to apply (3.14). By construction (3.14.2) is satisfied. In order to show (3.14.3), let $Y = g^{-1}(w) \subset U$ be the scheme theoretic fiber of g.

$$\sum m_i [U'_i \times_{W'} p^{-1}(w)] \to p^{-1}(w)$$

is a well defined family of d-dimensional cycles of Y and $\dim Y = d$. The following result is obvious:

3.17.1 Lemma. *Let* $(g : U \to V)$ *be a well defined family of d-dimensional algebraic cycles of a d-dimensional scheme* Y. *Assume that* V *is geometrically connected. Then* $g^{[-1]}(v)$ *is essentially independent of* $v \in V$. □

Since W is normal, $p^{-1}(w)$ is connected. Thus (3.17.1) implies (3.14.3). □

3.18 Definition. Let $(g : U \to W)$ be a well defined family of algebraic cycles over W. Let $f : W^1 \to W$ be a morphism. Assume that W^1 is reduced and that

(3.18.1)
$g^{[-1]}(f(w^1))$ is defined over $k(f(w^1))$ for every generic point $w^1 \in W^1$.

(E.g. W is a scheme over a field of characteristic zero.)

Then one can define a well defined family of algebraic cycles over W^1 as follows.

Let U^1_j be the irreducible components of $\operatorname{red}(\operatorname{Supp} U \times_W W^1)$. Let w^1 be a generic point of W^1 and U^1_{j,w^1} the fiber of U^1_j over w^1. Set $w = f(w^1)$. Let $\sum a_i[V_i]$ be the cycle theoretic fiber of g at w and $[w^1 \times_w V_i] = \sum_j b_{i,j}[U^1_{j,w^1}]$. Set

$$m^1_j = m^1(U^1_j) := \sum_i a_i b_{i,j} \quad \text{and} \quad U^1 := \sum_j m^1_j U^1_j.$$

$(g^1 : U^1 \to W^1)$ is called the *Chow pull-back* of g and is denoted by

$$(f^{ch}g : f^{ch}U \to W^1).$$

3.19 Proposition (Notation as above).

(3.19.1) *$(f^{ch}g : f^{ch}U \to W^1)$ is a well defined family of algebraic cycles.*

(3.19.2) *Let $h : W^2 \to W^1$ be another morphism. Assume that $g^{[-1]}((f \circ h)(w^2))$ is defined over $k((f \circ h)(w^2))$ for every generic point $w^2 \in W^2$. Then*

$$((f \circ h)^{ch}g : (f \circ h)^{ch}U \to W^2) \cong (h^{ch}(f^{ch}g) : h^{ch}(f^{ch}U) \to W^2)\,.$$

Proof. In order to see the first part we need to show that $(f^{ch}g : f^{ch}U \to W^1)$ satisfies the conditions (3.10.1–4). The first three conditions are clearly satisfied. Let $p : W' \to W$ and $g'_i : U'_i \to W'$ be as in (3.14.4). Set

$$W'^1 = W^1 \times_W W', \; \tilde{U}^1 = W^1 \times_W U \quad \text{and} \quad U'^1_i = W^1 \times_U U'_i.$$

The commutative diagram

$$(3.19.3) \qquad \begin{array}{ccc} U'^1_i & \longrightarrow & \tilde{U}^1 \\ {\scriptstyle g'^1_i}\big\downarrow & & \big\downarrow{\scriptstyle \tilde{g}^1} \\ W'^1 & \xrightarrow{\;p\;} & W^1 \end{array}$$

has the same properties as the diagram (3.14.1). Thus $(f^{ch}g : f^{ch}U \to W^1)$ also satisfies the condition (3.10.4) by (3.14). Similarly one can construct

$$(3.19.4) \qquad \begin{array}{ccc} U'^2_i & \longrightarrow & \tilde{U}^2 \\ {\scriptstyle g'^2_i}\big\downarrow & & \big\downarrow{\scriptstyle \tilde{g}^2} \\ W'^2 & \xrightarrow{\;p\;} & W^2 \end{array}$$

and it does not matter whether we construct (3.19.4) directly using $f \circ h$ or by pulling back the diagram (3.19.3) via h. Therefore (3.14.5) implies (3.19.2). □

3.20 Definition (Characteristic zero only). Let X be a scheme over S. Let

$$Chow(X/S)(Z) = \left\{ \begin{matrix} \text{Well defined families of nonnegative,} \\ \text{proper, algebraic cycles of } X \times_S Z/Z \end{matrix} \right\}.$$

If in addition $\mathcal{O}(1)$ is relatively ample on X, then let

$$Chow_{d,d'}(X/S)(Z) = \left\{ \begin{matrix} \text{Well defined families of nonnegative,} \\ \text{proper, algebraic cycles of dimension} \\ d \text{ and degree } d' \text{ of } X \times_S Z/Z \end{matrix} \right\}.$$

If Z is connected, then $Chow(X/S)(Z) = \cup_{d,d'} Chow_{d,d'}(X/S)(Z)$, and the union is disjoint.

The following is the main theorem of this section.

3.21 Theorem. *Let X/S be a scheme, projective over S and $\mathcal{O}(1)$ relatively ample. Assume that S is over a field of characteristic zero. Then:*

(3.21.1) *The Chow pull back makes $Chow_{d,d'}(X/S)$ into a functor*

$$\{\textit{semi normal } S\textit{-schemes}\} \to \{\textit{sets}\}.$$

(3.21.2) *$Chow_{d,d'}(X/S)$ is represented by a well defined family of nonnegative, proper, algebraic cycles of dimension d and degree d':*

$$(u : \mathrm{Univ}_{d,d'}(X/S) \to \mathrm{Chow}_{d,d'}(X/S)).$$

(3.21.3) $\mathrm{Chow}_{d,d'}(X/S)$ *is semi normal and projective over S.*

3.21.4 Remark. The characteristic zero case has the following two crucial advantages:

(3.21.4.1) Every cycle has a smallest field of definition (3.5).

(3.21.4.2) The notions "semi normal" and "weakly normal" coincide (7.2).

Both of these fail in positive characteristic, and (3.21) needs to be replaced by weaker variants. Even their formulation requires some lengthy definitions which can be given more naturally during the proof. Therefore we start the proof without assuming that the characteristic is zero and point out the difficulties when they arise. The results in positive characteristic are formulated and proved in the next section.

3.22 Exercise. Let X be a semi normal variety. Show that $\mathrm{Chow}_{0,m}(X) = S^m X$.

Proof of (3.21). (3.19) shows that $Chow_{d,d'}(X/S)$ is a functor.

In order to show representability, as in the case of *Hilb*, we first consider the special case when $X/S \cong \mathbb{P}/S$ is the projectivization of a rank $n+1$ vector bundle E.

3.23 Step 1. Construction of $Ch : Chow_{d,d'}(\mathbb{P}/S) \to CDiv$.

First we construct a natural transformation from $Chow_{d,d'}(\mathbb{P}/S)$ to a Chow functor parametrizing hypersurfaces in a smooth variety. Then we study Chow functors of hypersurfaces in (3.23.2). The results are summarized in (3.23.3). Several questions involving Chow forms over a field are left to the next step.

Let $\mathbb{P}^\vee$ be the dual projective space bundle and

$$I := \{(x, H_1, \ldots, H_{d+1}) \subset \mathbb{P} \times_S (\mathbb{P}^\vee)^{d+1} | x \in H_1 \cap \ldots \cap H_{d+1}\}$$

the *incidence correspondence*. Let $p_1 : I \to \mathbb{P}$ and $p_2 : I \to (\mathbb{P}^\vee)^{d+1}$ be the projections. p_1 is smooth and $p_1^{-1}(x) = (L_x)^{d+1}$, where $L_x \subset \mathbb{P}^\vee$ is the set of hyperplanes containing x.

Given $s : W \to S$ and $(g : U \to W)$, we get a diagram

$$\begin{array}{ccc} U \subset s^*\mathbb{P} & \xleftarrow{s^*p_1} & s^*I \\ & & \downarrow{\scriptstyle s^*p_2} \\ & & s^*(\mathbb{P}^\vee)^{d+1}. \end{array}$$

Set

$$[Ch(U)] := (s^*p_2)_*((s^*p_1)^*U) \quad \text{and} \quad Ch(U)_i := \operatorname{red}(s^*p_2)_*((s^*p_1)^*U_i).$$

(Note that in general $[Ch(U)_i] \neq [Ch(U_i)]$, though they agree in characteristic zero.) Let $Ch(g) : \operatorname{Supp}[Ch(U)] \to W$ be the natural projection.

3.23.1 Main Lemma. *Let $(g : U \to W)$ be a well defined family of algebraic cycles of $\mathbb{P}/S$. Then*

(3.23.1.1) *$(Ch(g) : [Ch(U)] \to W)$ is a well defined family of hypersurfaces in $(\mathbb{P}^\vee)^{d+1}$.*

(3.23.1.2) *$[Ch(U)] = \sum m_i k_i [Ch(U)_i]$ for some integers k_i. If every geometric generic fiber of $g|\operatorname{Supp} U$ is reduced, then $k_i = 1$ for every i. In particular, this holds if the characteristic is zero. (In general k_i is a power of the characteristic.)*

(3.23.1.3) *Let $f : W^1 \to W$ be a morphism. Then*

$$(Ch(f^{ch}g) : [Ch(f^{ch}U)] \to W^1) \cong (f^{ch}Ch(g) : f^{ch}[Ch(U)] \to W^1).$$

(3.23.1.4) *There is a unique subscheme $Ch(U) \subset (\mathbb{P}^\vee)^{d+1}$ such that $Ch(U)$ is purely one codimensional, has no embedded points and its fundamental cycle is $[Ch(U)]$.*

Proof. First we check that $(Ch(g) : [Ch(U)] \to W)$ satisfies the conditions (3.10.1–3). Since s^*p_1 is smooth, $(s^*p_1)^{-1}U_i \subset s^*I$ is reduced, irreducible and has dimension $d + (d+1)(n-1)$. For every $w \in W$ and $x \in U_{i,w}$, we can find $(H_1, \ldots, H_{d+1}) \in (\mathbb{P}^\vee_w)^{d+1}$ such that

(3.23.1.5) $$U_{i,w} \cap H_1 \cap \ldots \cap H_{d+1} = \{x\} \quad \text{(as sets)}.$$

Observe that by construction,

$$U_{i,w} \cap H_1 \cap \ldots \cap H_{d+1} = (s^*p_2)^{-1}(H_1, \ldots, H_{d+1}) \cap (s^*p_1)^{-1}(U_{i,w}).$$

In particular, $s^*p_2 : (s^*p_1)^{-1}(U_{i,w}) \to Ch(U_{i,w})$ is generically injective, thus

$$\dim Ch(U_{i,w}) = d + (d+1)(n-1) = (d+1)n - 1.$$

Therefore every $Ch(U_{i,w})$ is a hypersurface in $(\mathbb{P}^\vee_w)^{d+1}$.

Set $k_i = \deg((s^*p_1)^*U_i/Ch(U)_i)$. $s^*p_2 : (s^*p_1)^*(U_i) \to Ch(U)_i$ is a generically injective morphism between reduced schemes, thus k_i is a power of the characteristic of the residue field of the generic point of U_i. If g has reduced geometric generic fibers, then the equality in (3.23.1.5) also holds scheme theoretically, hence again $k_i = 1$. This proves (3.23.1.2).

Comparing with the definition of push forward of cycles (3.1.2), we see that

$$\sum m_i k_i [Ch(U)_i] = (s^*p_2)_*(s^*p_1)^*\Big(\sum_i m_i[U_i]\Big) = [Ch(U)]. \tag{3.23.1.6}$$

This shows (3.23.1.2).

Let T be the spectrum of a DVR with closed point 0, generic point t_g, and $h : T \to W$ a morphism. As in (3.9), we have flat morphisms $h_i : (U_i)'_T \to T$ such that $\lim_{h\to w}(U/W) \overset{ess}{=\!=} \sum m_i[h_i^{-1}(0)]$. We prove in (3.24.2) that this implies that

$$Ch(\lim_{h\to w}(U/W)) \overset{ess}{=\!=} \sum m_i Ch([h_i^{-1}(0)]).$$

Furthermore, by (3.2), we see that

$$\sum m_i Ch([h_i^{-1}(0)]) = \sum m_i([Ch(h_i)^{-1}(0)]).$$

$Ch(h_i) : [Ch((U_i)'_T)] \to T$ is flat by construction and its generic fiber is $Ch(h^*U_i|T_g) = h^*(Ch(U_i))|t_g$. Thus $[Ch((U_i)'_T)] \to T$ can be used to compute $\lim_{h\to w} Ch(U_i/W)$. Therefore,

$$Ch(\lim_{h\to w}(U/W)) \overset{ess}{=\!=} \lim_{h\to w} Ch(U/W), \tag{3.23.1.7}$$

hence $\lim_{h\to w} Ch(U/W)$ is essentially independent of h. Therefore $Ch(U)$ satisfies the condition (3.10.4), proving (3.23.1.1).

In order to see (3.23.1.3), it is sufficient to show that the two families are isomorphic over the generic points of W^1. This reduces the problem to the special case when $W^1 = \operatorname{Spec} F$ for some field F. In this case, the isomorphism follows from (3.1.8) and (3.23.1.7).

Finally (3.23.1.4) follows from (3.4.2). □

For later reference we formulate a result which was observed during the proof:

3.23.1.8 Corollary (Notation as above). *Assume that* $W = \operatorname{Spec} k$ *for a field* k. *Then* $s^*p_2 : (s^*p_1)^*(U_i) \to Ch(U)_i$ *is a local isomorphism at* $(x, H_1, \ldots, H_{d+1})$ *iff*

$$U_i \cap H_1 \cap \ldots \cap H_{d+1} = \{x\} \quad \textit{(as schemes)}. \qquad \square$$

3.23.2 Lemma. *Let* $Y \to S$ *be a smooth and projective morphism of relative dimension* N. *Let* $s : W \to S$ *be a morphism and* $(g : U \to W)$ *a well defined family of* $(N-1)$*-dimensional algebraic cycles of* Y/S.

Let $w_g \in W$ *be a generic point. By* (3.4.2) *there is a unique Cartier divisor* $C(w_g)$ *on* Y_{w_g} *whose fundamental class is* $g^{[-1]}(w_g)$. *Let* $W' \subset \operatorname{CDiv}(Y/S)$ *be the closure of the points* $[C(w_g)] \in \operatorname{CDiv}(Y/S)$ *and* $p : W' \to W$ *the projection. Then*

(3.23.2.1) *W' is dominated by the weak normalization of W.*

(3.23.2.2) *Assume that W is semi normal. Then p is an isomorphism iff $g^{[-1]}(w)$ is defined over $k(w)$ for every $w \in W$.*

(3.23.2.3) *If W is weakly normal, then g is flat.*

Proof. Let $g' : U' \to W'$ be the induced family of Cartier divisors over W'. p is proper by (1.13.2) and g' is flat by construction.

For $w \in W$, pick $w' \in W'$ such that $p(w') = w$. ${g'}^{-1}(w') \subset Y_{s'(w')}$ is a Cartier divisor, thus the fundamental cycle $[{g'}^{-1}(w')]$ determines ${g'}^{-1}(w')$ uniquely (3.4.2). Therefore $p : W' \to W$ is one-to-one. By construction, p is an isomorphism over the generic points of W. This proves (3.23.2.1), which implies (3.23.2.3).

By (3.4.2), $k(w')$ is the smallest field of definition of ${g'}^{-1}(w')$ containing $k(p(w')) = k(w)$. This shows (3.23.2.2). $\square$

3.23.3 Corollary. *Assume that* S *is over a field of characteristic zero. Let* $q : (\mathbb{P}^\vee)^{d+1} \to S$ *be the projection and*

$$\mathbb{H}/S := \operatorname{Proj}_S q_*(\mathcal{O}_{(\mathbb{P}^\vee)^{d+1}}(d', \ldots, d'))$$

the space parametrizing hypersurfaces of multidegree $(d', \ldots, d')$ *in* $(\mathbb{P}^\vee)^{d+1}$. *Then the above construction yields a transformation of functors*

$$Ch : Chow_{d,d'}(\mathbb{P}/S)(*) \to \operatorname{Hom}_S(*, \mathbb{H}).$$

Proof. Given a family $U \in Chow_{d,d'}(\mathbb{P}/S)(W)$, we obtain by (3.23.1) a family of hypersurfaces $Ch(U)/W$. By (3.23.2), $Ch(U)/W$ is flat, hence corresponds to a morphism $W \to \operatorname{Hilb}_S((\mathbb{P}^\vee)^{d+1})$. We prove in (3.24.2.4) that $\mathbb{H}$ is the component of $\operatorname{Hilb}_S((\mathbb{P}^\vee)^{d+1})$ containing the image. $\square$

3.23.4 Exercise. There is another way of defining a transformation $Chow \to CDiv$ which may seem more natural. At the set theoretic level, this can be described as follows.

Let $Z \subset \mathbb{P}^n$ be a cycle of dimension d. Let $\mathrm{Grass}(n-d, n+1)$ denote the Grassmann variety of $(n-d-1)$-dimensional linear subspaces in $\mathbb{P}^n$. Let

$$Ca(Z) := \{L \in \mathrm{Grass}(n-d, n+1) | Z \cap L \neq \emptyset\}.$$

Work out the correct definition which takes into account the multiplicities. Compare $Ch(Z)$ and $Ca(Z)$ to see that the two variants lead to equivalent theories.

3.24 Step 2. Chow or Cayley forms over fields.

3.24.1 Definition. Let $K \supset k$ be a field extension and $Z = \sum a_i[V_i] \in Z_d(\mathbb{P}_K)$ a cycle. $Ch(Z) \subset (\mathbb{P}^\vee_K)^{d+1}$ is a hypersurface, therefore it is defined by a multihomogeneous polynomial $ch(Z)$ which is unique up to a multiplicative constant. This polynomial is called the *Chow form* or *Cayley form* of the cycle Z.

The smallest field of definition of $Ch(Z)$ (over k) is called the *Chow field* of the cycle Z. It is denoted by $k^{ch}(Z)$. This field was constructed explicitly in (1.18).

3.24.2 Lemma. *Let $Z_K = \sum a_i[V_i] \in Z_d(\mathbb{P}_K)$ be a cycle.*

(3.24.2.1) *If $E \supset K$ is a field extension, then $ch(Z_K) = ch(Z_E)$.*

(3.24.2.2) *There is a one-to-one correspondence*

$$\{\textit{irreducible components of } Z_K\} \leftrightarrow \{\textit{irreducible divisors of } ch(Z_K)\}.$$

(3.24.2.3) $ch(Z) = \prod ch([V_i])^{a_i}$.

(3.24.2.4) *Let $d' = \sum a_i \deg V_i$. Then $ch(Z_K)$ has multidegree $(d', \dots, d')$.*

Proof. The first three claims are clear from the definition. To see the last one it is permissible to assume that k is algebraically closed and $Z = V_1$ is irreducible. Let us fix general hyperplanes $H_1, \dots, H_d$. Then $V_1 \cap H_1 \cap \dots \cap H_d$ consists of $\deg V_1$ distinct points $\{p_j\}$. Thus

$$ch(V_1)(H_1, \dots, H_d, *) = \prod_{j=1}^{\deg V_1} L_j(*),$$

where $(L_j(*) = 0) \subset \mathbb{P}^\vee$ is the set of hyperplanes containing the point p_j. □

We also define the inverse of Ch. At the moment, this can be done only set theoretically.

3.24.3 Definition. For a hypersurface $H \subset (\mathbb{P}^\vee)^{d+1}$, let

$$\operatorname{Supp} Ch^{-1}(H) := \{x \in \mathbb{P} | p_1^{-1}(x) \subset p_2^{-1}(H)\}.$$

3.24.4 Proposition. (3.24.4.1) $\dim \operatorname{Supp} Ch^{-1}(H) \leq d$.

(3.24.4.2) *Assume that H is irreducible, reduced and $k^p = k$. If* $\dim \operatorname{Supp} Ch^{-1}(H) = d$, *then* $\operatorname{Supp} Ch^{-1}(H)$ *is irreducible of dimension d and* $Ch(\operatorname{Supp} Ch^{-1}(H)) = H$.

Proof. If $\dim \operatorname{Supp} Ch^{-1}(H) \geq d+1$, then every codimension $d+1$ linear space intersects it. This gives the contradiction:

$$H \supset p_2(p_1^*(\operatorname{Supp} Ch^{-1}(H))) = (\mathbb{P}^\vee)^{d+1}.$$

Let $V \subset \operatorname{Supp} Ch^{-1}(H)$ be an irreducible component of dimension d. $\operatorname{Supp} Ch(V) \subset H$, hence $\operatorname{Supp} Ch(V) = H$ since H is irreducible. If $x \notin V$, then there are hypersurfaces H_i such that

$$x \in H_1 \cap \ldots \cap H_{d+1} \quad \text{and} \quad V \cap H_1 \cap \ldots \cap H_{d+1} \neq \emptyset.$$

Therefore $x \notin \operatorname{Supp} Ch^{-1}(Ch(V)) = \operatorname{Supp} Ch^{-1}(H)$. Since $k^p = k$, every algebraic extension of k is separable. Hence a reduced k-variety is geometrically reduced (cf. (7.1.1.1)). Thus by (3.23.1.2) $Ch(V)$ is reduced, hence $Ch(V) = H$. □

3.24.5 Corollary. *Assume that $k^p = k$. Then Ch gives a one-to-one correspondence:*

$$\left\{\begin{matrix} d\text{-dimensional cycles} \\ \text{of degree } d' \text{ in } \mathbb{P}_k \end{matrix}\right\} \leftrightarrow \left\{\begin{matrix} \text{Hypersurfaces } H \text{ of degree } (d', \ldots, d') \text{ in} \\ (\mathbb{P}_k^\vee)^{d+1} \text{ such that } \dim Ch^{-1}(H_j) = d \text{ for} \\ \text{every irreducible component } H_j \subset H \end{matrix}\right\}.$$

Proof. For a hypersurface H as above, let $[H] = \sum a_i [H_i]$ and define

$$Ch^{-1}(H) := \sum_i a_i [Ch^{-1}(H_i)].$$

By (3.24.4), Ch^{-1} is the inverse of Ch. □

3.24.6 Exercises.

(3.24.6.1) Give the equations for the incidence correspondence.

(3.24.6.2) Write down $ch(Z)$ in terms of the equations of Z in the following cases: Z is a linear subspace, Z is a hypersurface, Z is a codimension two complete intersection.

(3.24.6.3) Let $Z \subset \mathbb{P}$ be an irreducible variety with ideal sheaf I_Z. Let $H = Ch(Z)$. Show that $p_1^{-1}(\operatorname{Spec} \mathcal{O}_{\mathbb{P}}/I_Z^{(m)}) \subset p_2^{-1}(mH)$.

This shows that the natural scheme theoretic version of (3.24.3) gives too high multiplicity, unless Z is a hypersurface.

3.25 Step 3. Construction of $(u : \operatorname{Univ}_{d,d'}(\mathbb{P}/S) \to \operatorname{Chow}_{d,d'}(\mathbb{P}/S))$.

As a preliminary step, let

$$\begin{aligned}\mathrm{Chow}'_{d,d'}(\mathbb{P}/S) &:= \{H \in \mathbb{H}/S \mid \dim Ch^{-1}(H_j) = d \\ &= \text{for every component } H_j \subset H\}.\end{aligned}$$

3.25.1 Claim. $\mathrm{Chow}'_{d,d'}(\mathbb{P}/S)$ *is a closed algebraic subset of* $\mathbb{H}/S$.

Proof. Let $h : \mathbb{H} \to S$ be the projection and $\mathbb{U} \subset \mathbb{H} \times_S (\mathbb{P}^\vee)^{d+1}$ the universal hypersurface. We have the diagram

$$\begin{array}{ccc} h^* & \xleftarrow{h^* p_1} & h^* I \\ & & \downarrow{\scriptstyle h^* p_2} \\ & & \mathbb{U} \subset h^*(\mathbb{P}^\vee)^{d+1}. \end{array}$$

Let

$$\mathrm{Supp}\, Ch^{-1}(\mathbb{U}) := \{x \in h^*\mathbb{P} \mid (h^*p_1)^{-1}(x) \subset (h^*p_2)^{-1}(\mathbb{U})\}.$$

By (3.24.4), every fiber of $h^*q : \mathrm{Supp}\, Ch^{-1}(\mathbb{U}) \to \mathbb{H}$ has dimension at most d. By definition,

$$\begin{aligned}\mathrm{Chow}'_{d,d'}(\mathbb{P}/S) = \{H \in \mathbb{H} \mid &\text{ every component of } (h^*q)^{-1}(H) \\ &\text{has dimension } d\}.\end{aligned}$$

By the upper semicontinuity of fiber dimension, $\mathrm{Chow}'_{d,d'}(\mathbb{P}/S)$ is a closed algebraic subset. □

3.25.2 Definition (Notation as above). Let W be a semi normal scheme and $s : W \to S$ a morphism. Let $\mathbf{H} \subset s^*(\mathbb{P}^\vee)^{d+1}$ be a flat family of hypersurfaces over W such that for every field F and every morphism $\mathrm{Spec}\, F \to W$, the hypersurface $\mathbf{H} \times_W \mathrm{Spec}\, F$ is the Chow form of a cycle over F. Let

$$\mathrm{Supp}\, Ch^{-1}(\mathbf{H}) := \{x \in s^*\mathbb{P} \mid (s^*p_1)^{-1}(x) \subset (s^*p_2)^{-1}(\mathbf{H})\}.$$

We will see in (3.27.1) that every irreducible component of $\mathrm{Supp}\, Ch^{-1}(\mathbf{H})$ dominates an irreducible component of W.

In order to define $Ch^{-1}(\mathbf{H})$, we need to assign multiplicities to the irreducible components. To do this, we may localize at a generic point w_g of W. By assumption, there is a unique cycle over $k(w_g)$ whose Chow form is $\mathbf{H} \times_W w_g$. This assigns the multiplicities.

3.25.3 Definition. Let $\mathrm{Chow}_{d,d'}(\mathbb{P}/S)$ be the semi normalization of $\mathrm{Chow}'_{d,d'}$ $(\mathbb{P}/S)$. In particular, $\mathrm{Chow}_{d,d'}(\mathbb{P}/S)$ is semi normal and projective over S. Let

$$\mathrm{Univ}_{d,d'}(\mathbb{P}/S) := Ch^{-1}(\mathbb{U} \times_{\mathbb{H}} \mathrm{Chow}_{d,d'}(\mathbb{P}/S)).$$

3.26 Step 4. Construction of $(u : \mathrm{Univ}_{d,d'}(X/S) \to \mathrm{Chow}_{d,d'}(X/S))$.

Choose an embedding $X/S \hookrightarrow \mathbb{P}/S$. We already constructed

$$(u : \mathrm{Univ}_{d,d'}(\mathbb{P}/S) \to \mathrm{Chow}_{d,d'}(\mathbb{P}/S)).$$

Let

$$\mathrm{Chow}'_{d,d'}(X/S) := \{z \in \mathrm{Chow}_{d,d'}(\mathbb{P}/S) | u^{-1}(z) \subset X \times_S z\}.$$

Using the upper semicontinuity of fiber dimension, we see that $\mathrm{Chow}'_{d,d'}(X/S)$ is a closed algebraic subset. Let $\mathrm{Chow}_{d,d'}(X/S)$ be the semi normalization of $\mathrm{Chow}'_{d,d'}(X/S)$. Define $\mathrm{Univ}_{d,d'}(X/S)$ to be the Chow pull back of $\mathrm{Univ}_{d,d'}(\mathbb{P}/S)$. (In positive characteristic, this does not exist in some cases.)

It is clear that $(u : \mathrm{Univ}_{d,d'}(X/S) \to \mathrm{Chow}_{d,d'}(X/S))$ satisfies the properties (3.10.1–2). Also, $\mathrm{Chow}_{d,d'}(X/S)$ is semi normal and projective.

3.27 Step 5. Completion of the proof.

The main result needed is the following:

3.27.1 Proposition (Notation and assumptions as in (3.25.2)). *In particular* $\mathbf{H}/W$ *is flat.*

(3.27.1.1) $Ch^{-1}(\mathbf{H})$ *is a well defined family of algebraic cycles.*

(3.27.1.2) *If* W' *is a semi normal scheme over a field of characteristic zero and* $f : W' \to W$ *a morphism, then*

$$Ch^{-1}(f^*\mathbf{H}) \cong f^{ch}(Ch^{-1}(\mathbf{H})).$$

Proof. It is clear that $Ch^{-1}(\mathbf{H})$ satisfies the conditions (3.10.1–2). (3.10.3–4) is a consequence of the following stronger result:

Let T be the spectrum of a DVR with closed point $0 \in T$, generic point $t_g \in T$ and $h : T \to W$ a morphism such that $h(0) = w \in W$ and $h(t_g)$ is a generic point of W. Then

$$\lim_{h \to w} (Ch^{-1}(\mathbf{H})/W) \stackrel{ess}{=\!=} Ch^{-1}(\mathbf{H}_w). \tag{3.27.1.3}$$

To see this, let $\mathrm{Supp}\, U_T \subset \mathrm{Supp}\, Ch^{-1}(\mathbf{H}_T)$ be the closure of the generic fiber. Assign multiplicties to the irreducible components of $\mathrm{Supp}\, U_T$ as in (3.25.2) to obtain $(u : U_T \to T)$. $u : \mathrm{Supp}\, U_T \to T$ is flat, therefore it is a well defined family of algebraic cycles over T by (3.15). Thus $Ch(U_T) \to T$ is a flat family of hypersurfaces over T which agrees with $\mathbf{H}_T$ over the generic point of T. Therefore $Ch(U_T) = \mathbf{H}_T$, and hence $Ch^{-1}(\mathbf{H}_0) = U_0$. Thus $U_T = Ch^{-1}(\mathbf{H}_T)$. Furthermore,

$$Ch(\lim_{h \to w} (Ch^{-1}(\mathbf{H})/W)) = Ch(Ch^{-1}(\mathbf{H}_T)) \times_T T_0 = \mathbf{H}_T \times_T T_0 \stackrel{ess}{=\!=} \mathbf{H}_w.$$

By (3.24.5), this implies (3.27.1.3).

In order to prove (3.27.1.2), observe that it is sufficient to prove the case when $W' = \operatorname{Spec} F$ is the spectrum of a field by (3.11). Let $w = f(\operatorname{Spec} F)$. Choose $h : T \to W$ as above. By construction of f^{ch},

$$\begin{aligned} f^{ch}((Ch^{-1}(\mathbf{H})/W)) &= \lim_{h \to w} (Ch^{-1}(\mathbf{H})/W) \times_W \operatorname{Spec} F \\ &= Ch^{-1}(\mathbf{H} \times_W \operatorname{Spec} F) = Ch^{-1}(f^*\mathbf{H}). \end{aligned}$$

This was to be proved.

In positive characteristic $f^*\mathbf{H}$ need not satisfy the assumptions of (3.25.2), and therefore $Ch^{-1}(f^*\mathbf{H})$ may not exist. □

3.27.2 Corollary. *Let $(g : U \to W)$ be a well defined family of algebraic cycles. Assume that W is weakly normal. Then*

$$Ch^{-1}(Ch(U)) \cong U.$$

Proof. By (3.23.1) and (3.23.2), $Ch(U)$ is a flat family of hypersurfaces. Therefore by (3.27.1), $Ch^{-1}(Ch(U))$ is a well defined family of algebraic cycles. Thus the required isomorphism needs to be checked only at the generic points of W. This was done in (3.24.5). □

3.27.3. Let X/S be projective over S, W a weakly normal scheme, $s : W \to S$ a morphism and $(g : U \to W)$ a well defined family of algebraic cycles of X/S. Choose a projective embedding $X/S \hookrightarrow \mathbb{P}/S$. We can view $g : U \to W$ as a well defined family of algebraic cycles of $\mathbb{P}/S$. By step, 1 we get a morphism $\bar{s} : W \to \mathbb{H}/S$ and, by (3.26),

$$\operatorname{im} \bar{s} \subset \operatorname{im}[\operatorname{Chow}_{d,d'}(X/S) \to \mathbb{H}/S].$$

By (7.2.3.3) we can lift s to a morphism $\tilde{s} : W \to \operatorname{Chow}_{d,d'}(X/S)$. By (3.27.1.2) and (3.23.2)

$$\tilde{s}^{ch}(\operatorname{Univ}_{d,d'}(X/S)) = Ch^{-1}(s^*\mathbb{U}) = Ch^{-1}(Ch(U)) \cong U.$$

This means that $Chow_{d,d'}$ is represented by

$$(u : \operatorname{Univ}_{d,d'}(X/S) \to \operatorname{Chow}_{d,d'}(X/S)).$$ □

3.28 Exercise. The aim of this exercise is to establish an upper bound for the number of irreducible components of $\operatorname{Chow}_{k,d}(\mathbb{P}^n)$.

(3.28.1) Let $Z \subset \mathbb{P}^n$ be a purely k-dimensional reduced subvariety of degree d. Show by induction on k that

$$h^0(Z, \mathcal{O}_Z(m)) \le d\binom{m+k-1}{k} + \binom{m+k-1}{k-1} =: C(d,k,m).$$

(3.28.2) Let $f_0, \ldots, f_n \in H^0(\mathbb{P}^n, I_Z(d))$ be general sections. Then $Z = (f_0 = \ldots = f_n = 0)$ (set theoretically).

(3.28.3) One can encode $n+1$ polynomials of degree d as a point in the space

$$\mathbb{P}^N, \quad \text{where} \quad N = (n+1)\binom{n+d}{n} - 1.$$

(Usually, it is more natural to use instead products of projective spaces, but this choice turns out to be more convenient.) Let

$$U := \{(f_0, \ldots, f_n) \in \mathbb{P}^N \,|\, \dim(f_0 = \ldots = f_n = 0) \geq k\}.$$

Show that U is closed in $\mathbb{P}^N$.

(3.28.4) Show that

$$\begin{aligned}&\#\{\text{irreducible components of } U \text{ of codimension at most } (n+1)C(d,k,d)\}\\ &\geq \#\{\text{irreducible components of } \mathrm{Chow}_{k,d}(\mathbb{P}^n)\}.\end{aligned}$$

(3.28.5) Assume that so far everything was over a field K. Let ξ_{ij} be new variables and $h_i = \sum_j \xi_{ij}x_j$ linear forms for $i = 1, \ldots, k$. Show that

$$\begin{aligned}&\dim(f_0 = \ldots = f_n = 0) \geq k \quad \text{as a subset of } \mathbb{P}^n_K \quad \Leftrightarrow\\ &\dim(f_0 = \ldots = f_n = h_1 = \ldots = h_k = 0) \geq 0 \quad \text{as a subset of } \mathbb{P}^n_{K(\xi_{ij})} \quad \Leftrightarrow\\ &(f_0, \ldots, f_n, h_1, \ldots, h_k) \subset K(\xi_{ij})[x_0, \ldots, x_n] \quad \text{does not contain}\\ &\quad (x_0, \ldots, x_n)^{(n+1)d-n}.\end{aligned}$$

(To see the last part you may need (7.4.4.3).)

(3.28.6) Express the last condition as

$$\operatorname{rank} M < \binom{nd+d}{n},$$

where M is a certain matrix each of whose entries is either a coefficient in one of the polynomials $f_0, \ldots, f_n, h_1, \ldots, h_k$ or zero.

(3.28.7) Conclude that U can be defined by equations of degree at most $\binom{nd+d}{n}$.

(3.28.8) Let $S \subset \mathbb{P}^N$ be a subset which is (set theoretically) definable by equations of degree at most D. Then

$$\#\{\text{irreducible components of } S \text{ of codimension at most } K\} \leq K^D.$$

(3.28.9) Show that

$$\#\{\text{irreducible components of } \mathrm{Chow}_{k,d}(\mathbb{P}^n)\} \leq \binom{nd+d}{n}^{(n+1)C(d,k,d)}$$

(3.28.10) If $X \subset \mathbb{P}^m$ is a smooth projective variety of dimension k then it can be isomorphically projected to $\mathbb{P}^{2k+1}$.

(3.28.11) Give an explicit upper bound for the number of different irreducible familes of pairs (X, H), where X is a smooth projective variety of dimension k and H a very ample divisor with $(H^k) = d$.

(3.28.12) [Catanese92] Give an explicit upper bound for the number of different irreducible familes of smooth surfaces of general type with fixed K^2.

(3.28.13) Let $X \subset \mathbb{P}^n$ be an irreducible subvariety of dimension k' and degree d'. Give an explicit upper bound for the number of irreducible components of $\mathrm{Chow}_{k,d}(X)$.

I.4 Chow Varieties in Positive Characteristic

The aim of this section is to study the Chow functor in general. We start with several examples illustrating what can go wrong in positive characteristic. The most sophisticated example (4.2) is due to [Nagata55]. This shows that the semi normalization in (3.25.3) is probably indispensable in positive characteristic. The main new result is (4.5) which asserts that the Chow field of a cycle is the intersection of all fields of definition of the cycle. I could not find this result in the classical literature. This, however, does not take care of all the problems. There are still at least three natural ways of defining "good" families of algebraic cycles (4.11). The main result of the section is (4.13) which shows that $\mathrm{Chow}(X/S)$ (as constructed in Sect. 3) coarsely represents the Chow functor. It turns out that a universal family over $\mathrm{Chow}(X/S)$ exists in all the interesting cases (4.14).

4.1 Examples. (4.1.1) This example shows that there is no smallest field of definition for an arbitrary cycle.

Let L be an algebraically closed field of characteristic $p > 0$, $k = L(u, v)$ and $K = L(u^{p^{-1}}, v^{p^{-1}})$, where u, v are algebraically independent over L. Let $X = \mathbb{A}^2_k$ and $P = (x - u^{p^{-1}} = y - v^{p^{-1}} = 0) \subset \mathbb{A}^2_K$. Then $p[P]$ is defined over an algebraic extension $F \supset k$ iff F/k is not separable. In particular, $p[P]$ is defined over $L(u^{p^{-1}}, v)$ and also over $L(u, v^{p^{-1}})$ but it is not defined over $L(u, v)$.

Let $V \subset \mathbb{A}^2_F$ such that $[V] = p[P]$. Clearly, $(x^p - u, y^p - v) \subset I_V$. If F/k is separable, then

$$F[x, y]/(x^p - u, y^p - v) \cong k[x, y]/(x^p - u, y^p - v) \otimes_k F \cong K \otimes_k F$$

is a sum of fields by (7.1.1.1) and is local by (7.1.1.2). Therefore it is a field. Hence $I_V = (x^p - u, y^p - v)$ and $[V] = p^2[P]$, a contradiction.

If $f \in F$ is such that $f^p \in k$, then there is an $\bar{f} \in k[x, y]/(x^p - u, y^p - v)$ such that $f^p = \bar{f}^p$. Thus

$$\bar{f} \otimes 1 - 1 \otimes f \in k[x, y]/(x^p - u, y^p - v) \otimes_k F$$

is nilpotent of order p. Let $V' = \operatorname{red}\operatorname{Spec} F[x,y]/(x^p - u, y^p - v)$. Then $[V']$ is either $[P]$ or $p[P]$.

Some explicit examples are the following:

$$(x^p - u = y - v^{p^{-1}} = 0) \subset \mathbb{A}^2_{L(u,v^{p^{-1}})}$$

and

$$(x - u^{p^{-1}} = y^p - v = 0) \subset \mathbb{A}^2_{L(u^{p^{-1}},v)}$$

both define $p[P]$.

Observe also that if w is transcendental over k, then $p[P]$ is defined over $k(w, u^{p^{-1}} + v^{p^{-1}}w)$ but k is algebraically closed in $k(w, u^{p^{-1}} + v^{p^{-1}}w)$.

(4.1.2) This example shows that even for the simplest families $(g : U \to V)$, the cycle theoretic fiber $g^{[-1]}(w)$ is not always defined over $k(w)$.

We want to parametrize 0-cycles of degree two in the affine plane $\mathbb{A}^2_k$. The desired universal family should be

$$g : \mathbb{A}^2 \times \mathbb{A}^2 \to S^2(\mathbb{A}^2).$$

Let (x_1, y_1) and (x_2, y_2) be the coordinates on the two copies of $\mathbb{A}^2$. Then g^* is given by

$$k[x_1 + x_2, x_1x_2, y_1 + y_2, y_1y_2, x_1y_1 + x_2y_2] \hookrightarrow k[x_1, x_2, y_1, y_2].$$

The problem arises in characteristic two along the diagonal Δ. One can choose $u = x_1$ and $v = y_1$ as coordinates on the diagonal and the corresponding map becomes

$$g^*_\Delta : k[u^2, v^2] \hookrightarrow k[u, v].$$

This corresponds to a family of 0-cycles of degree 4 over $\operatorname{Spec} k[u^2, v^2]$. One can define a degree two family of 0-cycles over $\operatorname{Spec} k[u^2, v^2]$ which could be called the restriction, but it cannot be done in any canonical way.

Similar problems arise if we want to parametrize degree p cycles in characteristic p.

4.2 Example. This example, due to [Nagata55], shows that the classical definition of Chow varieties depends on the projective embedding chosen.

Let X be a quasi projective variety and choose a projective embedding $X \hookrightarrow \mathbb{P}$. We have natural morphisms

$$\operatorname{Chow}(X) \to \operatorname{Chow}(\mathbb{P}) \to \mathbb{H},$$

where $\mathbb{H}$ is defined in (3.23.3). The image of $\operatorname{Chow}(X)$ in $\mathbb{H}$ has been classically called the Chow variety of $X \subset \mathbb{P}$. We denote it by $\operatorname{Chow}(X \subset \mathbb{P})$. By our construction $\operatorname{Chow}(X)$ is the semi normalization of $\operatorname{Chow}(X \subset \mathbb{P})$.

[Nagata55] investigated the dependence of the image on the projective embedding $X \hookrightarrow \mathbb{P}$. His main result is that $\operatorname{Chow}_0(X \subset \mathbb{P})$ is independent of the embedding in characteristic zero, but it depends on the embedding in positive characteristic.

The results of [Barlet75] imply that over algebraically closed fields of characteristic zero $\mathrm{Chow}(X \subset \mathbb{P})$ is independent of the embedding.

Below, I reproduce the computations of [Nagata55] in the special case of degree two 0-cycles on affine varieties. The general case requires more complicated notation but is otherwise very similar.

On $\mathbb{A}^n$, choose coordinates $x_1, \dots, x_n$ with the convention that $x_0 = 1$ if it appears in a formula. Let $x^1 = (x_i^1)$ and $x^2 = (x_i^2)$ be two points in $\mathbb{A}^n$. Let $\mathbb{P}^\vee$ be the dual projective space with coordinates $u_0, \dots, u_n$. Then

$$ch(x^1 + x^2) = \Big(\sum_{i=0}^n u_i x_i^1\Big)\Big(\sum_{i=0}^n u_i x_i^2\Big) = u_0^2 + \sum_{i=1}^n (x_i^1 + x_i^2) u_i u_0$$
$$+ \sum_{i=1}^n (x_i^1 x_i^2) u_i u_i + \sum_{1 \le i < j \le n} (x_i^1 x_j^2 + x_j^1 x_i^2) u_i u_j.$$

Thus $\mathrm{Chow}_{0,2}(\mathbb{A}^n \subset \mathbb{P}^n)$ is obtained as the image of

$$p : \mathbb{A}^n \times \mathbb{A}^n \to \mathbb{A}^{(n^2+3n)/2},$$

where p is given by the coefficients of $ch(x^1 + x^2)$.

If $R = k[r_1, \dots, r_n]$ is a finitely generated k-algebra, then set

$$S^2_{r_1,\dots,r_n}(R) := k[r_i \otimes 1 + 1 \otimes r_i, r_i \otimes r_i, r_i \otimes r_j + r_j \otimes r_i : 1 \le i < j \le n],$$

considered as a subring of $R \otimes_k R$. If R is reduced, then $\operatorname{Spec} S^2_{r_1,\dots,r_n}(R)$ is exactly $\mathrm{Chow}_{0,2}(\operatorname{Spec} R \subset \mathbb{P}^n)$, where the embedding $\operatorname{Spec} R \subset \mathbb{A}^n \subset \mathbb{P}^n$ is given by the generators $r_1, \dots, r_n$.

On $R \otimes_k R$, there is a natural involution $\tau : a \otimes b \mapsto b \otimes a$. Let

$$S^2 R := \text{the ring of } \tau\text{-invariant elements of } R \otimes_k R.$$

If $\{r_\lambda : \lambda \in \Lambda\}$ is a k-basis of R, then it is easy to see that

$$S^2 R = k[r_\lambda \otimes 1 + 1 \otimes r_\lambda, r_\lambda \otimes r_\lambda, r_\lambda \otimes r_\mu + r_\mu \otimes r_\lambda : \lambda \neq \mu \in \Lambda]$$

4.2.1 Claim (Notation as above). *If* $\operatorname{char} k \neq 2$, *then*

$$S^2_{r_1,\dots,r_n}(R) = S^2 R.$$

Proof. The elements

$$x_i = r_i \otimes 1 + 1 \otimes r_i \quad \text{and} \quad y_i = r_i \otimes 1 - 1 \otimes r_i \quad 1 \le i \le n$$

generate $R \otimes_k R$ and, as in (1.19), one can choose a k-basis of $R \otimes_k R$ consisting of monomials in the x_i and y_i. These monomials are τ-eigenvectors, thus $S^2 R$

is the vector space spanned by the τ-invariant monomials in the x_i and y_i. Thus

$$S^2R = k[x_i, y_iy_j : 1 \leq i \leq j \leq n].$$

By explicit computation,

$$S^2_{r_1,\dots,r_n}(R) \supset k[x_i, y_iy_j : 1 \leq i \leq j \leq n],$$

and we are done. □

4.2.2 Claim. *If* char $k = 2$, *then:*

(4.2.2.1) *For a suitable choice of the generators* r_i,

$$S^2_{r_1,\dots,r_n}(R) = S^2R.$$

(4.2.2.2) *In general, the above equality does not hold.*

Proof. The ring S^2R is finitely generated, thus it has a finite generating set of the form

$$r_\lambda \otimes 1 + 1 \otimes r_\lambda, r_\lambda \otimes r_\lambda, r_\lambda \otimes r_\mu + r_\mu \otimes r_\lambda.$$

Pick every r_λ occurring in the above generating set to form a generating set of R. This shows (4.2.2.1).

For the second part, consider the example where $R = k[r_1, r_2, r_3]$ is a polynomial ring in three variables. In $R \otimes_k R$, a convenient generating set is

$$u_i = r_i \otimes 1 + 1 \otimes r_i \quad \text{and} \quad v_i = r_i \otimes 1.$$

The τ-action is

$$\tau(u_i) = u_i \quad \text{and} \quad \tau(v_i) = v_i + u_i.$$

Then

$$S^2_{r_1,r_2,r_3}(R) = k[u_i, v_i^2 + v_iu_i, v_iu_j + v_ju_i].$$

Observe that a degree three polynomial in $S^2_{r_1,r_2,r_3}(R)$ cannot contain $v_iv_ju_k$ for $i \neq j$. On the other hand

$$v_1v_2v_3 + (v_1 + u_1)(v_2 + u_2)(v_3 + u_3) \in S^2R$$

contains the monomial $v_1v_2u_3$. □

4.3 Proposition. *Let* $\mathbb{P}$ *be a projective space over a field* k *of characteristic* p *and* $\bar{K} \supset k$ *an algebraically closed field. Let* $V_j \subset \mathbb{P}_{\bar{K}}$ *be distinct, irreducible and reduced subschemes of dimension* d, $Z = \sum_{j \in J} a_j[V_j]$ *a cycle. For every* r *let* $J_r = \{j | a_j = r\}$ *and* $p^{\nu(r)}$ *the largest* p*-power dividing* r.

(4.3.1) *If* $(p, a_j) = 1$ *for every* j, *then* $k^{ch}(Z)$ *is the smallest field of definition of* Z *(cf. (3.4)).*

(4.3.2) *Let* F_r *be the smallest field of definition of* $\sum_{j \in J_r}[V_j]$. *Then*

$$k^{ch}(Z) = k(\sum_r F_r^{p^{\nu(r)}}).$$

Proof. Let us prove (4.3.1) in the special case when $a_j = 1$ for every j. By (3.3) this implies the general case of (4.3.1). Let F be the smallest field of definition of $\sum[V_j]$ and $E = k^{ch}(\sum[V_j])$. By construction, $ch(\sum[V_j])$ is defined over F, thus $F \supset E$.

We use the notation of (3.23). In order to prove $F \subset E$, consider $p_2^{-1}(Ch(\sum[V_j])) \subset I$. Applying (1.6) to the projection $p_1 : p_2^{-1}(Ch(\sum[V_j])) \to \mathbb{P}_E$, we obtain a locally closed subscheme $R \subset \mathbb{P}_E$ which is the largest subscheme such that p_1 is flat over R with fiber dimension $(d+1)(n-1)$.

We claim that $\sum V_i \subset R$ and equality holds at every generic point of R. Observe that we do not yet know that $\sum V_i$ is defined over E. It is, however, sufficient to prove the claim after a suitable base extension. As a consequence, we obtain that $\sum V_i$ (as the d-dimensional primary component of R) is defined over E.

Set theoretically, $\sum V_i = R$ by (3.24.4). Let $v \in V_i$ be any smooth point of $\sum V_i$. We show that

$$T_v V_i = T_v R.$$

Let $(v, H_1, \ldots, H_{d+1}) \in p_2^{-1}(Ch(\sum[V_j]))$. Then

$$\begin{aligned} &T_{(v,H_1,\ldots,H_{d+1})}(p_2^{-1}Ch(\textstyle\sum[V_j])) \\ &\quad = T_v V_i + T_{(v,H_1,\ldots,H_{d+1})}(p_1^{-1}(v)) + T_v(H_1 \cap \ldots \cap H_{d+1}). \end{aligned}$$

Thus

$$(p_1)_* T_{(v,H_1,\ldots,H_{d+1})}(p_2^{-1}Ch(\textstyle\sum[V_j])) = T_v V_i + T_v(H_1 \cap \ldots \cap H_{d+1}).$$

By assumption $T_v V_i$ is a d-dimensional linear space and depending upon the choice of $H_1, \ldots, H_{d+1}$, the other summand $T_v(H_1 \cap \ldots \cap H_{d+1})$ can be an arbitrary linear space of codimension $d+1$. In particular,

$$T_v R \subset \cap_{\{(H_1,\ldots,H_{d+1})\}} (T_v V_i + T_v(H_1 \cap \ldots \cap H_{d+1})) = T_v V_i.$$

Thus $\sum V_i = R$ in a neighborhood of v. This shows (4.3.1).

In a fixed k-coordinate system of $\mathbb{P}$, let ch_r be a defining equation of $Ch(\sum_{j \in J_r}[V_j])$. If $\{b_\lambda : \lambda \in \Lambda\}$ are the nonzero coefficients of ch_r, then by (1.18) $F_r = k(b_\lambda/b_\mu : \lambda, \mu \in \Lambda)$. Furthermore, $ch_r^{p^a}$ is a defining equation of $Ch(p^a \sum_{j \in J_r}[V_j])$, thus

$$k^{ch}(p^a \sum_{j \in J_r} [V_j]) = k(F_r^{p^a}).$$

By (3.24.2.3),

$$[Ch(Z)] = \sum_r r[Ch(\sum_{j \in J_r} [V_j])],$$

hence, by (3.3), the smallest field of definition of $[Ch(Z)]$ is the composite of the fields

$$k^{ch}(p^{\nu(r)} \sum_{j \in J_r} [V_j]) = F_r^{p^{\nu(r)}}.$$

This shows (4.3.2). □

4.4 Proposition – Definition. *Let X be a projective scheme defined over a field k and $i : X \hookrightarrow \mathbb{P}_k$ a projective embedding. For a cycle $Z \in Z_d(X_K)$, the Chow field $k^{ch}(i_* Z)$ is independent of the projective embedding i. It is called the Chow-field of Z and it is denoted by $k^{ch}(Z)$.*

Proof. Let $\bar{K}$ be the algebraic closure of K and $Z_{\bar{K}} = \sum a_i[V_i]$, where the V_i are irreducible and reduced. We use the notation of (4.3). Let E_r be the smallest field of definition of $\sum_{j \in J_r} V_j$. By (1.16),

$$E_r = \text{ the smallest field of definition of } \sum_{j \in J_r} i_*(V_j).$$

Thus, by (4.3.2), we obtain

$$k^{ch}(Z) = k(\sum_r E_r^{p^{\nu(r)}}).$$ □

4.5 Theorem. *Let $\mathbb{P}$ be a projective space over a field k of characteristic p and $\bar{K}$ an algebraically closed field containing k. Let $Z = \sum a_i[V_i]$ be a cycle over $\bar{K}$. The following three fields are the same:*

(4.5.1) $k^{ch}(Z)$.

(4.5.2) *The intersection of all fields of definition of Z.*

(4.5.3) *The intersection of all fields of definition of Z which are purely inseparable over $k^{ch}(Z)$.*

Proof. First we show that $k^{ch}(Z)$ is contained in any field of definition of Z. Assume that $Z \subset \mathbb{P}_{\bar{K}}$ is defined over a field E by a scheme Y_E. Let $[Y_E] = \sum n_j[Y_j]$ be the primary decomposition over E. (Over $\bar{K}$, the Y_j need not be irreducible or reduced.) Then $Ch(Z) = Ch([Y_E]) = \sum n_j k_j[Ch(Y_j)]$ and therefore $Ch(Z)$ is defined over E.

It remains to be proved that $k^{ch}(Z)$ is the intersection of all fields of definition of Z which are purely inseparable over $k^{ch}(Z)$.

4.5.4 Lemma. *Let $E \supset F$ be a purely inseparable field extension and $[E : F] = q$. Assume that a cycle $Z \in Z_d(X)$ is defined over E. Then qZ is defined over F.*

Proof. The morphism $g : X_E \to X_F$ is finite of degree q and is a homeomorphism. Thus $g^*(g_*(Z)) = qZ$, hence $g_*(Z)$ defines qZ over F. □

4.5.5 Main Lemma. *Let $E \supset F$ be a purely inseparable field extension such that $E^q \subset F$ for some $q = p^k$. Let $\bar{K}$ be an algebraic closure of E and*

$v \in \bar{K} - F$ such that $z := v^p \in F$. Let L be a maximal purely inseparable extension of F which does not contain v.

Then $EL \subset L(z^{q^{-1}})$.

Proof. Even stronger, I claim that $L(z^{q^{-1}})$ is the unique degree q purely inseparable extension of L. The proof is by induction on k.

Pick $u \notin L$ such that $u^p \in L$. Since L is maximal, $L(v) \subset L(u)$. They have the same degree, thus $L(v) = L(u)$. This shows the claim for $k = 1$.

In general, pick $w \in \bar{K}$ such that $w^{p^k} \in L$. We know by induction that $w^p \in L(z^{p^{1-k}})$. Thus we can write

$$w^p = \sum_{i=0}^{p^{k-1}-1} a_i (z^{p^{1-k}})^i \quad \text{hence} \quad w = \sum_{i=0}^{p^{k-1}-1} a_i^{p^{-1}} z^{ip^{-k}}, \qquad \text{where } a_i \in L.$$

By the $k = 1$ case, $a_i^{p^{-1}} \in L(z^{p^{-1}})$ for every i, hence $w \in L(z^{p^{-k}})$.

It is clear that a suitable finite subextension of L also has the required property. □

From now on, we use the notation of (4.3). As we proved,

$$k^{ch}(Z) = k\Big(\sum_r F_r^{p^{\nu(r)}}\Big).$$

Let $M \supset k^{ch}(Z)$ be the intersection of all fields of definition of Z which are purely inseparable over $k^{ch}(Z)$. If $M \neq k^{ch}(Z)$, then there is a $v \in M - k^{ch}(Z)$ such that $z := v^p \in k^{ch}(Z)$. Let L be a maximal purely inseparable extension of $k^{ch}(Z)$ which does not contain v. By (4.5.5),

$$F_r L \subset L(z^{p^{-\nu(r)}}).$$

In particular $[F_r L : L] \leq p^{\nu(r)}$. Since $\sum_{j\in J_r} [V_j]$ is defined over F_r, it is also defined over $F_r L$. Hence, by (4.5.4), $p^{\nu(r)} \sum_{j\in J_r} [V_j]$ is defined over L. Therefore Z is also defined over L, hence $M \subset L$. This is a contradiction since $v \in M - L$. □

4.6 Example. It is tempting to believe that in (4.5), it is sufficient to consider those fields of definition of Z which lie in between

$$k^{ch}(Z) = k\Big(\sum_r F_r^{p^{\nu(r)}}\Big)$$

and the field of definition of red Z, which is

$$k\Big(\sum_r F_r\Big).$$

This is, however, not the case, as shown by the following example.

Let $k = F(x, y, z)$, where F is arbitrary and the x, y, z are algebraically independent over F. Let

$$P = (s,t) := (x^{p^{-2}}, y^{p^{-1}} + z^{p^{-1}} x^{p^{-2}}) \in \mathbb{A}^2_k.$$

Then

$$\begin{aligned} ch(p^2[P]) &= (x^{p^{-2}} X + (y^{p^{-1}} + z^{p^{-1}} x^{p^{-2}})Y + Z)^{p^2} \\ &= xX^{p^2} + (y^p + z^p x)Y^{p^2} + Z^{p^2}, \end{aligned}$$

where $(X : Y : Z)$ are homogeneous coordinates on the dual projective plane. Thus $k^{ch}(p^2[P]) = k$.

$p^2[P]$ is not defined over k since

$$(u - s, v - t)k(s,t)[u,v] \cap k[u,v] = (u^{p^2} - x, v^p - (y + zu^p))$$

is a maximal ideal which defines $p^3[P]$.

Next, we show that the intersection of all subfields of $k(x^{p^{-2}}, y^{p^{-1}} + z^{p^{-1}} x^{p^{-2}})$ which properly contain k is $k(x^{p^{-1}})$.

4.6.1 Claim. *If $r \in k(s,t)$ then $r^p \in k \Leftrightarrow r \in k(s^p)$.*

Proof. Taking into account $t^p = y + zs^p$, we can assume that r is of the form

$$r = \sum_{i=0}^{p-1} t^i f_i(s, y, z),$$

for some rational functions f_i with coefficients in F. Thus

$$r^p = \sum_{i=0}^{p-1} (y + zx^{p^{-1}})^i f_i^{(p)}(x^{p^{-1}}, y^p, z^p) \in F(x, y, z).$$

Consider the derivation D on $F(x^{p^{-1}}, y, z)$ given by

$$D(x^{p^{-1}}) = 1, \; D(y) = 0, \; D(z) = 0.$$

Then, for $g \in F(x^{p^{-1}}, y, z)$, we have

$$D(g) = 0 \Leftrightarrow D(g) \in F(x, y, z) \Leftrightarrow g \in F(x, y, z). \tag{4.6.2}$$

Applying it to r^p, we obtain

$$0 = \sum_{i=0}^{p-1} (y + zx^{p^{-1}})^i \left[D(f_i^{(p)}) + (i+1) z f_{i+1}^{(p)} \right].$$

Observe that $y + zx^{p^{-1}} \notin F(x^{p^{-1}}, y^p, z)$ but $(y + zx^{p^{-1}})^p \in F(x^{p^{-1}}, y^p, z)$. Thus

$$(y + zx^{p^{-1}})^i : i = 0, \dots, p-1 \tag{4.6.3}$$

are linearly independent over $F(x^{p^{-1}}, y^p, z)$. Hence

$$D(f_i^{(p)}) + (i+1)zf_{i+1}^{(p)} = 0 \quad \text{for} \quad i = 0, \dots, p-1.$$

Starting with $i = p-1$ and using descending induction we obtain that

$$f_i^{(p)}(x^{p^{-1}}, y^p, z^p) \in F(x, y, z) \quad \text{for every } i.$$

The elements (4.6.3) are also independent over over k, therefore $r^p \in k$ implies that $r = f_0(s, y, z)$. Thus $r^p \in k$ iff $r \in k(s^p)$. □

4.7 Definition. Let $(g : U \to W)$ be a well defined family of algebraic cycles over W.

(4.7.1) We say that it satisfies the *Chow field condition* if

$$k(w)^{ch}([g^{[-1]}(w)]) = k(w) \quad \text{for every } w \in W.$$

(4.7.2) We say that it satisfies the *field of definition condition* if

$$g^{[-1]}(w) \quad \text{is defined over } k(w) \text{ for every } w \in W.$$

By the definition of Ch, we obtain:

4.8 Corollary. *Let $(g : U \to W)$ be a well defined family of algebraic cycles over W. The following are equivalent:*

(4.8.1) *$(g : U \to W)$ satisfies the Chow-field condition.*
(4.8.2) *$(Ch(g) : Ch(U) \to W)$ satisfies the field of definition condition.*

□

4.9 Corollary. *Let $(g : U \to W)$ be a well defined family of codimension one algebraic cycles of $\mathbb{P}/S$. Assume that it satisfies the Chow-field condition and that W is semi normal. Then $Ch(g) : Ch(U) \to W$ is flat.*

Proof. Combine (4.8) and (3.23.2.2). □

4.10 Corollary. *Let $(g : U \to W)$ be a well defined family of algebraic cycles of $\mathbb{P}/S$. Assume that W is weakly normal. Then $(g : U \to W)$ satisfies the Chow-field condition.*

Proof. Combine (4.8) and (3.23.2.1). □

There are at least three reasonable definitions of $Chow(X/S)(Z)$ in positive characteristic. They are the following:

4.11 Definition. Let X be a scheme over S. Let

$$Chow^{\text{big}}(X/S)(Z) = \left\{\begin{matrix}\text{Well defined proper algebraic}\\ \text{families of nonnegative cycles}\\ \text{of } X \times_S Z\end{matrix}\right\};$$

$$Chow(X/S)(Z) = \left\{\begin{matrix}\text{Well defined proper algebraic}\\ \text{families of nonnegative cycles}\\ \text{of } X \times_S Z \text{ which satisfy the}\\ \text{Chow-field condition}\end{matrix}\right\};$$

$$Chow^{\text{small}}(X/S)(Z) = \left\{\begin{matrix}\text{Well defined proper algebraic}\\ \text{families of nonnegative cycles}\\ \text{of } X \times_S Z \text{ which satisfy the}\\ \text{field of definition condition}\end{matrix}\right\}.$$

As in (3.20), one can also define $Chow^{\text{big}}_{d,d'}, Chow_{d,d'}, Chow^{\text{small}}_{d,d'}$.

4.12 Comments.

(4.12.1) The three definitions agree in characteristic zero.

(4.12.2) If Z is weakly normal, then $Chow^{\text{big}}(X/S)(Z) = Chow(X/S)(Z)$. Thus we do not loose much by considering $Chow$ instead of $Chow^{\text{big}}$.

(4.12.3) $Chow^{\text{big}}$ and $Chow$ do not define functors in general, only partial functors since a family cannot always be pulled back.

(4.12.4) $Chow^{\text{small}}$ is a well defined functor on the category of semi normal schemes (3.19). Unfortunately, many naturally occurring families of cycles are not in $Chow^{\text{small}}$. In particular, the "universal" family over Chow is usually not in $Chow^{\text{small}}$ (4.2).

(4.12.5) In general, Chow is not weakly normal. Therefore it is not reasonable to restrict ourselves to weakly normal schemes when considering families of cycles.

4.13 Theorem. *Let X/S be a scheme, projective over S and $\mathcal{O}(1)$ relatively ample. Then:*

(4.13.1) *The Chow pull back makes $Chow^{\text{small}}_{d,d'}(X/S)$ into a functor*

$$\{\textit{semi normal } S\textit{-schemes}\} \to \{\textit{sets}\}.$$

(4.13.2) *The Chow pull back makes $Chow_{d,d'}(X/S)$ into a partial functor*

$$\{\textit{semi normal } S\textit{-schemes}\} \to \{\textit{sets}\}.$$

(4.13.3) *$Chow^{\text{small}}_{d,d'}(X/S)$ and $Chow_{d,d'}(X/S)$ are both coarsely represented by the scheme* $\mathrm{Chow}_{d,d'}(X/S)$.

(4.13.4) $\mathrm{Chow}_{d,d'}(X/S)$ *is semi normal and projective over S.*

(4.13.5) $\mathrm{Chow}(X/S)$ *is independent of the choice of $\mathcal{O}(1)$.*

Proof. Most of the work was already done in the previous section. In particular (4.13.1–2) follow directly from (3.19). In order to prove (4.13.3), note that by (4.9) there is a natural transformation

$$Chow^{\text{small}}_{d,d'}(X/S)(\ *\ ) \subset Chow_{d,d'}(X/S)(\ *\ ) \xrightarrow{Ch} \mathrm{Hom}_S(\ *\ ,\mathrm{Chow}_{d,d'}(X/S)).$$

By (3.24.5), if $\bar{K}$ is an algebraically closed field, then

$$Chow^{\text{small}}_{d,d'}(X/S)(\bar{K}) = Chow_{d,d'}(X/S)(\bar{K}) \xrightarrow{\cong} \mathrm{Hom}_S(\bar{K},\mathrm{Chow}_{d,d'}(X/S)).$$

We still need to show that if

$$Chow^{\text{small}}_{d,d'}(X/S)(\ *\ ) \xrightarrow{Tr} \mathrm{Hom}_S(\ *\ ,\mathbf{Y}/S)$$

is a natural transformation, then there is a morphism $t : \mathrm{Chow}_{d,d'}(X/S) \to \mathbf{Y}$ such that $Tr = t \circ Ch$. This implies the similar statement for *Chow*.

We construct the graph of t as follows. For every $x \in \mathrm{Chow}_{d,d'}(X/S)$ let $k(v)^i$ be the purely inseparable closure of its residue field. There is a unique cycle Z_v defined over $k(v)^i$ which corresponds to v. This gives a morphism

$$\mathrm{Spec}\, k(v)^i \xrightarrow{(Ch,Tr)} \mathrm{Chow}_{d,d'}(X/S) \times_S \mathbf{Y}.$$

Let

$$\Gamma \subset \mathrm{Chow}_{d,d'}(X/S) \times_S \mathbf{Y}$$

be the union of all these images.

4.13.6 Claim. *Γ is a closed algebraic subset and the natural projection*

$$\pi_1 : \Gamma \to \mathrm{Chow}_{d,d'}(X/S)$$

is an isomorphism.

Proof. π_1 is one-to-one by construction. Pick a generic point v_g of $\mathrm{Chow}_{d,d'}$ (X/S) and let v_0 be any specialization of v_g. Let T be the spectrum of a DVR and $h : T \to \mathrm{Chow}_{d,d'}(X/S)$ any morphism such that $h(T_g) = v_g$ and $h(T_0) = v_0$. Over T, there is a flat family of algebraic cycles $(g : U \to T)$ such that $Ch(g) = h \in \mathrm{Hom}(T, \mathrm{Chow}_{d,d'}(X/S))$. Thus we obtain a morphism

$$\bar{h} : T \to \mathrm{Chow}_{d,d'}(X/S) \times_S \mathbf{Y}$$

such that $\bar{h}(T_g) = \pi_1^{-1}(v_g)$ and $\bar{h}(T_0) = \pi_1^{-1}(v_0)$. Thus Γ consists of the points $\pi_1^{-1}(v_{g_i})$ and their specializations where v_{g_i} are all generic points of $\mathrm{Chow}_{d,d'}(X/S)$. Therefore Γ is a closed algebraic subset.

For $v \in \mathrm{Chow}_{d,d'}(X/S)$, let Z_v be the cycle corresponding to v (over $k(v)^i$) and $\bar{v} = \pi_1^{-1}(v)$. If $E \supset k(v)$ is a field of definition of Z_v, then we obtain a morphism $\mathrm{Spec}\, E \to \bar{v}$. Therefore $k(\bar{v}) \supset k(v) = k^{ch}(Z_v)$ is contained in all fields of definition of Z_v. Thus by (4.5), $k(\bar{v}) = k(v)$. Since $\mathrm{Chow}_{d,d'}(X/S)$ is semi normal, this implies that π_1 is an isomorphism. □

This completes the proof of (4.13.3).

(4.13.4) is true by construction. Finally, note that $\mathrm{Chow}(X/S)$ is the coarse moduli space for the functor $Chow(X/S)$ which is defined without reference to any ample line bundle. Thus $\mathrm{Chow}(X/S)$ is independent of the choice of $\mathcal{O}(1)$. □

Since Chow does not represent *Chow* in general, there is no universal family over Chow, at least not in $Chow^{\mathrm{small}}(\mathrm{Chow})$. However in most cases there is a universal family which is in $Chow(\mathrm{Chow})$. We state the result in the case when $S = \operatorname{Spec} k$ for a field k.

4.14 Theorem. *Let X be a projective scheme over a field k and V an irreducible component of $\mathrm{Chow}_{d,d'}(X)$. Let*

$$\mathbb{H}_V \subset (\mathbb{P}^{\vee}_{k(V)})^{d+1}$$

be the Chow form over the generic point $v_g = \operatorname{Spec} F$ of V.

(4.14.1) *If $\mathbb{H}_V$ is geometrically reduced, then*

$$\mathrm{Univ}_{d,d'}(X) \in Chow_{d,d'}(X)(V)$$

exists.

(4.14.2) *The geometrically nonreduced part of $\mathbb{H}_V$ is defined over a finite extension of k.*

One can informally formulate the above result as follows: A universal family exists except possibly for constant families.

Proof. If $\mathbb{H}_V$ is geometrically reduced, then $Ch^{-1}(\mathbb{H}_V)$ is defined over F by (4.3.1).

In order to see the second part, it is sufficient to consider the special case when $\mathbb{H}_V$ is irreducible over F but not geometrically reduced. Let $V \subset \mathrm{Chow}(X)$ be the closure of v_g. We need to show that V is zero dimensional. Assume the contrary. There is a quasi finite and dominant morphism $q : V' \to V$ and a flat family of geometrically reduced cycles $(g' : U' \to V')$ such that

$$[{g'}^{-1}(v')] = p^a(\text{ the cycle corresponding to } q(v')\ ).$$

Let $W = V' \times \cdots \times V'$ (p^a-times) with cordinate projections p_i. Let $U_W = \sum p_i^* U'$. We obtain a family of cycles $(U_W \to W)$ whose generic fiber is geometrically reduced. By construction, the closure of the image of the morphism $W \to \mathrm{Chow}(X)$ properly contains V. This is a contradiction. □

I.5 Related Moduli Problems

In the previous sections we constructed and investigated the Hilbert and Chow schemes of a projective morphism $f : X \to S$. The aim of this section is to give a short overview (mostly without proofs) of the known results in the case when f is a morphism in some other category of spaces.

5.1 Cases to be considered.

(5.1.1) quasi projective morphisms;
(5.1.2) arbitrary morphisms of schemes;
(5.1.3) morphisms of algebraic spaces;
(5.1.4) morphisms of complex analytic spaces;
(5.1.5) Quot-schemes.

5.2 Definition. Let $f : X \to S$ be a morphism in one of the above categories. The functor

$$Hilb(X/S) : \begin{Bmatrix} \text{schemes over } S, \text{ resp.} \\ \text{algebraic spaces over } S, \text{ resp.} \\ \text{complex spaces over } S \end{Bmatrix} \to \{\text{sets}\}$$

is defined as follows:

$$Hilb(X/S)(Z) = \begin{Bmatrix} \text{Subschemes (resp. sub-} \\ \text{spaces) } V \subset X \times_S Z \\ \text{which are proper and} \\ \text{flat over } Z \end{Bmatrix} = \begin{Bmatrix} \text{Quotient sheaves } F \text{ of the} \\ \text{structure sheaf of } X \times_S Z \\ \text{which are flat and have} \\ \text{proper support over } Z \end{Bmatrix}.$$

5.3 Definition. Let $f : X \to S$ be a morphism in one of the above categories. The functor

$$Chow(X/S) : \begin{Bmatrix} \text{schemes over } S, \text{ resp.} \\ \text{algebraic spaces over } S, \text{ resp.} \\ \text{complex spaces over } S \end{Bmatrix} \to \{\text{sets}\}$$

is defined as follows:

$$Chow(X/S)(Z) = \begin{Bmatrix} \text{Well defined proper algebraic} \\ \text{(resp. analytic) families of} \\ \text{nonnegative cycles of } X \times_S Z \end{Bmatrix}.$$

The existence problem for quasiprojective morphisms can easily be reduced to the projective one:

5.4 Theorem. *Let $\bar{f} : \bar{X} \to S$ be a projective morphism, $X \subset \bar{X}$ an open subscheme. Then $Hilb(X/S)$ (resp. $Chow(X/S)$) is representable (resp. representable in characteristic zero and coarsely representable in general) by an open subscheme*

$$\operatorname{Hilb}(X/S) \subset \operatorname{Hilb}(\bar{X}/S) \quad \textit{resp.} \quad \operatorname{Chow}(X/S) \subset \operatorname{Chow}(\bar{X}/S). \qquad \square$$

In the remaining cases, the existence results for Hilb and Chow require very different methods that lie completely outside the scope of this book. See [Campana-Peternell94] for an overview of the complex analytic case.

It turns out that the existence theorem for *Hilb* is not easier for schemes than for algebraic spaces. One reason is that even for a proper scheme over a field, the Hilbert functor is usually represented by an algebraic space which is not a scheme (5.5.1).

5.5 Theorem [Artin69b]. *Let $f : X \to S$ be a morphism of algebraic spaces which is locally of finite presentation. Then $Hilb(X/S)$ is representable by an algebraic space* $\mathrm{Hilb}(X/S)/S$ *which is locally of finite presentation. If f is separated, then* $\mathrm{Hilb}(X/S) \to S$ *is also separated.* □

5.5.1 Exercise. Using the construction of Hironaka explained in [Hartshorne77, App. B, 3.4.1–2] find a smooth proper scheme X of dimension 3 over $\mathbb{C}$ with a fixed point free automorphism τ of order 2 such that X/τ is not a scheme. Conclude from this that $Hilb(X)$ cannot be represented by a scheme.

5.6 Remark. [Angéniol81] constructs Chow varieties for smooth schemes. As far as I know there are no other results about $Chow(X/S)$ for schemes or algebraic spaces.

5.7 Theorem [Douady66; Pourcin69]. *Let $f : X \to S$ be a morphism of complex analytic spaces. Then $Hilb(X/S)$ is representable by a complex analytic space* $\mathrm{Hilb}(X/S)$. *(Usually it is called the Douady space.)* □

5.8 Theorem [Barlet75]. *Let X be a complex analytic space. Then $Chow(X)$ is representable by a complex analytic space* $\mathrm{Chow}(X)$. *(Usually it is called the Barlet space.)* □

5.9 Remark. The result of Barlet is even stronger. His definition of Chow is more refined, than the one I use. He does not have to restrict himself to semi normal spaces.

If X is a proper variety, then the connected components of $\mathrm{Hilb}(X)$ are usually not of finite type. There can be infinitely many irreducible components in a single connected component (5.13.1). The irreducible components themselves are, however, of finite type, even compact under some mild conditions.

5.10 Theorem [Fujiki84]. *Let X be a compact complex analytic space. Assume that X is bimeromorphic to a Kähler space. Then every irreducible component of* $\mathrm{Hilb}(X)$ *or of* $\mathrm{Chow}(X)$ *is compact.* □

5.11 Remark. [Fujiki84] proves the relative version of the above result as well.

5.12 Exercise [Blanchard56]. Let C be a smooth projective curve over $\mathbb{C}$ and M a line bundle of positive degree over C which is generated by two sections s and t. The four sections

$$(s,t),(is,-it),(t,-s),(it,is) \in \Gamma(M+M)$$

are independent over $\mathbb{R}$ in every fiber of $M+M$. Therefore they generate a lattice bundle $L \subset M+M$ and the quotient $q: M+M \to X = M+M/L$ is a compact complex threefold. There is a natural projection $f: X \to C$ and every fiber is a compact complex torus. (In fact, any fiber is isomorphic to $\mathbb{C}/(\mathbb{Z}+i\mathbb{Z}) + \mathbb{C}/(\mathbb{Z}+i\mathbb{Z})$, and is therefore projective.)

Let $H \subset \operatorname{Hilb}(X)$ be the subset parametrizing sections $s: C \to X$. Show that $H \subset \operatorname{Hilb}(X)$ is closed.

Assume that $C \cong \mathbb{P}^1$. Fix a point p in the zero section of $M+M$ and let $H_p \subset H$ be the set of those sections of f that contain $q(p)$. Show that H_p is closed in H and $H_p \cong \mathbb{C}^{2\deg M}$.

Conclude that H is a noncompact connected component of $\operatorname{Hilb}(X)$.

5.13 Theorem. *Let $f: X \to S$ be a proper morphism of algebraic spaces. Then every irreducible component of* $\operatorname{Hilb}(X/S)$ *is proper over S.*

Proof. Step 5 of (1.8) applies verbatim to show that if $f: X \to S$ is a proper morphism of algebraic spaces, then $\operatorname{Hilb}(X/S)$ satisfies the valuative criterion of properness. Thus if an irreducible component of Hilb is of finite type, then it is also proper. This observation is very useful, however it does not seem to be helpful in proving properness.

The actual proof is a simple special case of the arguments in [Fujiki82]. Let $Y \subset \operatorname{Hilb}(X/S)$ be closed and irreducible. Let $Y^0/S \subset Y/S$ be an open affine subset and $u^0: U^0 \to Y^0$ the universal family. Let $Y^0/S \subset \bar{Y}/S$ be an open immersion such that $\bar{Y}$ is proper over S. Let $\bar{U} \subset X \times_S \bar{Y}$ be the closure of $U^0 \subset X \times_S Y^0 \subset X \times_S \bar{Y}$. By [Raynaud-Gruson71, 5.7.9], there is a proper modification $Y' \to \bar{Y}$ which is an isomorphism over Y^0 such that the closure U' of $U^0 \subset X \times_S Y'$ is flat over Y'. Thus we obtain a morphism $h: Y' \to \operatorname{Hilb}(X/S)$. Since Y'/S is proper and $\operatorname{Hilb}(X/S)$ is separated, $h(Y') \subset \operatorname{Hilb}(X/S)$ is proper and closed. By construction $Y^0 \subset h(Y')$, hence $Y \subset h(Y')$. This shows that Y is proper over S. □

5.13.1 Exercise. Use the construction of Hironaka explained in [Hartshorne77, App. B, 3.4.1–2] to find a smooth proper algebraic space X of dimension 3 over $\mathbb{C}$ and a connected component $V \subset \operatorname{Hilb}(X)$ such that V has infinitely many irreducible components.

The infinitesimal computations done in Sect. 2 did not use any particular properties of the projective category. All the proofs can be applied without change in any of the above categories as well. In particular, we obtain:

5.14 Theorem. (5.14.1) *Let $f : Y \to S$ be a flat morphism in any of the above categories. Then* (2.15) *holds.*

(5.14.2) *Let $f : Y \to S$ be a flat morphism and $g : X \to S$ a flat and proper morphism in any of the above categories. Then* (2.17) *holds.* □

Finally we consider the Quot-schemes introduced by [Grothendieck62, 221]. For projective morphisms the proofs are very close to the proofs given in Sect. 1 and 2. These generalizations are left as an exercise. Analogous results hold for algebraic spaces and for complex analytic spaces.

5.15 Definition. Let X/S be a scheme (or an algebraic space) and E a coherent sheaf on X. The *Quot functor*

$$\mathit{Quot}(E/S) : \{\text{schemes over } S\} \to \{\text{sets}\}$$

is defined as follows. For a morphsim $p : Z \to S$ let $p_X : X \times_S Z \to X$ denote its pull back. Then,

$$\mathit{Quot}(E/S)(Z) = \left\{ \begin{matrix} \text{Quotient sheaves } F \text{ of } p_X^* E \text{ which are} \\ \text{flat and have proper support over } Z \end{matrix} \right\}.$$

If $\mathcal{O}(1)$ is a relatively ample line bundle and P a polynomial, then, as in (1.3), one can define $\mathit{Quot}_P(E/S)$.

5.16 Theorem. *Let X/S be a scheme (or an algebraic space), E a coherent sheaf on X. The functor $\mathit{Quot}(E/S)$ is represented by an algebraic space* $\mathrm{Quot}(E/S)$.

If $\mathcal{O}(1)$ is a relatively ample line bundle and P a polynomial then Quot_P (E/S) *is projective over S.*

5.17 Theorem. *Let X/S be a scheme (or an algebraic space), E a coherent sheaf on X, flat over S and $s \in S$ a point. Let F be a field and $p : \operatorname{Spec} F \to s \in S$ a morphism. Let $X^F = X \times_S \operatorname{Spec} F$ be the fiber over $\operatorname{Spec} F$ and and E^F the pull back of E to X^F. Let*

$$0 \to I_G \to E^F \to G \to 0$$

be an exact sequence, $\bar{p} : \operatorname{Spec} F \to \mathrm{Quot}(E/S)$ the corresponding morphism and set $[G] = \operatorname{im} \bar{p}$. Assume that S is equidimensional at s. Then

(5.17.1) *The dimension of every irreducible component of* $\mathrm{Quot}(E/S)$ *at $[G]$ is at least*

$$\dim_F \operatorname{Hom}_{X^F}(I_G, G) - \dim_F \operatorname{Ext}^1_{X^F}(I_G, G) + \dim_s S.$$

(5.17.2) *If equality holds, then $\mathrm{Quot}(E/S) \to S$ is a local complete intersection morphism at $[G]$. In particular, it is flat at $[G]$.*

I.6 Comparing Chow Varieties and Hilbert Schemes

The aim of this section is to describe some of the natural maps between the various schemes Hilb, Hom and Chow which are not obvious from the definitions. We see that in some important cases Univ → Chow is flat (6.5–6). Also, Chow turns out to be a covariant functor on the category of S-schemes (6.8). This property completely fails for Hilb.

6.1 Proposition. *Let $f : X/S \to Y/S$ be a morphism, $[W] \in \mathrm{Hilb}(X/S)$ a point and $W \subset X$ the corresponding subscheme. Assume that $f|W : W \to Y$ is a closed immersion. Then there is an open neighborhood $[W] \in U \subset \mathrm{Hilb}(X/S)$ such that*

$$\bar{f} : \mathrm{Univ}(X/S)|U \to X \times_S U \xrightarrow{f} Y \times_S U$$

is a closed immersion. In particular we obtain a morphism $f_* : U \to \mathrm{Hilb}(Y/S)$.

Proof. f gives a $\mathrm{Hilb}(X/S)$-morphism

$$\mathrm{Univ}(X/S) \to Y \times_S \mathrm{Hilb}(X/S),$$

which is a closed immersion on the fiber over $[W]$. By (1.10.1) it is a closed immersion over an open neighborhood $U \ni [W]$. □

6.2 Corollary. *Let $[f_0] \in \mathrm{Hom}_S(X,Y)$ be a point. Assume that the corresponding morphism $f_0 : X_0 \to Y_0$ is a closed immersion. Then there is an open neighborhood $[W] \in U \subset \mathrm{Hom}_S(X,Y)$ such that the natural morphism*

$$X \times_S U \to Y \times_S U$$

is a closed immersion. In particular we obtain a morphism $U \to \mathrm{Hilb}(Y/S)$.

Proof. By construction

$$\mathrm{Hom}_S(X,Y) \subset \mathrm{Hilb}(X \times_S Y).$$

Let W be the graph of f_0. Apply (6.1) to the second projection. □

6.3 Theorem. *Let X be a scheme over S.*

(6.3.1) *Taking the fundamental class of a subscheme* (3.1.3) *gives a natural morphism from the semi normalization of* Hilb *to* Chow*:*

$$FC : \mathrm{Hilb}_d^{sn}(X/S) \to \mathrm{Chow}_d(X/S).$$

(6.3.2) *If $V \subset X$ is geometrically reduced and has pure dimension d, then FC is injective near V.*

Proof. Let $\mathrm{Univ}_d \to \mathrm{Hilb}_d$ be the universal family. By (3.15)

$$[\mathrm{Univ}_d \times_{\mathrm{Hilb}} \mathrm{Hilb}_d^{sn}] \to \mathrm{Hilb}_d^{sn}$$

is a well defined family of algebraic cycles. This gives the required morphism $\mathrm{Hilb}^{sn} \to \mathrm{Chow}$. Since we defined Chow to be semi normal, in general FC does not descend to a morphism Hilb → Chow.

If V is reduced and has pure dimension d, then the same holds in an open neighborhood of V in Hilb. FC is clearly injective in this neighborhood. □

Next we show that under some conditions Univ → Chow is flat. This is nearly equivalent to the question: when is FC a local isomorphism.

6.4 Definition. Let $W = \sum w_i[W_i]$ be an algebraic cycle over a field k.

(6.4.1) We say that $x \in \cup W_i$ is a *normal point* (resp. a *smooth point*) of W if there is exactly one W_j such that $x \in W_j$; moreover W_j is normal (resp. smooth) at x and $w_j = 1$.

(6.4.2) We say that $x \in \cup W_i$ is a *reduced point* of W if $x \in W_j$ implies that $w_j = 1$.

(6.4.3) We say that $x \in \cup W_i$ is *geometrically* normal (resp. smooth or reduced) if x is a normal (resp. smooth or reduced) point of W_K where $K \supset k$ is an algebraic closure.

(6.4.4) Let $(g : U \to V)$ be a well defined family of proper algebraic cycles. If $u \in U$ is a reduced point of U, then we say that g is *flat* (resp. *smooth*) at u if $g : \mathrm{Supp}\, U \to V$ is flat (resp. smooth).

6.5 Theorem. *Let $(g : U \to V)$ be a well defined family of proper algebraic cycles. Assume that V is semi normal and g satisfies the Chow field condition. (By* (4.10) *this is always the case if V is weakly normal.) Let $v \in V$ and $u \in g^{[-1]}(v)$ a smooth point of $g^{[-1]}(v)$. Then g is smooth at u.*

Proof. Let $[U] = \sum m_i U_i$. If $u \in U_i$, then $m_i = 1$, even after an inseparable base change. Thus $U_i \to V$ is generically smooth.

We can view g as a family of algebraic cycles in a projective space $\mathbb{P}_V$. We use the notation of (3.23) and consider the diagram:

$$\begin{array}{ccc} U \subset \mathbb{P}_V & \xleftarrow{p_1} & I \\ & & \downarrow{\scriptstyle p_2} \\ & & (\mathbb{P}_V^\vee)^{d+1}. \end{array}$$

Let $I_U = p_1^{-1}(U) \subset I$. Since $u \in g^{[-1]}(v)$ is smooth, there are hyperplanes $H_1, \dots, H_{d+1} \in \mathbb{P}^\vee_{k(v,u)}$ such that

$$g^{[-1]}(v) \cap H_1 \cap \dots \cap H_{d+1} = \{u\} \quad \text{(as schemes)}.$$

Set $\bar{u} = (u, H_1, \dots, H_{d+1}) \in I_U$. We have a commutative diagram

$$\begin{array}{ccc} U & \stackrel{p_1}{\longleftarrow} & I_U \\ {\scriptstyle g}\downarrow & & \downarrow{\scriptstyle p_2} \\ V & \stackrel{Ch(g)}{\longleftarrow} & Ch(U). \end{array}$$

We prove that p_1 is smooth at $\bar{u}$, p_2 is a local isomorphism at $\bar{u}$ and $Ch(g)$ is smooth at $p_2(\bar{u})$. By (7.4.3) these imply that g is smooth at u.

$I_U \to U$ is always smooth. $Ch(g) : Ch(U) \to V$ is flat by (4.9), hence smooth at $p_2(\bar{u})$ iff its fiber is smooth at $p_2(\bar{u})$. To see the latter, consider the maps

$$g^{[-1]}(v) \stackrel{p_1}{\longleftarrow} I_{g^{[-1]}(v)} \stackrel{p_2}{\longrightarrow} Ch(g^{[-1]}(v)).$$

By construction p_1 is smooth. By (3.23.1.8)

$$p_2 : I_{g^{[-1]}(v)} \to Ch(g^{[-1]}(v))$$

is a local isomorphism at $\bar{u} = (u, H_1, \dots, H_{d+1})$. This implies that $Ch(g^{[-1]}(v))$ is smooth at $p_2(\bar{u})$. Thus $Ch(g)$ is smooth at $p_2(\bar{u})$. By (7.2.6) this implies that $Ch(U)$ is semi normal at $p_2(\bar{u})$.

$I_U \to Ch(U)$ is birational over z by (3.23.1.2). $p_2^{-1}(z) \cap I_U = \bar{u}$ (as sets) thus $I_U \to Ch(U)$ is finite over a neighborhood of z.

Since $Ch(U)$ is semi normal at $p_2(\bar{u})$, $p_2 : I_U \to Ch(U)$ is a local isomorphism near $p_2(\bar{u})$ iff it induces an isomorphism on the residue fields near $p_2(\bar{u})$. This condition can be checked on the nearby fibers of $Ch(U) \to V$.

$Ch(g)$ is smooth, hence it has smooth fibers. A finite and birational map is an isomorphism over a smooth point. This shows that p_2 is a local isomorphism near $p_2(\bar{u})$. □

6.5.1 Exercise. In many cases there is a simpler geometric proof of (6.5).

Claim. Let $f : X \to Y$ be a generically smooth projective morphism between normal schemes such that every fiber has dimension 1. Let H be a relatively ample line bundle. Assume that $y \mapsto H \cdot \operatorname{red} f^{-1}(y)$ is constant on Y. Let $x \in \operatorname{red} f^{-1}(y)$ be a smooth point such that $k(x) = k(y)$. Show that f is smooth at x.

Hints. We may assume that Y is the spectrum of a complete local ring and y the closed point. Let $D \subset X$ be a hyperplane section which intersects $\operatorname{red} f^{-1}(y)$ transversally at x. Show that D intersects the generic fiber in a single point. Thus $D \to Y$ is finite and birational, hence an isomorphism. Use [Matsumura86, 22.1] to conclude that f is flat at x.

More generally, discuss the case when Y is seminormal and $k(x) \supset k(y)$ is separable.

The following result is a stronger version of a conjecture of Catanese. The global case is also discussed in [Catanese92].

6.6 Theorem. *Let $(g : U \to V)$ be a well defined family of proper algebraic cycles. Assume that V is semi normal and g satisfies the Chow field condition.*

(By (4.10) this is always the case if V is weakly normal.) Let $v \in V$ and $u \in g^{[-1]}(v)$ a geometrically normal point of $g^{[-1]}(v)$. Then g is flat at u.

Proof. By definition there is exactly one irreducible component W_j of $\sum w_i W_i := g^{[-1]}(v)_K$ which contains u and for that component $w_j = 1$. Therefore g is smooth at the smooth points of W_j by (6.5). If $u \in U_i$ then also $W_j \subset U_i$, hence $a_i = 1$. This shows that $u \in U$ is a reduced point. $\mathcal{O}_{v,V} \to \mathcal{O}_{u,\operatorname{Supp} U}$ satisfies the conditions of (7.3.3), hence g is flat at u. □

6.6.1 Corollary (Notation as in (6.3)). *If $V \subset X$ is geometrically reduced, normal and has pure dimension d then FC is a local isomorphism at V.*

Proof. By (6.6), Univ → Chow is flat at $[V]$, thus $FC : \mathrm{Hilb}^{sn} \to \mathrm{Chow}$ has a local "semi" inverse Chow → Hilb at $[V]$. Since Chow is semi normal, we can lift this to an actual inverse $\mathrm{Chow} \to \mathrm{Hilb}^{sn}$ by (7.2.3.3). □

6.7 Definition. Let X/S be a scheme over S and $(g : U/S \to W/S)$ a well defined family of proper algebraic cycles of X. Let $f : X/S \to Y/S$ be a morphism, f_*U the push forward of the cycle U under the morphism

$$f : U/S \hookrightarrow X \times_S W \xrightarrow{(f,id)} Y \times_S W,$$

and $f_*g : f_*U/S \to W/S$ the natural projection. $(f_*g : f_*U/S \to W/S)$ is called the *push forward* of $(g : U/S \to W/S)$.

6.8 Theorem (Notation as above).

(6.8.1) *$(f_*g : f_*U/S \to W/S)$ is a well defined family of proper algebraic cycles of Y.*

(6.8.2) *If $(g : U/S \to W/S)$ satisfies the Chow field condition, then so does $(f_*g : f_*U/S \to W/S)$.*

(6.8.3) *The push forward induces a natural transformation of (partial) functors*

$$Chow_d(X/S) \to Chow_d(Y/S) \quad \text{and} \quad Chow_d^{\mathrm{small}}(X/S) \to Chow_d^{\mathrm{small}}(Y/S).$$

(6.8.4) *The above natural transformations induce a morphism*

$$f_* : \mathrm{Chow}_d(X/S) \to \mathrm{Chow}_d(Y/S),$$

which makes Chow_d into a covariant functor

$$\{\, S\text{-schemes} \,\} \to \{\, \text{semi normal } S\text{-schemes} \,\}.$$

Proof. First we need to check the conditions (3.10–11). Only (3.10.3–4) need some arguments. Let $g_i : U_i \to W_i$ be an irreducible component. We have a commutative diagram

$$\begin{array}{ccc} U_i & \xrightarrow{f} & f(U_i) \\ {\scriptstyle g_i}\downarrow & & \downarrow{\scriptstyle f_* g_i} \\ W & = & W. \end{array}$$

Every fiber of g_i has dimension d. Taking into account that $g_i = (f_* g_i) \circ f_*$ and the upper semicontinuity of the fiber dimension of a morphism, we conclude that every fiber of $f_* g_i$ has the same dimension $e \leq d$. If $e < d$, then by definition $f_*[U_i] = \emptyset$. Here it is very convenient that the empty family is considered to have arbitrary dimension. Otherwise let $a_i = \deg(U_i/f(U_i))$.

(3.10.4) is implied by the following formula:

$$\begin{aligned} (f_* g)^{[-1]}(w) &\stackrel{ess}{=\!=} \sum m_i a_i \lim_{h \to w} (f(U_i)/W) \\ &= \sum m_i \lim_{h \to w} (f_* U_i/W) \\ &= f_* \left(\sum m_i \lim_{h \to w} (U_i/W) \right) \\ &\stackrel{ess}{=\!=} f_* \left(g^{[-1]}(w) \right). \end{aligned}$$

Since

$$(f_* g)^{[-1]}(w) \stackrel{ess}{=\!=} f_* \left(g^{[-1]}(w) \right),$$

if $g^{[-1]}(w)$ is defined over a field E, so is $(f_* g)^{[-1]}(w)$. Hence by (4.5)

$$k^{ch}\left((f_* g)^{[-1]}(w) \right) \subset k^{ch}\left(g^{[-1]}(w) \right).$$

This shows (6.8.2).

(6.8.3) is an immediate consequence of (6.8.1–2) and of the definition of Chow pull back (3.18). Thus there is a natural transformation

$$Chow_d^{\text{small}}(X/S)(\,*\,) \to \text{Hom}_S(\,*\,, \text{Chow}_d(Y/S)),$$

which yields the required morphism in (6.8.4). □

6.9 Corollary. *Let X, Y be projective schemes over S. Then there is a natural morphism*

$$\text{Hom}_S(X, Y)^{sn} \to \text{Chow}(Y/S).$$

Proof. We have natural morphisms

$$\text{Hom}_S(X, Y)^{sn} \to \text{Hilb}(X \times_S Y/S) \to \text{Chow}(X \times_S Y/S) \to \text{Chow}(Y/S). \ \square$$

6.10 Exercise. The aim of this exercise is to prove a smoothability criterion for curves.

(6.10.1) Let Y be a smooth projective variety and $X \subset Y$ a reduced subvariety. Show that there is an exact sequence

$$0 \to Hom(\Omega^1_X, \mathcal{O}_X) \to T_Y|X \to Hom(I_X/I_X^2, \mathcal{O}_X) \to 0.$$

(6.10.2) Let C be a reduced curve which is embedded into a projective space $\mathbb{P}$. Show that if $H^1(C, \mathcal{O}_C(1)) = 0$, then $H^1(C, Hom(I_C/I_C^2, \mathcal{O}_C)) = 0$.

(6.10.3) (For those who are familiar with the deformation theory of singularities, e.g. [Artin76].) Let $P_i \in C$ be the singular points and $Def(P_i \in C)$ the miniversal deformation spaces. Show that there is a neighborhood $[C] \in U \subset \mathrm{Hilb}(\mathbb{P})$ and a morphism $r : U \to \prod_i Def(P_i \in C)$. Show that r is smooth at $[C]$ if $H^1(C, Hom(I_C/I_C^2, \mathcal{O}_C)) = 0$.

(6.10.4) Conclude that a reduced and proper curve is smoothable iff its singularities are smoothable.

I.7 Appendix: Some Results from Commutative Algebra

7.1 Field Extensions and Artin Algebras

(R, m) (or (R_i, m_i) etc.) always stands for an Artin local k-algebra with maximal ideal m. $l(R)$ denotes the length of R over R/m.

7.1.1 Proposition. *Let $L \supset k$ be a finite algebraic field extension.*

(7.1.1.1) *If $L \supset k$ is separable, then*

$$(R, m) \otimes_k L \cong \sum_i (R_i, m_i), \quad \textit{and} \quad l(R) = l(R_i)\ \forall i.$$

(7.1.1.2) *If $L \supset k$ is purely inseparable, then*

$$(R, m) \otimes_k L \cong (R', m').$$

(7.1.1.3) *If $L \supset k$ is normal algebraic, then*

$$(R, m) \otimes_k L \cong \sum_i (R_i, m_i), \quad \textit{and} \quad l(R_i) = p^a \cdot l(R)\ \forall i, \ (p = \mathrm{char}\, k, a \geq 0).$$

(7.1.1.4) *If $L \supset k$ is algebraic, then*

$$(R, m) \otimes_k L \cong \sum_i (R_i, m_i) \quad \textit{and} \quad [L : k] \cdot l(R) = \sum [R_i/m_i : R/m] \cdot l(R_i).$$

Proof. Let $K = R/m$. Since R is Artin, there is a field of representatives, thus we can write (noncanonically) $R = K + m$.

If (L_i, m_i) is a local K-algebra, then $R \otimes_K L_i = L_i + m \otimes_K L_i$ is local with maximal ideal $m_i + m \otimes_K L_i$. Thus it has residue field L_i/m_i and $l(R \otimes_K L_i) = l(R) \cdot l(L_i)$.

Let $K \otimes_k L \cong \sum(L_i, m_i)$. Then

$$R \otimes_k L \cong R \otimes_K K \otimes_k L \cong \sum_i R \otimes_K L_i.$$

This shows that it is sufficient to prove (7.1.1) in the special case when $R = K$ is a field.

It is sufficient to prove the first two claims for extensions L/k generated by one element. If $L = k[x]/f(x)$, then $k[x]/f(x) \otimes_k K \cong K[x]/f(x)$. If f is separable then over K it decomposes into relatively prime factors $f = \prod f_i$ and $K[x]/f(x) \cong \sum K[x]/f_i(x)$ is a sum of fields. If f is purely inseparable of degree $p = \operatorname{char} k$ over k, then either f stays irreducible over K, thus $K[x]/f(x)$ is a field or f becomes a pth-power and, then $K[x]/f(x) \cong K[x]/x^p$.

If $L \supset k$ is normal, then the Galois group of L/k acts on $L \otimes_k K$ and the action is transitive on the maximal ideals.

Finally, (7.1.1.4) is a combination of (7.1.1.1–2). □

7.2 Semi and Weak Normalization

The following definitons and their basic properties are taken from [Andreotti-Noguet67; Andreotti-Bombieri69; Traverso70]. The notions of semi and weak normalization coincide in characteristic zero.

7.2.1 Definition. Let $f : Z \to X$ be a finite and surjective morphism of reduced schemes.

(7.2.1.1) The *semi normalization* (resp. *weak normalization*) of X in Z is a scheme X' together with a finite and surjective morphism $m : X' \to X$ such that

(7.2.1.1.1) for every $x \in X$ there is a unique $x' \in X'$ such that $m(x') = x$;

(7.2.1.1.2^{semi}) for every $x \in X$ the field extension $k(x') \supset k(x)$ is an isomorphism;

(7.2.1.1.2^{weak}) for every $x \in X$ the field extension $k(x') \supset k(x)$ is purely inseparable and an isomorphism if $x \in X$ is a generic point;

(7.2.1.1.3) there is a factorization $f : Z \to X' \to X$;

(7.2.1.1.4) if $g : Y \to X$ satisfies the above three properties, then there is a unique factorization

$$m : X' \to Y \xrightarrow{g} X.$$

(7.2.1.2) X is called *semi normal in Z* (resp. *weakly normal in Z*) if it is isomorphic to its semi (resp. weak) normalization in Z.

(7.2.1.3) X is called *semi normal* (resp. *weakly normal*) if it is isomorphic to its semi (resp. weak) normalization in its normalization $Z = \bar{X}$.

(7.2.1.4) We use the notation $X^{Z/sn}$ resp. $X^{Z/wn}$ to denote the semi normalization (resp. weak normalization) of X in Z. If $Z = \bar{X}$ is the normalization of X, then we use the shorter versions X^{sn} resp. X^{wn}.

(7.2.1.5) For an arbitrary (nonreduced) scheme Y, set $Y^{sn} := (\operatorname{red} Y)^{sn}$ resp. $Y^{wn} := (\operatorname{red} Y)^{wn}$.

7.2.2 Exercise. (7.2.2.1) Let R be a complete one dimensional equicharacteristic local ring with algebraically closed residue field k. Then $\operatorname{Spec} R$ is semi normal iff

$$R \cong k[[x_1, \dots, x_n]]/(x_i x_j : 1 \le i < j \le n).$$

I.e. $\operatorname{Spec} R$ is the completion of the coordinate ring of the n coordinate axes in k^n.

(7.2.2.2) Let k be a field of characteristic $p > 0$. For every $n > 0$, the ring

$$R_n = k[x^{p^n}, yx^i : 0 \le i \le p^n - 1] \subset k[x, y]$$

is semi normal but not weakly normal. The weak normalization is $k[x, y]$.

(7.2.2.3) Let $X_1 = \operatorname{Spec} k[x^2, x^3, y, xy]$ and $X_2 = \operatorname{Spec} k[u, v]$. Let $C_1 \subset X_1$ resp. $C_2 \subset X_2$ be the cuspidal cubics given by the ideals (y, xy) resp. $(u^2 - v^3)$. Let X be obtained by gluing together X_1 and X_2 along $C_1 \cong C_2$. Then X is semi normal, but X_1 is not.

7.2.3 Proposition. *Let $f : Z \to X$ be a finite and surjective morphism of reduced schemes. Then:*

(7.2.3.1) *X has a semi (resp. weak) normalization $X^{Z/sn}$ (resp. $X^{Z/wn}$) in Z.*

(7.2.3.2) *A normal scheme is weakly normal. A weakly normal scheme is semi normal.*

(7.2.3.3) *Let $g : Y \to X$ be a morphism. Assume that Y is semi normal. Then g can be factored, in a unique way, as*

$$g : Y \to X^{sn} \to X.$$

(7.2.3.4) *We have a natural morphism $f^{sn} : Z^{sn} \to X^{sn}$. Thus $\mathcal{O}_Z$ and $\mathcal{O}_{X^{sn}}$ can be viewed as subsheaves of $\mathcal{O}_{Z^{sn}}$. Then*

$$\mathcal{O}_{X^{Z/sn}} = \mathcal{O}_Z \cap \mathcal{O}_{X^{sn}} \subset \mathcal{O}_{Z^{sn}}.$$

Proof. For a point $x \in X$, let $\bar{x} = f^{-1}(x) \subset Z$ with reduced scheme structure. $k(\bar{x})$ is the direct sum of the residue fields of the points of $\bar{x}$, $i_x : k(x) \to k(\bar{x})$ the natural injection. For an open affine $U \subset X$, let

$$\mathcal{O}^{sn}(U) := \{h \in \mathcal{O}_{n^{-1}(U)} | \forall x \in U, h(\bar{x}) \in \operatorname{im} i_x\}, \quad \text{and}$$

$$\mathcal{O}^{wn}(U) := \left\{ h \in \mathcal{O}_{n^{-1}(U)} | \forall x \in U, h(\bar{x})^q \in \operatorname{im} i_x, \quad \begin{matrix} \text{where } q \text{ is a power of} \\ \operatorname{char} k(x), \text{ and } q = 1 \\ \text{if } x \text{ is a generic point.} \end{matrix} \right\}.$$

These define sheaves of rings $\mathcal{O}^{sn}$ and $\mathcal{O}^{wn}$ over X. Set

$$X^{Z/sn} = \operatorname{Spec}_X \mathcal{O}^{sn} \quad \text{and} \quad X^{Z/wn} = \operatorname{Spec}_X \mathcal{O}^{wn}.$$

If $g : Y \to X$ satisfies the conditions (7.2.1.1–3), then there is a factorization

$$f : Z \to Y \xrightarrow{g} X.$$

It is also clear that $g_*\mathcal{O}_Y$ is a subsheaf of $\mathcal{O}^{sn}$ (resp. $\mathcal{O}^{wn}$). This shows (7.2.3.1) while (7.2.3.2) is clear.

(7.2.3.3) is also obvious. Observe that the analog of (7.2.3.3) is false for the weak normalization (and for the normalization). An example is provided by $\operatorname{Spec} k[x^{p^n}] \to \operatorname{Spec} R_n$ where R_n is defined in (7.2.2.2).

Finally, it is clear that $\operatorname{Spec}_X(\mathcal{O}_Z \cap \mathcal{O}_{X^{sn}}) \to X$ satisfies the requirements (7.2.1.1.1–3), thus $\mathcal{O}_{X^{Z/sn}} \supset \mathcal{O}_Z \cap \mathcal{O}_{X^{sn}}$. Conversely, there are natural morphisms $Z \to X^{Z/sn}$ and $X^{sn} \to X^{Z/sn}$. These show that $\mathcal{O}_{X^{Z/sn}} \subset \mathcal{O}_Z \cap \mathcal{O}_{X^{sn}}$. □

7.2.4 Notation. Let $f : Z \to X$ be a finite and surjective morphism of reduced schemes which is an isomorphism over every generic point of X (for instance, Z is the normalization of a reduced scheme X). Let $C_X \subset X$ be the closed subset where f is not a local isomorphism and $C_Z = f^{-1}(C_X)$ (with reduced scheme structure). The corresponding ideal sheaves are denoted by I_X resp. I_Z.

7.2.5 Proposition. *Assumptions as in* (7.2.4). *X is semi normal in Z iff $I_X = I_Z$ and C_X is semi normal in C_Z.*

Proof. The question is local on X, thus we may assume that X and Z are affine. Let $h \in \mathcal{O}_Z$. Since f is an isomorphism over $X - C_X$, we see from the construction (7.2.3) that $h \in \mathcal{O}^{sn}(X)$ iff $h(\bar{x}) \in \operatorname{im} i_x$ for every $x \in C_X$.

Assume that X is semi normal in Z. If $h \in I_Z$ then h vanishes along C_Z. Hence $h \in \mathcal{O}^{sn}(X)$ and so $h \in I_X$. Thus $I_X = I_Z$. Let $h' \in \mathcal{O}_{C_Z}$ be such that h' is in the semi normalization of C_X in C_Z. Extend h' to a function h on Z. Then $h \in \mathcal{O}^{sn}(X)$, hence $h' \in \mathcal{O}_{C_X}$. Thus C_X is semi normal in C_Z.

Conversely, assume that $I_X = I_Z$ and C_X is semi normal in C_Z. This means that if $h \in \mathcal{O}_Z$ is any function and $h|C_Z$ is in the semi normalization of C_X in C_Z then $h \in \mathcal{O}_X$. Thus X is semi normal in Z. □

7.2.6 Proposition. *Let $f : Y \to X$ be a flat morphism with geometrically normal fibers. Then*

$$Y^{sn} = Y \times_X X^{sn}.$$

Proof. It is clear that we have a morphism $Y^{sn} \to Y \times_X X^{sn}$. Thus it is sufficient to prove (7.2.6) in case X is semi nomal.

The proof is by induction on $\dim X$. There is nothing to prove if $\dim X = 0$.

Let $\bar{X} \to X$ be the normalization. Then $Y \times_X \bar{X}$ is flat over $\bar{X}$ with geometrically normal fibers, thus $Y \times_X \bar{X}$ is normal by [Matsumura80, 21.E] and $Y \times_X \bar{X} = \bar{Y}$.

Let $C_X \subset X$, $C_{\bar{X}} \subset \bar{X}$, $C_Y \subset Y$, $C_{\bar{Y}} \subset \bar{Y}$ be as in (7.2.4). By [Matsumura80, 21.E], $Y \times_X C_X$ and $Y \times_X C_{\bar{X}}$ are reduced, thus $C_Y = Y \times_X C_X$ and $C_{\bar{Y}} = Y \times_X C_{\bar{X}}$. By (7.2.5) $I_X = I_{\bar{X}}$, hence also $I_Y = I_{\bar{Y}}$.

By (7.2.5) C_X is semi normal in $C_{\bar{X}}$. We need to show that C_Y is semi normal in $C_{\bar{Y}}$. By (7.2.3.4),

$$\mathcal{O}_{C_X} = \mathcal{O}_{C_{\bar{X}}} \cap \mathcal{O}_{(C_X)^{sn}} \subset \mathcal{O}_{(C_{\bar{X}})^{sn}}.$$

By induction on the dimension,

$$(C_Y)^{sn} = Y \times_X (C_X)^{sn} \quad \text{and} \quad (C_{\bar{Y}})^{sn} = Y \times_X (C_{\bar{X}})^{sn}.$$

Intersection of subsheaves commutes with flat base change, thus

$$\mathcal{O}_{C_Y} = \mathcal{O}_{C_{\bar{Y}}} \cap \mathcal{O}_{(C_Y)^{sn}} \subset \mathcal{O}_{(C_{\bar{Y}})^{sn}}.$$

This shows that C_Y is semi normal in $C_{\bar{Y}}$, thus Y is semi normal in $\bar{Y}$. □

7.2.7 Corollary. *Let $f : Y \to X$ be a faithfully flat morphism with geometrically normal fibers. Then X is semi normal iff Y is.* □

7.2.8 Remark. I do not know how to prove (7.2.6–7) in the weakly normal case.

7.3 A Flatness Theorem

For ease of reference, we restate the following special case of [Kollár95b].

7.3.1 Theorem. *Let $f : X \to Y$ be a morphism of pure relative dimension n. Assume for simplicity that X and Y are of finite type. Let $Z = f^{-1}(y)$. Assume that*

(7.3.1.1) *Z is reduced at its generic points;*
(7.3.1.2) red Z *is normal (or at least S_2);*
(7.3.1.3) *Y is reduced;*
(7.3.1.4) *X has no embedded points;*
(7.3.1.5) *f is flat at the generic points of Z.*

Then Z is reduced and f is flat along Z.

7.4 Miscellaneous

7.4.1 Proposition. *Let (A, m) be a local ring and B an A-algebra. Let M, N be finite B-modules. Assume that N is flat over A and M is separated in the m-adic topology. Let $g : M \to N$ be a morphism and set $Q = \operatorname{coker} g$.*

(7.4.1.1) *The following are equivalent:*

(7.4.1.1.1) *g is injective and Q is flat over A;*
(7.4.1.1.2) *$g \otimes (A/m)$ is injective.*

(7.4.1.2) *g is an isomorphism iff $g \otimes (A/m)$ is an isomorphism.*

Proof. [Matsumura86, 22.5] states (7.4.1.1) in case B is also local, but the proof uses only that M is separated in the m-adic topology.

Only one direction of the second part is nontrivial. If $g \otimes_A (A/m)$ is an isomorphism, then g is injective by the first part. By the Nakayama lemma g is also surjective. □

7.4.2 Proposition [EGA, 0.10.3.1]. *Let A be a Noetherian local ring with residue field k, $F \supset k$ a field extension. Then there is a Noetherian local A-algebra B such that B is faithfully flat and unramified over A and*

$$[A/m_A \hookrightarrow B/m_B] \cong [k \hookrightarrow F].$$

Proof. We prove the case when $F \supset k$ is finitely generated. The general case uses the same construction and a study of the resulting direct limit.

By induction, it is sufficient to consider the case when $F = k(x)$ for a single $x \in F$. If x is transcendental over k, then let

$$B = A[t]_{(m,t)}.$$

If x is algebraic over k, then let $x^m + \sum_0^{m-1} a_i x^i = 0$ be its minimal polynomial. Let $\bar{a}_i \in A$ be any lifting of a_i. Set

$$B = A[t]/(t^m + \sum_0^{m-1} \bar{a}_i t^i). \qquad \square$$

7.4.3 Proposition. *Let $X \xrightarrow{f} Y \xrightarrow{g} Z$ be morphisms of schemes. The following are equivalent:*

(7.4.3.1) *f and $g \circ f$ are flat (resp. smooth) at $x \in X$;*
(7.4.3.2) *f is flat (resp. smooth) at $x \in X$ and g is flat (resp. smooth) at $y = f(x) \in Y$.*

Proof. (7.4.3.2)$\Rightarrow$ (7.4.3.1) is in [Hartshorne77, III.9.2 and 10.1]. Since $\mathcal{O}_{x,X}$ is faithfully flat over $\mathcal{O}_{y,Y}$, the reverse implication is easy for flatness.

Next, consider smoothness. We already know that g is flat at y, thus g is smooth at y iff $g^{-1}(g(y))$ is smooth at y. Thus we are reduced to the case when $Z = \operatorname{Spec} k$ for a field $k = k(g(y))$. By [Matsumura80, 21.D] Y is regular at y, furthermore it stays regular after an arbitrary field extension $K \supset k$. Thus Y is smooth at y. $\square$

7.4.4 Theorem (Macaulay). *Let $R = k[x_0, \dots, x_n]$ be the polynomial ring in $n+1$ variables and $m = (x_0, \dots, x_n)$. Let $f_0, \dots, f_n \in R$ be homogeneous polynomials. Assume that $(f_0, \dots, f_n)$ is m-primary. Let $M^{(d)}$ denote the degree d piece of a graded R-module M. Then*

(7.4.4.1) $\dim_k (f_0, \dots, f_n)^{(d)}$ *depends only on* $\deg f_0, \dots, \deg f_n$,

(7.4.4.2) $(f_0, \dots, f_n) \supset m^{1+\sum(\deg f_i - 1)}$.

Proof. The following proof was explained to me by M. Green.

Le us look at the Koszul complex of the f_i [Matsumura80, 18.D]:

$$\cdots \wedge^2 \Big(\sum_i R(-\deg f_i)\Big) \xrightarrow{\wedge^2(f_0,\dots,f_n)} \sum R(-\deg f_i) \xrightarrow{(f_0,\dots,f_n)} I \to 0.$$

This is exact, thus $\dim_k I^{(d)}$ can be computed as

$$\dim_k I^{(d)} = \sum (-1)^{r-1} \dim_k \Big(\wedge^r \Big(\sum_i R(-\deg f_i)\Big)\Big)^{(d)},$$

which clearly depends only on $\deg f_0, \dots, \deg f_n$.

Using (7.4.4.1), we see that it is sufficient to prove (7.4.4.2) for a particular choice of the f_i. Set $f_i' = x_i^{\deg f_i}$. It is clear that $(f_0', \dots, f_n') \supset m^{1+\sum(\deg f_i - 1)}$. $\square$

7.4.4.3 Corollary (Notation as above). *Let $g_0, \dots, g_s \in R$ be homogeneous and assume that* $\deg g_0 \geq \dots \geq \deg g_s$. *Then $(g_0, \dots, g_s)$ is m-primary iff*

$$(g_0, \dots, g_s) \supset m^{\deg g_0 + \dots + \deg g_n - n}.$$

Proof. We may assume that k is infinite. The condition is clearly sufficient. Conversely assume that $(g_0, \dots, g_s)$ is m-primary. Let $g_i' = g_i + \sum_{j>i} h_{ij} g_j$, where the h_{ij} are general such that $\deg h_{ij} = \deg g_i - \deg g_j$. Then $(g_0', \dots, g_n')$ is m-primary and by (7.4.4),

$$(g_0, \dots, g_s) \supset (g_0', \dots, g_n') \supset m^{\deg g_0 + \dots + \deg g_n - n}. \qquad \square$$

Chapter II. Curves on Varieties

Chapter I discussed the general theory of parametrizing subvarieties of a given variety and morphisms between varieties. In this chapter, these general results are used to study curves on varieties and morphisms from curves to varieties.

General deformation theory tells us that the tangent space to Hilb or Hom is given by H^0 of a certain sheaf and the obstructions are in H^1 of the same sheaf. In the curve case $H^0 - H^1$ is the Euler characteristic, which can usually be computed easily. Thus a simple Euler characteristic computation gives a lower bound for the dimension of the deformation space. This is the content of (II.1.2), which is used repeatedly later. The existence of this lower bound is the main reason why the deformation theory of curves is especially manageable.

Section 1 contains general results about the Hilbert scheme of curves on a scheme and about morphisms from curves to schemes. We also consider deformations with base conditions. In the cases that are needed later, it is possible to reduce this question to the already studied deformation problem. Deformations of curves are also considered briefly. We only need a lower bound for the dimension of the miniversal deformation space of a curve; this can be obtained by the method of Hilbert schemes rather easily.

Chow varieties parametrize curves on a variety. In particular, a suitable subvariety of the Chow variety parametrizes rational curves. One can also view rational curves as the images of maps from $\mathbb{P}^1$ and it is desirable to have a parametrization of rational curves in this form as well. These questions are considered in Sect. 2. The main existence result (2.11) is not very difficult in characteristic zero; some more work is needed in positive characteristic. An alternative way to get the same result is to start with $\mathrm{Hom}(\mathbb{P}^1, X)$ and take the quotient by $\mathrm{Aut}(\mathbb{P}^1)$. This approach is somewhat more technical; see [Mori79] for details.

In Sect. 3 we consider morphisms from curves to varieties where there are no obstructions to deformation in a strong sense. These morphisms behave very well. There are few such morphisms from higher genus curves, but many from $\mathbb{P}^1$. These are called free morphisms and they play a central role in Chap. IV. The most important result of the section (3.11) says that if X is a smooth projective variety over a field of characteristic zero and $x \in X$ a very general point, then every rational curve through x is free. Despite the ease of its proof, this is a rather powerful result. In practice, it is very hard to determine which points are "very general". The answer is not known even for simple examples like surfaces in $\mathbb{P}^3$.

The cone of curves is introduced and studied through some examples in Sect. 4. This cone first appears in [Hironaka60, Kleiman66], but the main impetus for its study comes from [Mori82]. One of the basic results is the so called Theorem on the Base of Néron–Severi (4.5), which says that the vector space of 1-cycles modulo numerical equivalence is finite dimensional. Over $\mathbb{C}$, this follows easily from the finite dimensionality of the second homology group. In positive characteristic, the argument is more complicated and the only proof I know relies on the finite generation of Mordell–Weil groups over function fields. I would very much like to see a more elementary approach. The cone of curves is another example where explicit computations are very difficult. Examples show the complexity of the cone even for surfaces (4.15–18).

The first major application of the previous techniques is in Sect. 5. This is the bend–and–break technique of [Mori79]. The original form and its later development [Miyaoka-Mori86] are proved in (5.4–5). The main consequence (5.8) asserts that if X is a smooth projective variety and X contains a curve C which has negative intersection number with the canonical class, then X contains such a rational curve as well. Moreover, it gives rather precise information about the resulting rational curve. Mori's original proof (which is still the only one known) proceeds through positive characteristic, even for varieties defined over $\mathbb{C}$. The method is very efficient and powerful, but a more differential geometric approach would be quite interesting to see.

Section 6 is more of an aside. It turns out that the bend-and-break method of producing rational curves sometimes predicts the existence of a rational curve which should not exist. Through a construction of [Ekedahl88] this can be used to prove vanishing results in positive characteristic. The special cases (6.4–5) are the most transparent. The method is rather nice and provides an interesting application of the bend–and–break technique. These results are used occasionally later, but are not essential.

From the technical point of view Sect. 7 is the most involved. Given a morphism from a reducible curve to a variety X we would like to obtain it as a limit of morphisms from smooth curves to X. The most powerful result is (7.9). This is the main reason why the general theory of Hilbert schemes had to be developed in the relative setting. The main point of these results becomes clear through the applications in Chap. IV.

II.1 Deformations of Morphisms of Curves

The aim of this section is to translate the general results of Chap. I to the special case of the Hilbert scheme of curves. The main special feature of this case is that $h^0 - h^1$ is an Euler characteristic, thus readily computable in many cases. The first main result is (1.2) which is a direct consequence of (I.2.16). A more sophisticated variant is (1.3).

We frequently use deformations which are trivial on a subscheme. These behave predictably and (1.7) is a perfect analogue of (I.2.16).

(1.8) is a rather straightforward computation describing when a general deformation of a morphism is an immersion or an embedding.

The Hilbert schemes give information about the deformations of curves which are not a priori embedded. The curve case is again much simpler due to the fact that any flat deformation of a curve is relatively projective. Finally (1.13–17) contain some lower bounds on the dimension of Hilb and Chow which do not readily follow from (1.2).

1.1 Notation. Let $f : C \to X$ be a morphism from a proper curve to a scheme and L a line bundle or Cartier divisor on X. We use any of the following notation to denote the intersection number of C and L:

$$C \cdot L = C \cdot_f L = \deg_C f^*L = f_*[C] \cdot L.$$

(See (I.3.1.5) for a definition of the last expression.)

1.2 Theorem. *Let C be a proper algebraic curve without embedded points and $f : C \to Y$ a morphism to a smooth variety Y of pure dimension n. Then*

$$\dim_{[f]} \operatorname{Hom}(C, Y) \geq -K_Y \cdot C + n\chi(\mathcal{O}_C),$$

and equality holds if $H^1(C, f^*T_Y) = 0$. In fact, every irreducible component of $\operatorname{Hom}(C, Y)$ *at $[f]$ has dimension at least $-K_Y \cdot C + n\chi(\mathcal{O}_C)$.*

1.2.1 Remark. Using the results from (I.5) we see that the same estimate holds for algebraic or complex analytic spaces as well.

Proof. If F is a locally free sheaf on a scheme Z, then $\operatorname{Ext}^i_Z(F, \mathcal{O}_Z) = H^i(Z, F^*)$ (cf. [Hartshorne77, III.6.3 and 7]). Thus by (I.2.16) we have

$$\begin{aligned}\dim_{[f]} \operatorname{Hom}(C, Y) &\geq \operatorname{Hom}_C(f^*\Omega^1_Y, \mathcal{O}_C) - \operatorname{Ext}^1_C(f^*\Omega^1_Y, \mathcal{O}_C)\\ &= h^0(C, f^*T_Y) - h^1(C, f^*T_Y) = \chi(C, f^*T_Y)\\ &= \deg f^*T_Y + n\chi(\mathcal{O}_C) = -K_Y \cdot C + n\chi(\mathcal{O}_C). \qquad \square\end{aligned}$$

It is very desirable to have similar formulas for singular Y as well. Unfortunately, very little is known. The following result is a weak step in this direction.

1.3 Theorem. *Let C be a proper algebraic curve without embedded points and $f : C \to Y$ a morphism to a quasi projective variety Y of pure dimension n. Assume that the image of every irreducible component of C intersects the smooth locus of Y. Assume furthermore that Y is a local complete intersection. Then*

$$\begin{aligned}\dim_{[f]} \operatorname{Hom}(C,Y) &\geq \dim \operatorname{Hom}(f^*\Omega^1_Y, \mathcal{O}_C) - \dim \operatorname{Ext}^1(f^*\Omega^1_Y, \mathcal{O}_C)\\ &= -K_Y \cdot C + n\chi(\mathcal{O}_C).\end{aligned}$$

In fact, every irreducible component of $\operatorname{Hom}(C,Y)$ *at* $[f]$ *has dimension at least* $-K_Y \cdot C + n\chi(\mathcal{O}_C)$.

Proof. Y can be embedded into a smooth variety $Y \subset V$. Let J be the ideal sheaf of Y in V. Since Y is a local complete intersection we have an exact sequence

$$0 \to J/J^2 \to \Omega^1_V|Y \to \Omega^1_Y \to 0.$$

The sequence splits locally at smooth points of Y. Pull it back to obtain

$$f^*(J/J^2) \xrightarrow{d} f^*\Omega^1_V \to f^*\Omega^1_Y \to 0.$$

$f^*(J/J^2)$ is locally free and the sequence splits generically on each component of C. Therefore d is generically injective on each component of C, hence it is injective. Thus

$$0 \to f^*(J/J^2) \xrightarrow{d} f^*\Omega^1_V \to f^*\Omega^1_Y \to 0 \tag{1.3.1}$$

is exact. We obtain the long exact sequence

$$\begin{aligned}0 \longrightarrow \operatorname{Hom}(f^*\Omega^1_Y, \mathcal{O}_C) &\longrightarrow \operatorname{Hom}(f^*\Omega^1_V, \mathcal{O}_C) \longrightarrow \operatorname{Hom}(f^*(J/J^2), \mathcal{O}_C)\\ \longrightarrow \operatorname{Ext}^1(f^*\Omega^1_Y, \mathcal{O}_C) &\longrightarrow \operatorname{Ext}^1(f^*\Omega^1_V, \mathcal{O}_C) \xrightarrow{r} \operatorname{Ext}^1(f^*(J/J^2), \mathcal{O}_C)\end{aligned}$$

The dual of r is the morphism

$$H^0(C, \omega_C \otimes f^*(J/J^2)) \to H^0(C, \omega_C \otimes f^*\Omega^1_V)$$

which is injective. Thus r is surjective and we obtain that

$$\begin{aligned}&\dim \operatorname{Hom}(f^*\Omega^1_Y, \mathcal{O}_C) - \dim \operatorname{Ext}^1(f^*\Omega^1_Y, \mathcal{O}_C)\\ &= \chi(C, (f^*\Omega^1_V)^*) - \chi(C, f^*(J/J^2)^*)\\ &= c_1((\det f^*\Omega^1_V)^{-1} \otimes \det f^*(J/J^2)) + n\chi(\mathcal{O}_C)\\ &= -\deg_C f^*K_Y + n\chi(\mathcal{O}_C).\end{aligned} \tag{1.3.2}$$

By (I.2.16) this implies the result. □

1.3.3 Remark. (1.3) also holds if Y is a scheme, algebraic space or a complex analytic space. The proof needs to be modified as follows. (We use scheme terminology.)

Let $\{Y_i\}$ be an affine cover of Y. Let $C_i = f^{-1}(Y_i)$ and $f_i : C_i \to Y_i$ the restriction. Each Y_i can be embedded into a smooth scheme $Y_i \subset V_i$. Let J_i be the ideal sheaf of Y_i in V_i. Since Y_i is a local complete intersection we have an exact sequence

$$0 \to J_i/J_i^2 \to \Omega^1_{V_i}|Y_i \to \Omega^1_{Y_i} \to 0.$$

As in the proof of (1.3) we see that the sequence

$$0 \to f_i^*(J_i/J_i^2) \xrightarrow{d} f_i^*\Omega^1_{V_i} \to f_i^*\Omega^1_{Y_i} \to 0.$$

is exact.

By the above considerations $f^*\Omega^1_Y$ has projective dimension one. Since C is projective, $f^*\Omega^1_Y$ admits a global projective resolution of length two

$$0 \to E \to F \to f^*\Omega^1_Y \to 0$$

with locally free sheaves E and F. As in (1.3.2) we get that

$$\begin{aligned} &\dim \operatorname{Hom}(f^*\Omega^1_Y, \mathcal{O}_C) - \dim \operatorname{Ext}^1(f^*\Omega^1_Y, \mathcal{O}_C) \\ &= \chi(C, F^*) - \chi(C, E^*) \\ &= c_1(\det F^{-1} \otimes \det E) + n\chi(\mathcal{O}_C). \end{aligned} \tag{1.3.3.1}$$

To compute the last term observe that there are canonical local isomorphisms

$$\det E^{-1} \otimes \det F|C_i \cong \det f_i^*(J_i/J_i^2)^{-1} \otimes \det(f_i^*\Omega^1_{V_i}) \cong f_i^*\omega_{Y_i}.$$

Thus we have a global isomorphism

$$\det E^{-1} \otimes \det F \cong f^*\omega_Y.$$

By (I.5.14) we are done. □

1.3.4 Exercise. Let S be a normal surface, C a smooth curve and $f : C \to S$ a morphism. Prove that

$$\dim_{[f]} \operatorname{Hom}(C, S) \geq -K_S \cdot C + 2\chi(\mathcal{O}_C).$$

Hint: Use the minimal resolution $S' \to S$.

1.4 Definition. Let X/S and Y/S be schemes, $B \subset X$ a subscheme, proper over S and $g : B \to Y$ a morphism. $Hom(X, Y, g)$ is the functor

$$Hom(X,Y,g)(T) = \left\{ \begin{matrix} T\text{-morphisms } f : X \times_S T \to Y \times_S T \\ \text{such that } f|B \times_S T = g \times_S id_T. \end{matrix} \right\}.$$

If $B = \{b_i\}$ is finite and $y_i = g(b_i)$, then we sometimes use the notation

$$Hom(X, Y, b_i \mapsto y_i).$$

It is possible to develop the infinitesimal theory of the functors $Hom(X, Y, g)$ along the lines of (I.2) (see, for instance, [Mori79]). We need only some special cases when the necessary results can easily be established by indirect methods. Representability of $Hom(X, Y, g)$ is reduced to the existence of $\mathrm{Hom}(X, Y)$ by the following obvious construction:

1.5 Proposition (Notation as above). *Assume that X/S and B/S are flat. Let*

$$R : \mathrm{Hom}(X, Y) \to \mathrm{Hom}(B, Y)$$

be the restriction morphism. $g : B \to Y$ gives a section $\bar{g} : S \to \mathrm{Hom}(B, Y)$. Then

$$\mathrm{Hom}(X, Y, g) := R^{-1}(\bar{g}(S)) \subset \mathrm{Hom}(X, Y)$$

represents $Hom(X, Y, g)$. □

1.6 Corollary (Notation as above). *Assume that X/S is flat, B/S is flat and finite of degree k and Y/S is smooth of relative dimension n. Then*

$$\mathrm{Hom}(X, Y, g) \subset \mathrm{Hom}(X, Y)$$

is locally defined by kn equations. In particular,

$$\mathrm{codim}(\mathrm{Hom}(X, Y, g), \mathrm{Hom}(X, Y)) \leq kn.$$

Proof. Let $p : B \to S$ be the projection. By (I.2.16) $\mathrm{Hom}(B, Y)$ is smooth over S of relative dimension $\mathrm{rank}\, p_*(g^*T_{Y/S}) = kn$. Thus $\bar{g}(S) \subset \mathrm{Hom}(B, Y)$ is locally defined by kn equations. Pulling back these equations by R we obtain local defining equations for $\mathrm{Hom}(X, Y, g)$. □

The following is the main theorem describing $\mathrm{Hom}(X, Y, g)$.

1.7 Theorem. *Let C/S be a flat projective curve without embedded points and Y/S a smooth scheme over S. Let $B/S \subset C/S$ be a closed subscheme finite and flat over S. Assume that C/S is smooth along B. Let $G : B/S \to Y/S$ be a morphism. Let $s \in S$ be a point and $f_s \in Hom(C_s, Y_s, G_s)(k(s))$ a morphism. Then:*

(1.7.1) *The Zariski tangent space of* $\mathrm{Hom}(C_s, Y_s, G_s)$ *at* $[f_s]$ *is isomorphic to*

$$H^0(C_s, f^*T_{Y_s} \otimes I_{B_s}).$$

(1.7.2) *Assume that S is equidimensional at s. The dimension of every irreducible component of* $\mathrm{Hom}(C, Y, G)$ *at* $[f_s]$ *is at least*

$$h^0(C_s, f^*T_{Y_s} \otimes I_{B_s}) - h^1(C_s, f^*T_{Y_s} \otimes I_{B_s}) + \dim_s S.$$

(1.7.3) *If the dimension of* $\mathrm{Hom}(C, Y, G)$ *at* $[f_s]$ *equals the lower bound given in* (1.7.2), *then* $\mathrm{Hom}(C, Y, G) \to S$ *is a local complete intersection morphism at* $[f_s]$. *In particular it is flat at* $[f_s]$.

Proof. Let $I_B \subset \mathcal{O}_C$ be the ideal sheaf of B. We prove the special case when there are sections $p_i : S \to C$ such that $I_B = \mathcal{O}_C(-\sum_{i=1}^m p_i(S))$, where we allow repetitions among the p_i. This is the only case that we use. The general case is outlined in [Mori79, Sect. 1].

Let $X_0 = C \times_S Y$ and $\gamma_0 : C_s \cong \Gamma_0 \subset X_0$ be the graph of f_s. If $\gamma_i : C_s \cong \Gamma_i \subset X_i$ is already defined, then let $\pi_{i+1} : X_{i+1} \to X_i$ be the blow up of X_i along $p_i(S)$ and Γ_{i+1} the birational transform of Γ_i.

Let $I_i \subset \mathcal{O}_{X_i}$ be the ideal sheaf of $\Gamma_i \subset X_i \times_S \{s\}$. By construction

$$\gamma_0^*(I_0/I_0^2) \cong f^*\Omega^1_{Y_s} \quad \text{and} \quad \gamma_{i+1}^*(I_{i+1}/I_{i+1}^2) \cong \gamma_i^*(I_i/I_i^2) \otimes \mathcal{O}_{C_s}(-p_{i+1}(s)).$$

Therefore

$$\gamma_m^*(I_m/I_m^2) \cong f^*\Omega^1_{Y_s} \otimes I_{B_s}.$$

1.7.4 Claim. *There is an open neighborhood*

$$[\Gamma_m] \in U \subset \operatorname{Hilb}(X_m/S)$$

such that $\operatorname{Hom}(C, Y, G) \cong U$.

Proof. Let $U \subset \operatorname{Hilb}(X_m/S)$ be the open set parametrizing those 1-cycles D for which the projection $D \to C$ is an isomorphism onto a fiber of C/S. By (I.1.10.1) U is open. The universal family over U is clearly in $\operatorname{Hom}(C, Y, G)(U)$.

Conversely let $p_0 : C \times_S R \to Y \times_S R$ be a family in $Hom(C, Y, G)(S)$ given by its graph $G_0 \subset X_0 \times_S R$. Assume that we already constructed

$$C \times_S R \cong G_i \subset X_i \times_S R.$$

By construction $\{p_{i+1}(S)\} \times_S R \subset G_i$ and $G_i \to R$ is smooth along $\{p_{i+1}(S)\} \times_S R$. Thus if G_{i+1} denotes the birational transform of G_i in $X_{i+1} \times_S R$, then $G_{i+1} \cong G_i \cong C \times_S R$. In particular it is flat over R.

Finally we obtain

$$C \times_S R \cong G_m \subset X_m \times_S R \quad \text{and} \quad G_m \in Hilb(X_m/S)(R).$$

This gives a natural transformation

$$\operatorname{Hom}(C, Y, G)(*) \to \operatorname{Hom}(*, U)$$

which is an isomorphism. □

Thus (I.2.15) implies (1.7). □

It is not hard to give conditions which imply that a general deformation of a morphism is an immersion or an embedding:

1.8 Theorem. *Let $f : C \to X$ be a morphism from a smooth projective curve to a smooth variety. Let $B \subset C$ be a subscheme and $g = f|B$.*

(1.8.1) *If* $H^1(C, f^*T_X \otimes I_B(-2p)) \le \dim X - 2$ *for every* $p \in C$*, then a general deformation of* f *over* g *is an immersion on* $C - B$*. Furthermore, if* g *is an immersion, then a general deformation of* f *over* g *is an immersion on* C*.*

(1.8.2) *If* $H^1(C, f^*T_X \otimes I_B(-p-q)) \le \dim X - 3$ *for every* $p, q \in C$*, then a general deformation of* f *over* g *is an embedding on* $C - B$*. Furthermore, if* g *is an embedding, then a general deformation of* f *over* g *is an embedding on* C*.*

Proof. Let $[f] \in U \subset \mathrm{Hom}(C, X, g)$ be an open neighborhood such that the estimates (1.8.1–2) are satisfied for every morphism in U. By (1.7.2)

$$\dim U \ge \chi(C, f^*T_X \otimes I_B).$$

If $[f_t] \in U$ is not an immersion at p and $f_t(p) = x$, then set $g_{2p}|B = g$ and $g_{2p}|2p = f_t|2p$. By (1.7.1)

$$\begin{aligned}\dim \mathrm{Hom}(C, X, g_{2p}) &\le h^0(C, f_t^*T_X \otimes I_B(-2p))\\ &= \chi(C, f_t^*T_X \otimes I_B(-2p)) + h^1(C, f_t^*T_X \otimes I_B(-2p))\\ &= \chi(C, f_t^*T_X \otimes I_B) - 2\dim X + h^1(C, f_t^*T_X \otimes I_B(-2p))\\ &\le \chi(C, f_t^*T_X \otimes I_B) - \dim X - 2.\end{aligned}$$

Since

$$\{[f_t] \in U | f_t \text{ is not an immersion on } (C-B)\} = \bigcup_{p\in C-B} \bigcup_{x\in X} \mathrm{Hom}(C, X, g_{2p}) \cap U,$$

we conclude that

$$\begin{aligned}&\mathrm{codim}(\{[f_t] \in U | f_t \text{ is not an immersion on } (C - B)\}, U)\\ &\qquad \ge \dim X + 2 - 1 - \dim X \ge 1.\end{aligned}$$

Assume next that g is an immersion. If $p \in B$ and f is an immersion at p, then every general deformation of f is an immersion at p. If f is not an immersion at p, then $\mathrm{length}_p B = 1$. Let $g_p = f|B \cup \{p\}$. By considering $\mathrm{Hom}(C, X, g_p)$ and arguing as above we see that

$$\mathrm{codim}(\{[f_t] \in U | f_t \text{ is not an immersion along } B\}, U) \ge 2.$$

Thus the proof of the first part is complete.

The proof of the second part is similar. □

1.8.3 Exercise. Give examples which show that (1.8) is sharp.

The following result can easily be established by elementary methods. We mostly include it for ease of reference.

1.9 Exercise. Let C be a smooth and irreducible curve. Let D be an arbitrary curve and D^n the normalization. Then

(1.9.1) Composition with $D^n \to D$ gives a natural isomorphism:

$$\{\text{nonconstant morphisms } C \to D^n\} \leftrightarrow \{\text{nonconstant morphisms } C \to D\}.$$

(1.9.2) Assume that C, B are irreducible, smooth and proper curves. Let $f : C \to B$ be a nonconstant morphism, $F \subset C$ a closed subscheme and $g = f|F$. Then

$$\dim_{[f]} \operatorname{Hom}(C, B, g) \le \begin{cases} 0 & \text{if } g(B) \ge 2; \\ \max\{0, 1 - \deg F\} & \text{if } g(B) = 1; \\ \max\{0, 2\deg f + 1 - \deg F\} & \text{if } g(B) = 0. \end{cases}$$

So far we have avoided considering the most classical deformation problem in algebraic geometry: deformations of a variety (or more generally of a scheme). We need this only for curves, but unfortunately for singular curves. The moduli problem is more delicate than the problem of parametrizing subschemes, because the resulting functor is not representable if the curves have automorphisms. The following construction can be used as a substitute of the "miniversal deformation space" of a curve. The result in fact computes the dimension of a component of the "miniversal deformation space" when it applies.

1.10 Definition. A proper curve C over a field k is called *smoothable* over k if there is an irreducible pointed k-variety $0 \in T$ and a proper flat family of curves $g : Z \to T$ such that $C \cong g^{-1}(0)$ and the generic fiber of g is smooth.

Any such $g : Z \to T$ is called a smoothing of C.

1.11 Theorem. *Let C be a proper smoothable curve without embedded points over an algebraically closed field k. Then there is a normal pointed k-variety $0 \in S$ and a flat and proper family of curves $f : X \to S$ such that:*

(1.11.1) $f^{-1}(0) \cong C$;

(1.11.2) *if V is an irreducible pointed k-variety and $t : (v \in V) \to (0 \in S)$ a morphism such that*

$$[V \times_S X \to V] \cong [V \times_{\operatorname{Spec} k} C \to V],$$

then $t(V) = \{0\}$;

(1.11.3) $\dim S = -3\chi(\mathcal{O}_C) + \dim \operatorname{Aut} C$.

Proof. Let $g : Z \to T$ be a smoothing and L a relatively ample line bundle on Z. A sufficiently high multiple L^m of L gives an embedding

$$h : Z \to \mathbb{P}_T = \mathbb{P}_k \times_{\operatorname{Spec} k} T,$$

where $\mathbb{P}_k = \mathbb{P}_k^N$ for some $N > 0$. Let $h_0 = h|C$. h gives a morphism $T \to \operatorname{Hilb}(\mathbb{P}_k)$. Let $W \subset \operatorname{Hilb}(\mathbb{P})$ be the irreducible component containing the image of T (we will see that there is only one such).

1.11.4 Claim. *If $m \gg 0$, then*

(1.11.4.1) $\dim W = (N+1)\deg_C L^m + (N-3)\chi(\mathcal{O}_C)$;

(1.11.4.2) $\dim_{[h_0]} \operatorname{Hom}(C, \mathbb{P}) = (N+1)\deg_C L^m + N\chi(\mathcal{O}_C)$.

Proof. Let D be a smooth fiber of $Z \to T$. We compute the dimension of W at $h(D)$. The normal bundle of $h(D)$ is given by a sequence

$$0 \to T_D \to T_{\mathbb{P}}|D \to N_{h(D),\mathbb{P}} \to 0.$$

The usual presentation of $T_{\mathbb{P}}$ gives surjections

$$\sum_{i=0}^{N+1} L^m|D \twoheadrightarrow T_{\mathbb{P}}|D \twoheadrightarrow N_{h(D),\mathbb{P}}.$$

If $m \gg 0$, then $H^1(D, L^m) = 0$, hence also $H^1(D, N_{h(D),\mathbb{P}}) = 0$. By (I.2.8.4) and (I.2.14.3) we obtain that W is smooth at $[h(D)]$ of dimension

$$\begin{aligned}\chi(D, N_{h(D),\mathbb{P}}) &= (N+1)\deg_D L^m + (2g(D)-2) + (N-1)\chi(\mathcal{O}_D)\\ &= (N+1)\deg_C L^m + (N-3)\chi(\mathcal{O}_C).\end{aligned}$$

Similarly, if $H^1(C, L^m) = 0$, then also $H^1(C, h_0^*T_{\mathbb{P}}) = 0$. Hence by (I.2.16.2)

$$\dim_{[h_0]} \operatorname{Hom}(C, \mathbb{P}) = (N+1)\deg_C L^m + N\chi(\mathcal{O}_C). \qquad \square$$

By (I.6.2) there is an open neighborhood $[h_0] \in U \subset \operatorname{Hom}(C, \mathbb{P})$ such that the natural morphism

$$C \times U \to \mathbb{P} \times \operatorname{Hilb}(\mathbb{P})$$

is an immersion. This gives a morphism $G : U \to \operatorname{Hilb}(\mathbb{P})$. The fibers of G are principal homogeneous spaces over $\operatorname{Aut} C$. Thus

$$\dim \operatorname{im} G = (N+1)\deg_C L^m + (N-3)\chi(\mathcal{O}_C) - \dim \operatorname{Aut} C,$$

and hence

$$\operatorname{codim}_{[h_0(C)]}(\operatorname{im} G \cap W, W) = -3\chi(\mathcal{O}_C) + \dim \operatorname{Aut} C.$$

Let S be the normalization of a sufficiently general linear space section of W containing $[h_0(C)]$. By construction S has all the required properties. $\square$

1.11.5 Remark. I do not know any example of a proper curve without embedded points for which (1.11) fails. This is very much connected to the question of existence of rigid curve singularities.

1.11.6 Exercise (Notation as in (1.11)). Assume that the singularities of C are k ordinary nodes. For $0 \le j \le k$ let $S_j \subset S$ be the set of curves with at least j nodes. Show that S_j is closed, nonempty and has codimension exactly j.

1.12 Example. In order to use (1.11) we need to know when a curve is smoothable. By (I.6.10) a reduced proper curve is smoothable iff it is locally smoothable. We use this only in some special cases when the smoothing can be constructed explicitly. Let $C = \sum_0^n C_i$ be a curve which is obtained from the smooth and irreducible curve C_0 by attaching trees of smooth rational curves to it (see (7.4) for the precise definition).

Then C is smoothable.

Proof. Let $b \in B$ be a smooth curve and set $g_0 : S_0 = C_0 \times B \to B$. If $g_j : S_j \to B$ is already constructed such that $g_j^{-1}(b) \cong \sum_0^j C_i$, then let S_{j+1} be the blow up of p_j in S_j and $g_{j+1} : S_{j+1} \to S_j \to B$ the natural projection.

Finally we obtain $g_n : S_n \to B$ whose fiber over b is C and every other fiber is isomorphic to C_0. □

It is quite interesting that one can use the above results about Hom to get information about the Hilbert scheme of curves.

1.13 Proposition. *Let C be a proper algebraic curve without embedded points. Let $h : C \to Y$ be a morphism to a quasi projective variety Y of pure dimension n. Assume that Y is a local complete intersection and that the image of every component of C intersects the smooth locus of Y. Assume that C is smoothable and let $f : X \to S$ be as in* (1.11). *Then*

$$\dim_{[h]} \operatorname{Hom}(X/S, Y \times S/S) \geq -K_Y \cdot C + (n-3)\chi(\mathcal{O}_C) + \dim \operatorname{Aut} C.$$

Proof. This is an immediate corollary of (1.3), (1.11) and (I.2.17.1). □

1.13.1 Exercise. Let Y be a smooth K3 surface, C a smooth curve of genus g and $h : C \to Y$ a nonconstant morphism. Let $f : X \to S$ be as in (1.11). (1.13) gives that

$$\dim_{[h]} \operatorname{Hom}(X/S, Y \times S/S) \geq g - 1 + \dim \operatorname{Aut} C.$$

Replace Y with a complex space $p : \mathbf{Y} \to \Delta$ where $\mathbf{Y}$ is a flat deformation of Y whose general fiber contains no algebraic curves (the existence of such a deformation is a nontrivial fact, see e.g. [BPV84, VIII]). Use $\mathbf{Y}$ to improve the above estimate to

$$\dim_{[h]} \operatorname{Hom}(X/S, Y \times S/S) \geq g + \dim \operatorname{Aut} C.$$

In particular, every (singular) elliptic curve on a K3 surface is contained in a 1-parameter family of (singular) elliptic curves. This is a special case of [Ran93, Cor. 4].

1.14 Theorem. *Let C be a proper algebraic curve without embedded points. Let $h : C \to Y$ be a closed immersion such that the image of every component* *of C intersects the smooth locus of Y. Assume that Y is a local complete intersection of pure dimension n and C is smoothable. Then*

$$\dim_{[C]} \mathrm{Hilb}(Y) \geq -K_Y \cdot C + (n-3)\chi(\mathcal{O}_C).$$

Proof. Let $\pi : \mathrm{Hom}(X/S, Y \times S/S) \to S$ be the projection. We have a natural morphism

$$u : X \times_S \mathrm{Hom}(X/S, Y \times S/S) \to Y \times \mathrm{Hom}(X/S, Y \times S/S),$$

which is a closed immersion over $[h]$. Thus by (I.6.2) u is a closed immersion over an open neighborhood $[h] \in U \subset \mathrm{Hom}(X/S, Y \times S/S)$. This gives a morphism

$$p : U \to \mathrm{Hilb}(Y).$$

Let

$$V = p^{-1}([h(C)]) \subset U \subset \mathrm{Hom}(X/S, Y \times S/S).$$

By construction,

$$u : V \times_S X \to V \times h(C)$$

is an isomorphism, hence by (1.11.3) $\pi(V) = \{0\} \in S$. Therefore

$$V = \mathrm{Hom}(C, h(C)) \cap U$$

has dimension $\dim \mathrm{Aut}\, C$. Thus

$$\begin{aligned} \dim_{[C]} \mathrm{Hilb}(Y) &\geq \dim \mathrm{im}[U \to \mathrm{Hilb}(Y)] \\ &\geq \dim \mathrm{Hom}(X/S, Y \times S/S) - \dim \mathrm{Aut}\, C \\ &= -K_Y \cdot C + (n-3)\chi(\mathcal{O}_C). \end{aligned}$$

□

1.15 Theorem. *Let Y be a local complete intersection variety of pure dimension n and $Z = \sum a_i[C_i]$ a 1-cycle in Y. Assume that none of the C_i is contained in the singular locus of Y. Let $h_i : D_i \to C_i$ be the normalization. Then*

$$\dim_Z \mathrm{Chow}(Y) \geq -\sum a_i(K_Y \cdot C_i) + (n-3)\sum a_i \chi(\mathcal{O}_{D_i}).$$

Proof. Let D be the union of a_i disjoint copies of D_i for every i. Let $h : D \to Y$ be the morphism which is h_i on every copy of D_i. The rest of the proof is the same as for (1.14). □

The above results are especially nice in dimension three since, then $\chi(\mathcal{O}_C)$ does not enter into the estimate. If Y is smooth, (1.14) can be strengthened a little:

1.16 Theorem [Ein86, Lemma 5]. *Let Y be a smooth projective threefold and $C \subset Y$ a purely one dimensional closed subscheme without embedded points. Then*

$$\dim_{[C]} \operatorname{Hilb}(Y) \geq -K_Y \cdot C.$$

Proof. Let $I \subset \mathcal{O}_X$ be the ideal sheaf of C. I has a length two resolution by locally free sheaves (see e.g. [Artin76, §5]):

$$0 \to F \to E \to I \to 0. \tag{1.16.1}$$

First Hom it into $\mathcal{O}_Y$ to get

$$0 \to Hom(I, \mathcal{O}_Y) \to E^* \to F^* \to Ext^1(I, \mathcal{O}_Y) \to 0, \tag{1.16.2}$$

and note that

$$Ext^1(I, \mathcal{O}_Y) \cong Ext^2(\mathcal{O}_C, \mathcal{O}_Y) \cong \omega_C \otimes \omega_Y^{-1}.$$

Next Hom (1.16.1) into $\mathcal{O}_C$ to get

$$0 \to Hom(I, \mathcal{O}_C) \to E^*|C \to F^*|C \to Q \to 0, \tag{1.16.3}$$

for some sheaf Q. Tensor (1.16.2) by $\mathcal{O}_C$ to get

$$E^*|C \to F^*|C \to \omega_C \otimes \omega_Y^{-1} \otimes \mathcal{O}_C \to 0. \tag{1.16.4}$$

Comparing (1.16.2–4) we see that $Q \cong \omega_C \otimes \omega_Y^{-1}$. $Hom(I, \mathcal{O}_C) \cong Hom(I/I^2, \mathcal{O}_C)$, thus (1.16.2) becomes

$$0 \to Hom(I/I^2, \mathcal{O}_C) \to E^*|C \to F^*|C \to \omega_C \otimes \omega_Y^{-1} \to 0. \tag{1.16.5}$$

Since I is the ideal sheaf of a curve on a threefold, its first Chern class is zero. Therefore $c_1(E) = c_1(F)$. Now we can calculate $\chi(Hom(I/I^2, \mathcal{O}_C))$ from the sequence (1.16.5) to obtain that

$$\chi(Hom(I/I^2, \mathcal{O}_C)) = -C \cdot K_Y.$$

A curve in a smooth threefold has no local deformation obstructions (I.2.12.3). Substituting into (I.2.15.3) we are done. □

1.16.6 Remark. If C is smoothable, then (1.16) is a consequence of (1.14). By (I.2.12.3), C is locally smoothable but it need not be smoothable if C is not reduced.

1.16.7 Exercise. Let $C \subset \mathbb{P}^3$ be an irreducible curve such that every small flat deformation of C is projectively equivalent to C. Prove that C is a smooth rational curve of degree at most 3.

1.17 Remark. Theorems (1.14–16) hold if Y is an algebraic space or a complex analytic space. The proof needs no changes for (1.14–15). See [Kollár91a, 5.1] for a fairly cumbersome proof for (1.16).

II.2 Families of Rational Curves

The aim of this section is to investigate families of rational curves in more detail. The first main result is (2.8) which asserts that any well defined family of irreducible and reduced rational curves is a $\mathbb{P}^1$-bundle. This is a rather special property of rational curves. Using this we construct universal families of rational curves

$$\mathrm{Univ}^{rc}(X/S) \to \mathrm{RatCurves}(X/S)$$

on any scheme (2.11). These are just quotients of $\mathrm{Hom}(\mathbb{P}^1, X)$ by $\mathrm{Aut}(\mathbb{P}^1)$, though the construction is different (2.15–16).

2.1 Definition. (2.1.1) A variety X over a field k is called *rational* (or *k-rational* or *rational over k*) if X is k-birational to $\mathbb{P}_k^{\dim X}$.

(2.1.2) A variety X over a field k is called *geometrically rational* if there is a field extension $K \supset k$ such that X_K is reduced, irreducible and K-birational to $\mathbb{P}_K^{\dim X}$.

(2.1.3) A scheme X over a field k is said to have *geometrically rational components* if there is a field extension $K \supset k$ such that every irreducible component of $\mathrm{red}\, X_K$ is rational over K.

See (IV.3.1.4) for a warning concerning this terminology.

2.2 Proposition. *Let $f : X \to Y$ be a proper morphism of relative dimension one. Assume that if T is the spectrum of a DVR and $h : T \to Y$ a morphism, then every irreducible component of $T \times_Y X$ has dimension two. (By* (I.3.16) *this is always the case if f is a well defined family of proper algebraic 1-cycles.) Then*

$$\{y \in Y |\ f^{-1}(y) \text{ has geometrically rational components}\} \quad \text{is closed in } Y.$$

Proof. Let $\bar{f} : \bar{X} \to Y$ be the normalization of X and $\bar{X}_g = \bar{f}^{-1}(y_g)$ a generic fiber of $\bar{f}$. There is a finite extension $K \supset k(y_g)$ such that the normalization of $\bar{X}_g \times_{y_g} \mathrm{Spec}\, K$ is smooth. (The field extension is necessary only in positive characteristic.) Let Y^0 be an open subset of the normalization of Y in K and $g : Y^0 \to Y$ the natural morphism. g is a quasifinite morphism with open image. If Y^0 is suitably small, then $f^0 : \mathrm{red}(\bar{X} \times_Y Y^0) \to Y^0$ is smooth and proper. The genus of the fibers of f^0 is locally constant. If there is a nonempty open set $U \subset Y$ such that the fibers of f^0 over $g^{-1}(U)$ have positive genus, then we can replace Y by $Y_1 = Y - U$ and apply Noetherian induction. Therefore it is sufficient to consider the case when for every generic point $y_g \in Y$ the fiber $f^{-1}(y_g)$ has only geometrically rational components.

Let $y \in Y$ be an arbitrary point, T the spectrum of a DVR and $h : T \to Y$ a morphism such that $h(T_0) = y$ and $h(T_g) = y_g$. By enlarging

T we may assume that every irreducible component of the generic fiber of $\operatorname{red}(T \times_Y X)$ is rational over the quotient field of T. Let Z be an irreducible component of the normalization of $T \times_Y X$ and $h : Z \to T$ the projection. By assumption the general fiber Z_g is rational, thus there is a birational morphism $n : \mathbb{P}^1_{T_g} \to Z_g$. Let $\Gamma \subset \mathbb{P}^1_T \times_T Z$ be the closure of the graph of n. The first projection p_1 is proper and birational. $\mathbb{P}^1_T$ is regular of dimension two, thus every positive dimensional fiber of p_1 is a rational curve (see (VI.1.2) or (2.2.1)) In particular, the special fiber of Γ/T consists of $\mathbb{P}^1_{T_0}$ and the positive dimensional fibers of p_1. Since Γ_0 dominates $T_0 \times_Y Z$ we conclude that every component of $h^{-1}(y)$ is rational. Finally every component of $f^{-1}(y) \subset X$ is dominated by a component of the special fiber of $Z \to T$ for some Z (here we use the assumption about f). Thus every component of $f^{-1}(y)$ is geometrically rational by the Lüroth theorem. □

2.2.1 Exercise. Let $f : X \to T$ be a morphism from a normal surface X to an irreducible curve T. Let g_{gen} be the genus of the normalization of the geometric general fiber of f. Let $0 \in T$ be a point and C_i the normalizations of the irreducible geometric components of $f^{-1}(0)$. Prove that $\sum g(C_i) \leq g_{gen}$.

Hint: First do the case when $f^{-1}(0)$ is reduced. The general case can be reduced to this by a suitable base change $T' \to T$.

2.3 Definition. (2.3.1) Let $(g : U \to V)$ be a family of proper algebraic cycles of X/S. The image of the natural morphism $U \to X$ is called the *locus* of $(g : U \to V)$. It is denoted by $\operatorname{Locus}(g : U \to V)$. We may also write $\operatorname{Locus}(g), \operatorname{Locus}(U)$ or $\operatorname{Locus}(V)$ if no confusion is likely. If $V \to S$ is proper, then the cycle map $U \to X$ is also proper. Hence in this case $\operatorname{Locus}(g : U \to V) \subset X$ is closed.

(2.3.2) Let $(g : U \to V)$ be a family of proper algebraic 1-cycles of X/S. Let $U' \subset U$ be the set of points $u \in U$ which are contained in a geometrically rational component of $g^{-1}(g(u))$. The image of the natural morphism $U' \to X$ is called the rational locus of $(g : U \to V)$. It is denoted by $\operatorname{RatLocus}(g : U \to V)$.

2.4 Corollary (Notation as in (2.3.2)). *Assume that $V \to S$ is proper. Then* $\operatorname{RatLocus}(g : U \to V)$ *is proper over S.*

Proof. It is sufficient to consider one irreducible component at a time, thus we may assume that V is irreducible. Let $U = \sum a_i U_i$. Again it is sufficient to consider every $(g_i : U_i \to V)$ separately. The generic fiber D_i of g_i is an irreducible curve. If D_i is rational, then by (2.2) every fiber of g_i has rational components. Thus

$$\operatorname{RatLocus}(g_i : U_i \to V) = \operatorname{im}[U_i \to X]$$

is proper over S. If D_i is not rational, then there is an open subset $\emptyset \neq W \subset V$ such that the fibers of g_i over W are irreducible and nonrational. Thus

$$\operatorname{RatLocus}(g_i : U_i \to V) = \operatorname{RatLocus}(g_i : g_i^{-1}(V - W) \to (V - W)),$$

and we can apply Noetherian induction. □

2.5 Definition. Let X be a scheme. We say that a morphism $g : Y \to X$ is a $\mathbb{P}^1$-*bundle* if g is smooth, proper and for every $x \in X$ the fiber $f^{-1}(x)$ is a rational curve. (It is easy to see that g is a locally trivial fiber bundle in the étale topology.)

2.6 Definition. Let $\operatorname{Hom}_{bir}(\mathbb{P}^1, X) \subset \operatorname{Hom}(\mathbb{P}^1, X)$ be the open subscheme corresponding to those morphisms $f : \mathbb{P}^1 \to X$ which are birational onto their image. f is birational onto its image iff f is an immersion at its generic point. Therefore by (I.1.10.1) this is an open condition.

By the Lüroth theorem, every morphism $g : \mathbb{P}^1 \to X$ is either constant or it can be written as

$$\mathbb{P}^1 \xrightarrow{h} \mathbb{P}^1 \xrightarrow{f} X$$

where f is birational onto its image. Thus, at least set theoretically, $\operatorname{Hom}_{bir}(\mathbb{P}^1, X)$ contains all information about $\operatorname{Hom}(\mathbb{P}^1, X)$. (However $\operatorname{Hom}_{bir}(\mathbb{P}^1, X)$ does not carry all the information about degenerations of morphisms.)

2.7 Comment. It is clear that the scheme $\operatorname{Hom}_{bir}(\mathbb{P}^1, X)$ is still too large for many purposes. If $f : \mathbb{P}^1 \to X$ is any morphism and $h \in \operatorname{Aut}(\mathbb{P}^1)$, then $f \circ h$ is counted as a different morphism. The group scheme $\operatorname{Aut}(\mathbb{P}^1) \cong PGL_2$ acts on $\operatorname{Hom}_{bir}(\mathbb{P}^1, X)$ and it is the quotient that "really parametrizes" morphisms of $\mathbb{P}^1$ into X. One can show (cf. [Mori79, Sect. 4]) that the morphism constructed in (I.6.9) $\operatorname{Hom}_{bir}(\mathbb{P}^1, X)^{sn} \to \operatorname{Chow}(X)$ is affine and $\operatorname{Aut}(\mathbb{P}^1)$-invariant. Thus by [Mumford-Fogarty82, p.148] we conclude that the quotient $\operatorname{Hom}_{bir}(\mathbb{P}^1, X)^{sn} / \operatorname{Aut}(\mathbb{P}^1)$ exists. Similarly one can show that there is a universal $\mathbb{P}^1$-bundle

$$\operatorname{Univ} \to \operatorname{Hom}_{bir}(\mathbb{P}^1, X)^{sn} / \operatorname{Aut}(\mathbb{P}^1)$$

and a natural morphism $\operatorname{Univ} \to X$.

In characteristic zero (and frequently also in characteristic p) one can construct the quotient using the Chow variety. To avoid technical complications, we construct only the normalization of the quotient. This is sufficient for the applications.

2.8 Theorem. *Let $f : X \to Y$ be a proper and surjective morphism between irreducible and normal schemes. Assume that the dimension of every fiber is one. (Thus by* (I.3.17 *and* I.4.10) *f is a well defined family of proper 1-cycles satisfying the Chow field condition.) Assume that*

(2.8.1) *For every $y \in Y$ the cycle theoretic fiber $f^{[-1]}(y)$ is an irreducible and reduced rational curve.*

(2.8.2) *The generic fiber is a smooth rational curve. (This assumption is automatic in characteristic zero.)*

Then f is a $\mathbb{P}^1$-bundle.

Proof. We need to show that f is smooth. By (I.6.5) f is smooth at the generic point of every fiber. For $y \in Y$ pick three different points $x_1, x_2, x_3 \in f^{-1}(y)$ such that f is smooth at x_i. Let $S_i \subset X$ be a Cartier divisor which intersects $f^{[-1]}(y)$ transversally at x_i (there may be other intersection points). $S_i \to Y$ is étale at x_i. Let

$$Z = S_1 \times_Y S_2 \times_Y S_3; \; z = (x_1, x_2, x_3) \in Z \quad \text{and} \quad X_Z = X \times_Y Z.$$

$Z \to Y$ is étale at z, thus by [Matsumura80, 21.E] X_Z is normal along $f_Z^{-1}(z)$ and f is smooth above y iff f_Z is smooth above z. Furthermore, f_Z has three sections $s_i : Z \to X_Z$ corresponding to the S_i. By shrinking Z we may assume that these sections are disjoint.

In $\mathbb{P}^1_Z \to Z$ we have three disjoint sections $p_i : Z \to \mathbb{P}^1_Z$ corresponding to $\{0, 1, \infty\}$. Our aim is the construct an isomorphism $q : \mathbb{P}^1_Z \to X_Z$ such that $q \circ p_i = s_i$. Let $h : \mathbb{P}^1_Z \times_Z X_Z \to Z$ be the projection. In order to construct the graph of q let $\Gamma \subset \operatorname{Chow}^1(\mathbb{P}^1_Z \times_Z X_Z/Z)$ be the closed subvariety parametrizing 1-cycles D with the following properties:

(2.8.3.1) $\deg \mathcal{O}_{\mathbb{P}^1}(1)|D = 1$;
(2.8.3.2) $\deg \mathcal{O}(s_1(Z))|D = 1$;
(2.8.3.3) $(p_i(h(D)), s_i(h(D))) \in D$ for $i = 1, 2, 3$.

Let $U_\Gamma \to \Gamma$ be the universal family. The following claim completes the proof:

2.8.4 Claim. *The natural projections $\pi_1 : U_\Gamma \to \mathbb{P}^1_Z$ and $\pi_2 : U_\Gamma \to X_Z$ are isomorphisms.*

Proof. For any $t \in Z$ consider $h^{-1}(t)$. By construction

$$\operatorname{red} h^{-1}(t) \cong \mathbb{P}^1_{k(t)} \times C_t,$$

where C_t is an irreducible geometrically rational curve, smooth for general t. D gives a 1-cycle on $\operatorname{red} h^{-1}(t)$ which has bidegree (1,1). Thus D is either the graph of a birational morphism $q_t : \mathbb{P}^1_{k(t)} \to C_t$ or the union of a vertical and of a horizontal section. In the latter case it can not contain all three points $(p_i(t), s_i(t))$. Hence D is the graph of the unique birational morphism q_t such that $q_t(p_i(t)) = s_i(t)$ (i=1,2,3). Thus π_i (i=1,2) are one-to-one. If C_t is smooth, then q_t is defined over $k(t)$, thus π_i (i=1,2) are isomorphisms over the generic point of Z. Since X_Z and $\mathbb{P}^1_Z$ are normal, this implies that π_i (i=1,2) are isomorphisms. □

2.8.5 Exercise. It is tempting to believe that if $f : X \to Y$ is a proper family of reduced 1-cycles and both X and Y are normal, then f is flat with reduced fibers. Unfortunately this is not true. Let X be any normal scheme which is not S_3 at a single point (e.g. a cone over an Abelian surface). Let $|H|$ be a general net of ample divisors with base points at smooth points only. This gives a morphism $f : BX \to \mathbb{P}^2$ where $BX \to X$ is obtained by blowing up the base points. For sufficiently general choices every fiber is a generically reduced and irreducible curve. The fiber containing the non-S_3 point is not reduced.

A cone over an Abelian threefold produces a similar example $f : BX \to \mathbb{P}^3$ where f is not even flat.

2.8.6 Exercise. The aim of this exercise is to give an alternate approach to study degenerations of rational curves. We assume that the characteristic is zero and use the following result:

(2.8.6.1) [Kollár86a, 7.1] Let $f : X \to Y$ be a morphism between smooth varieties whose general fiber is $\mathbb{P}^1$. Then $R^i f_* \mathcal{O}_X = 0$ for $i > 0$.

(2.8.6.2) Let $f : X \to Y$ be a morphism whose general fiber is $\mathbb{P}^1$, Y smooth and X normal. Show that $R^1 f_* \mathcal{O}_X = 0$.

(2.8.6.3) Let $f : X \to Y$ be a morphism whose general fiber is $\mathbb{P}^1$, Y smooth and X normal. Assume that the dimension of every fiber is one. Show that $C_y := \operatorname{red} f^{-1}(y)$ is a connected curve with $h^1(C_y, \mathcal{O}) = 0$ for every $y \in Y$. Thus if C_y is irreducible, then $C_y \cong \mathbb{P}^1$.

(2.8.6.4) Notation and assumptions as in (2.8). Assume in addition that Y is smooth. Use (I.6.5) to show that f is smooth.

In positive characteristic we use the following result to deal with the cases when (2.8.2) is not satisfied.

2.9 Theorem. *Let Z be a variety defined over a field k. Assume that Z is geometrically reduced (2.9.1). Then there is a unique smallest field extension $K \supset k$ such that the normalization $\bar{Z}_K$ of Z_K is geometrically normal. Furthermore, $K \supset k$ is purely inseparable.*

Proof. Let $F \supset k$ be an extension. Assume that the normalization $\bar{Z}_F \to Z_F$ is geometrically normal. Let

$$I_F := \operatorname{Hom}(\mathcal{O}_{\bar{Z}_F}, \mathcal{O}_{Z_F}) \subset \mathcal{O}_{Z_F}$$

be the conductor. If $L \supset F$ is another extension, then by [Matsumura80, 21.E] $\bar{Z}_L \cong (\bar{Z}_F)_L$, hence

$$I_L = I_F \otimes_F L.$$

Thus if $F \supset K \supset k$ is an extension such that $\bar{Z}_K$ is geometrically normal, then I_F is defined over K. Conversely, if I_K defines I_F, then consider the sheaf of finite $\mathcal{O}_{Z_K}$-algebras $\operatorname{Hom}(I_K, I_K)$. Let

$$Z'_K = \operatorname{Spec}_{Z_K} \operatorname{Hom}(I_K, I_K).$$

Since $\operatorname{Hom}(I_K, I_K)$ is finite over $\mathcal{O}_{Z_K}$,

$$\operatorname{Hom}(I_K, I_K) \otimes_K F \cong \operatorname{Hom}(I_F, I_F) \cong \mathcal{O}_{\bar{Z}_F}.$$

Thus Z'_K is geometrically normal.

By (I.1.16,19) I_F has a smallest field of definition $K \supset k$.

Over an algebraically closed field normal implies geometrically normal, thus $K \supset k$ is an algebraic extension. Also, $K \supset k$ is clearly a normal field extension. Thus there is a maximal purely inseparable subextension $K \supset K' \supset k$ such that $K \supset K'$ is separable. By [Matsumura80, 21.E]

$$\bar{Z}_{K'} \times_{\operatorname{Spec} K'} \operatorname{Spec} K$$

is normal, therefore it is isomorphic to $\bar{Z}_K$. Therefore $\bar{Z}_K$ is defined over K'. Since K is the minimal field of definition of $\bar{Z}_K$, $K = K'$ and K is purely inseparable over k. □

2.9.1 Definition. A scheme X defined over a field k is called *geometrically reduced* (resp. *irreducible*, resp. *normal*) if $X_{\bar{k}}$ is reduced (resp. irreducible, resp. normal) where $\bar{k} \supset k$ is the algebraic closure.

2.10 Exercise. The assumption that X be geometrically reduced is necessary in the above theorem. To see this let F be a field of characteristic $p \neq 2$. Note that if $k' \supset k$ is a finite field extension, then $\operatorname{Spec}_k k'$ is geometrically normal iff $k' \supset k$ is separable.

Let $k = F(x^p + y^p, x^p y^p)$ where x, y are algebraically independent over F. Let $Z = \operatorname{Spec}_k k(x)$. Show that x is separable over $k(x)$ and also over $k(x+y, xy) = F(x+y, xy)$ but it is not separable over

$$k = F(x^p + y^p, x^p y^p) = F(x^p + y^p, x^p y^p, x) \cap F(x+y, xy).$$

2.11 Definition – Proposition. (2.11.1) Let X/S be a scheme, projective over S. Let

$$\operatorname{Hom}_{bir}(\mathbb{P}^1, X)^{sn} = \bigcup_i W_i$$

be the decomposition into irreducible subschemes. In (I.6.9) we constructed a natural morphism $\operatorname{Hom}(\mathbb{P}^1, X)^{sn} \to \operatorname{Chow}(X)$. Let $\bar{V}_i \subset \operatorname{Chow}(X)$ be the closure of the image of W_i. By (2.2) $\bar{V}_i$ parametrizes 1-cycles with geometrically rational components, and the generic 1-cycle is irreducible. Let $V_i \subset \bar{V}_i$ be the open subset parametrizing irreducible 1-cycles.

Let $v_{i,g} \in V_i$ be the generic point and $C_{i,g}$ the corresponding curve. By (2.9) there is a smallest field extension $K_i \supset k(v_{i,g})$ such that the normalization of $C_{i,g} \times_{v_{i,g}} \operatorname{Spec} K_i$ is a smooth rational curve. Let V_i^m be the normalization of V_i in K_i. The *family of rational curves on* X is, by definition,

$$\operatorname{RatCurves}^n(X/S) = \bigcup_i V_i^m.$$

(The superscript ^{n} is a reminder that we normalized.) If $\mathcal{O}_X(1)$ is ample on X/S, then

$$\mathrm{RatCurves}^n(X/S) = \bigcup_d \mathrm{RatCurves}^n_d(X/S),$$

where $\mathrm{RatCurves}^n_d(X/S)$ parametrizes rational curves of degree d. $\mathrm{RatCurves}^n_d(X/S)$ is quasi projective over S for every d. Let

$$\begin{aligned}&\mathrm{Univ}^{rc}(X/S)\\ &\quad = \text{the normalization of } \mathrm{RatCurves}^n(X/S) \times_{\mathrm{Chow}(X/S)} \mathrm{Univ}^{\mathrm{chow}}(X/S).\end{aligned}$$

(2.11.2) Let X/S be a scheme, projective over S and $f : S \to X$ a section. We can view f as a morphism from the zero section of $\mathbb{P}^1_S/S$ to X. Let

$$\mathrm{Hom}_{bir}(\mathbb{P}^1, X, f)^{sn} = \bigcup_i W_i$$

be the decomposition into irreducible components. As before we can construct V_i^m and set

$$\mathrm{RatCurves}^n(f, X/S) = \bigcup_i V_i^m.$$

As before one can define $\mathrm{RatCurves}^n_d(f, X/S)$ and $\mathrm{Univ}^{rc}(f, X/S)$.

We refer to $\mathrm{RatCurves}^n(f, X/S)$ as the *family of rational curves passing through* $\operatorname{im} f$. If $S = \operatorname{Spec} k$ and $f(\operatorname{Spec} k) = x \in X$, then we usually use the notation $\mathrm{RatCurves}^n(x, X)$ instead of $\mathrm{RatCurves}^n(f, X)$.

From (2.8) we obtain:

2.12 Corollary (Notation as above). *Then*

$$\begin{aligned}&\mathrm{Univ}^{rc}(X/S) \to \mathrm{RatCurves}^n(X/S) \quad \text{and}\\ &\mathrm{Univ}^{rc}(x, X) \to \mathrm{RatCurves}^n(x, X)\end{aligned}$$

are $\mathbb{P}^1$*-bundles.* □

2.13 Example. It is interesting to analyze the family of rational plane cubics given by an equation

$$C(u, v, w) = (y^2 = x^3 - ux^2 + vx - w).$$

$C(u, v, w)$ is rational iff the right hand side of the equation has a double root. Let $\Delta \subset \mathbb{A}^3(u, v, w)$ be the space of rational cubics. If we let s be the double root and t the third root, then

$$\Delta = \operatorname{im}[(s, t) \mapsto (2s + t, s^2 + 2st, s^2t)].$$

The inverse, at least if char $\neq 2$, is given by

$$s = \frac{uv - 9w}{2u^2 - 6v}, \quad t = \frac{u^3 - 4uv + 9w}{u^2 - 3v}.$$

The three dimensional linear system $|L|$ spanned by $\{C(u,v,w)\}$ has six base conditions at infinity. Let $X \to \mathbb{P}^2$ be the surface obtained by blowing up the six (infinitely near) base points. $|L|$ is a complete linear system on X and $L \cdot K_X = -3$. Let $f : \mathbb{P}^1 \to X$ be the normalization of a nodal cubic in $|L|$. Then $\deg f^*T_X = 3$. Since f is an immersion,

$$\mathcal{O}(2) \cong T_{\mathbb{P}^1} \to f^*T_X$$

is locally a direct summand. Since $\operatorname{Ext}^1_{\mathbb{P}^1}(\mathcal{O}(1), \mathcal{O}(2)) = 0$, this implies that $f^*T_X \cong \mathcal{O}(2) + \mathcal{O}(1)$. By (1.2)

$$\dim_{[f]} \operatorname{Hom}(\mathbb{P}^1, X) = 5$$

and the automorphisms of $\mathbb{P}^1$ and the two dimensional family Δ account for all deformations of f. Let $\Delta' \subset \operatorname{RatCurves}^n(X)$ be the corresponding irreducible component of $\operatorname{RatCurves}^n(X)$.

If the characteristic is 2, then $\Delta = (uv + w = 0)$ is smooth and for u, v algebraically independent over k the cubic $C(u, v, uv)$ is regular over $k(u, v)$. It becomes singular over the field extension $k(u, \sqrt{v})$. Thus $p : \Delta' \to \Delta$ is inseparable of degree 2.

If the characteristic is 3, then $\Delta = (v^3 + wu^3 - u^2v^2 = 0)$. For $u \neq 0$ the singular point of the cubic is given by $(x, y) = (-v/u, 0)$. Thus for $u \neq 0$ the rational cubic

$$C(u, v, \frac{u^2v^2 - v^3}{u^3})$$

is singular over $k(u, v)$. Therefore $p : \Delta' \to \Delta$ is birational. Δ is singular along the locus $Z = (u = v = 0)$ and $p^{-1}(Z) \to Z$ is inseparable of degree 3.

Z parametrizes the family of cuspidal cubics $C(w) = (y^2 = x^3 - w)$. If w is transcendental over k, then $C(w)$ is regular over $k(w)$. It becomes singular over the field extension $k(\sqrt[3]{w})$. $k[\sqrt[3]{w}] \supset k[w]$ corresponds to the morphism $p^{-1}(Z) \to Z$.

2.14 Proposition (Notation as above).

(2.14.1) *Let* $m = \min\{d | \operatorname{RatCurves}^n_d(X/S) \neq \emptyset\}$. *Then* $\operatorname{RatCurves}^n_k(X/S)$ *is proper over* S *for* $k < 2m$.

(2.14.2) *For* $x \in X$ *let* $m(x) = \min\{d | \operatorname{RatCurves}^n_d(x, X) \neq \emptyset\}$. *Then* $\operatorname{RatCurves}^n_k(x, X)$ *is proper for* $k < m + m(x)$.

(2.14.3) *Let* $u : \mathbb{P}^1 \times \operatorname{RatCurves}^n_k(x, X) \to X$ *be the cycle map.* u *is quasifinite on* $(\mathbb{P}^1 - \{0\}) \times \operatorname{RatCurves}^n_k(x, X)$ *for* $k < m + m(x)$.

Proof. The proofs of (2.14.1–2) are identical; we consider the first part only. $\operatorname{Chow}_{1,k}(X/S)$ is proper over S, thus we need to show that $\operatorname{RatCurves}^n_k(X/S) \subset \operatorname{Chow}_{1,k}(X/S)$ is closed. Let $\sum a_i D_i$ be a 1-cycle in the closure of $\operatorname{RatCurves}_k(X/S)$. By (2.2) every D_i is rational and $\sum a_i \deg D_i = k < 2m$.

By assumption $\deg D_i \geq m$, thus $\sum a_i D_i$ is an irreducible and reduced rational curve.

In order to see (2.14.3) assume that there is an ireducible and proper curve

$$D \subset \mathbb{P}^1 \times \operatorname{RatCurves}^n_k(x, X), \quad D \not\subset \{0\} \times \operatorname{RatCurves}^n_k(x, X)$$

such that $u(D) = x' \in X$. Let $D' \to D$ be the normalization. We obtain a morphism

$$u' : D' \times \mathbb{P}^1 \xrightarrow{n} \mathbb{P}^1 \times \operatorname{RatCurves}^n_k(x, X) \xrightarrow{u} X$$

which contracts the sections $n^{-1}(D) \cong D'$ and $D' \times \{0\}$. By (5.5.2) we obtain a contradiction. □

2.15 Theorem (Notation as above). *Let* $\operatorname{Hom}^n_{bir}$ *be the normalization of* Hom_{bir}. *There is a natural commutative diagram*

$$\begin{array}{ccc} \mathbb{P}^1 \times \operatorname{Hom}^n_{bir}(\mathbb{P}^1_S, X/S) & \xrightarrow{U} & \operatorname{Univ}^{rc}(X/S) \\ \downarrow & & \downarrow \\ \operatorname{Hom}^n_{bir}(\mathbb{P}^1_S, X/S) & \xrightarrow{u} & \operatorname{RatCurves}^n(X/S). \end{array}$$

Furthermore, U and u are smooth of relative dimension three with connected fibers. (In fact both U and u are principal $\operatorname{Aut}(\mathbb{P}^1)$*-bundles.)*

Proof. Let

$$G = \operatorname{Hom}_{bir}(\mathbb{P}^1_S \times_S \operatorname{RatCurves}^n(X/S), \operatorname{Univ}^{rc}(X/S)/\operatorname{RatCurves}^n(X/S)).$$

If $C \subset X_F$ is a rational curve over an algebraically closed field F, then the fibers of the morphisms

$$\begin{aligned} \operatorname{Hom}^n_{bir}(\mathbb{P}^1_S, X/S) &\to \operatorname{Chow}(X/S), \quad \text{and} \\ G &\to \operatorname{RatCurves}^n(X/S) \end{aligned}$$

over $[C]$ are both in one-to-one correspondence with the set of birational morphisms $\mathbb{P}^1_F \to C$. We prove that in fact $G \cong \operatorname{Hom}^n_{bir}(\mathbb{P}^1_S, X/S)$.

The composition

$$\mathbb{P}^1_G \to \operatorname{Univ}^{rc}(X/S) \to X$$

induces a morphism

$$\rho : G \to \operatorname{Hom}^n(\mathbb{P}^1_S, X/S).$$

If $[g] \in G$, then the induced morphism

$$\mathbb{P}^1 \xrightarrow{g} \operatorname{Univ}^{rc}(X/S) \to X$$

is birational onto its image. Thus $\operatorname{im} \rho \subset \operatorname{Hom}^n_{bir}(\mathbb{P}^1_S, X/S)$. As we observed at the beginning of the proof, ρ is bijective.

In characteristic zero this implies that ρ is an isomorphism onto $\operatorname{Hom}^n_{bir}(\mathbb{P}^1_S, X/S)$.

In order to show that ρ is an ismorphism in positive characteristic it is sufficient to construct its inverse at the generic points of $\operatorname{Hom}^n_{bir}(\mathbb{P}^1_S, X/S)$.

Let $W \subset \operatorname{Hom}^n_{bir}(\mathbb{P}^1_S, X/S)$ be an irreducible component. Let

$$\pi : \operatorname{Hom}^n_{bir}(\mathbb{P}^1_S, X/S) \to \operatorname{Chow}(X/S)$$

be the morphism constructed in (I.6.9). Let $V = \pi(W)$, $w_g \in W$ the generic point of W and $v_g = \pi(w_g)$ with residue field L.

Let $K = k(W)$ be the residue field of w_g and $f_g : \mathbb{P}^1_K \to X$ the corresponding morphism. Set $C_g = \operatorname{im} f_g$. Then $[C_g] = v_g$ and L is the smallest field of definition of C_g. Let $V' \subset \operatorname{RatCurves}^n(X/S)$ be the component of $\operatorname{RatCurves}^n$ corresponding to V and $v'_g \in V'$ the generic point of V' with residue field L'.

Let $\bar{C}_g$ be the normalization of C_g over L and $\bar{C}'_g$ be the normalization of C_g over L'. L' is the unique smallest field such that C_g has smooth normalization over L'.

By definition

$$f_g : \mathbb{P}^1_K \to C_g \times_{\operatorname{Spec} L} \operatorname{Spec} K$$

is a birational morphism, thus the normalization of

$$C_g \times_{\operatorname{Spec} L} \operatorname{Spec} K$$

is smooth, hence $K \supset L'$. Therefore f_g can be lifted to

$$\bar{f}_g : \mathbb{P}^1_K \to \bar{C}'_g \times_{\operatorname{Spec} L'} \operatorname{Spec} K.$$

The map $f_g \mapsto \bar{f}_g$ is the inverse of ρ.

The smoothness of u and U and the connectivity of the fibers follow from the next lemma applied to

$$\operatorname{Univ}^{rc}(X/S) \to \operatorname{RatCurves}^n(X/S).$$

2.15.1 Lemma. *Let $h : Y \to Z$ be a $\mathbb{P}^1$-bundle. Then the morphisms*

$$\operatorname{Hom}_{bir}(\mathbb{P}^1_Z, Y) \xrightarrow{u} Z \quad \text{and}$$
$$\mathbb{P}^1_Z \times_Z \operatorname{Hom}_{bir}(\mathbb{P}^1_Z, Y) \xrightarrow{U} Y$$

are smooth of relative dimension three with connected fibers.

Proof. The result about u follows from (I.2.17). Both $\mathbb{P}^1_Z \times_Z \operatorname{Hom}_{bir}(\mathbb{P}^1_Z, Y)$ and U are smooth over Z, thus U is smooth iff it is smooth on the fibers over Z. Finally, over a field F

$$\mathbb{P}^1_F \times \operatorname{Aut}(\mathbb{P}^1_F) \to \mathbb{P}^1_F$$

is an $\operatorname{Aut}(\mathbb{P}^1_F)$ bundle, so smooth of relative dimension three. □

The same method as above also gives:

2.16 Theorem. *Let X be projective over a field k and $x \in X$ a k-point. Let* Hom^n_{bir} *be the normalization of* Hom_{bir}. *There is a natural commutative diagram*

$$\begin{array}{ccc} \mathbb{P}^1 \times \mathrm{Hom}^n_{bir}(\mathbb{P}^1, X, 0 \mapsto x) & \xrightarrow{U} & \mathrm{Univ}^{rc}(x, X) \\ \downarrow & & \downarrow \\ \mathrm{Hom}^n_{bir}(\mathbb{P}^1_S, X, 0 \mapsto x) & \xrightarrow{u} & \mathrm{RatCurves}^n(x, X). \end{array}$$

Furthermore, U and u are smooth of relative dimension two with connected fibers. (In fact both U and u are principal $\mathrm{Aut}(0, \mathbb{P}^1)$*-bundles.)* □

II.3 Free Morphisms of Curves

In this section we study those morphisms of curves whose deformation theory is unobstructed in a strong sense. The precise definition is given in (3.1). The notion is especially useful for morphisms of rational curves. Several simple properties are considered first. The most important result of the section is (3.11). This says that if X is a smooth variety in characteristic zero and $x \in X$ a very general point, then every morphism $\mathbb{P}^1 \to X$ whose image contains x is free. The proof is very simple but this principle is crucial in many applications.

Generic deformations of free morphisms behave especially well (3.7, 3.13–14). For instance, with a few exceptions, generic deformations of a free morphism are imbeddings. This is proved in (3.14).

Everything is over an algebraically closed field k.

3.1 Definition. Let C be a proper curve without embedded points, X a smooth variety and $f : C \to X$ a morphism. Let $B \subset C$ be a closed subscheme with ideal sheaf I_B and $g = f|B$.

f is called *free over* g if f is nonconstant and the following two equivalent conditions are satisfied:

(3.1.1) $H^1(C, f^*T_X \otimes J) = 0$ for every subsheaf $J \subset I_B$ such that $\mathrm{length}(I_B/J) = 1$. (If C is smooth, then this can be written as: $H^1(C, f^*T_X \otimes I_B(-p)) = 0$ for every $p \in C$.)

(3.1.2) $H^1(C, f^*T_X \otimes I_B) = 0$ and $f^*T_X \otimes I_B$ is generated by global sections.

When $B = \emptyset$ we say simply that that f is *free*.

3.1.3 Exercise. If $f : C \to X$ is free, then $C \cdot K_X < 0$. Use this and (5.8) to show that if there is any free morphism $f : C \to X$, then X is uniruled.

It is clear that being free is an open property:

3.2 Proposition. *Let $p : C \to S$ be a proper and flat family of curves and $q : X \to S$ a smooth morphism. Let $B/S \subset C/S$ be a closed subscheme, flat over S. Let $F : C/S \to X/S$ be a morphism and $G = F|B$. Assume that for some $s \in S$ the induced morphism $F_s : C_s \to X_s$ is free over G_s. Then there is an open neighborhood $s \in U \subset S$ such that $F_t : C_t \to X_t$ is free over G_t for every $t \in U$.* □

Free morphisms have especially pleasant differential properties. In order to see this, we recall the notation of (I.2.18–19).

3.3 Notation. Let C be a proper curve without embedded points and X a smooth variety. Let $B \subset C$ be a closed subscheme with ideal sheaf I_B and $g : B \to X$ a morphism. Let

$$F : C \times \operatorname{Hom}(C, X, g) \to X$$

be the universal morphism. For later applications we also consider the induced morphism

$$F^{(2)} : C \times C \times \operatorname{Hom}(C, X, g) \to X \times X.$$

Let $p, q \in C$ be closed points and $f : C \to X$ a morphism such that $f|B = g$. If $p, q \notin B$, then by (I.2.16) and (1.9)

$$T_{C\times \operatorname{Hom}(C,X,g)} \otimes k(p, [f]) = T_C \otimes k(p) + H^0(C, f^*T_X \otimes I_B) \quad \text{and}$$
$$T_{C\times C\times \operatorname{Hom}(C,X,g)} \otimes k(p, q, [f]) = T_C \otimes k(p) + T_C \otimes k(q) + H^0(C, f^*T_X \otimes I_B).$$

Let $df(s) : T_C \otimes k(s) \to T_X \otimes k(f(s))$ be the differential of f at $s \in C$,

$$\phi(p, f) : H^0(C, f^*T_X \otimes I_B) \to f^*T_X \otimes k(p) \quad \text{and}$$
$$\phi^{(2)}(p, q, f) : H^0(C, f^*T_X \otimes I_B) \to f^*T_X \otimes k(p) + f^*T_X \otimes k(q)$$

the evaluation maps.

The following is a reformulation of (I.2.19):

3.4 Proposition (Notation as above). *Then*

$$dF(p, [f]) = df(p) + \phi(p, f) \quad \textit{and}$$
$$dF^{(2)}(p, q, [f]) = (df(p) + \phi(p, f), df(q) + \phi(q, f)).$$ □

3.5 Proposition (Notation as above). *Assume in addition that C is smooth at the points p, q and $\operatorname{Hom}(C, X, g)$ is smooth at $[f]$.*

(3.5.1) *If $\phi(p, f)$ is surjective, then F is smooth at $(p, [f])$. The converse also holds if*

$$H^0(C, T_C \otimes I_B) \to T_C \otimes k(p)$$

is surjective.

(3.5.2) *If* $\phi^{(2)}(p,q,f)$ *is surjective, then* $F^{(2)}$ *is smooth at* $(p,q,[f])$. *The converse also holds if*

$$H^0(C, T_C \otimes I_B) \to T_C \otimes k(p) + T_C \otimes k(q)$$

is surjective.

Proof. In both parts the first implication follows from (3.4).

Assume conversely that F is smooth at $(p,[f])$. Then $\phi(p,f)$ is surjective as a map to

$$f^*T_X \otimes k(p)/\operatorname{im}[T_C \otimes k(p) \to f^*T_X \otimes k(p)].$$

The assumption in (3.5.1) implies that $T_C \otimes k(p)$ is in the image of $\phi(p,f)$. Thus $\phi(p,f)$ is surjective. The converse in (3.5.2) is proved analogously. □

3.5.3 Corollary. *If C is smooth and $f : C \to X$ is free over g, then*

$$F : C \times \operatorname{Hom}(C, X, g) \to X$$

is smooth along $(C - B) \times [f]$. □

The following special cases of (3.2) and of (3.5.3) will be used repeatedly:

3.5.4 Corollary. *Let X be a smooth variety and* $\operatorname{Hom}^{\text{free}}(\mathbb{P}^1, X) \subset \operatorname{Hom}(\mathbb{P}^1, X)$ *the set of free morphisms. Then:*

(3.5.4.1) $\operatorname{Hom}^{\text{free}}(\mathbb{P}^1, X)$ *is open in* $\operatorname{Hom}(\mathbb{P}^1, X)$, *and*
(3.5.4.2) $F : \mathbb{P}^1 \times \operatorname{Hom}^{\text{free}}(\mathbb{P}^1, X) \to X$ *is smooth.* □

3.6 Definition (Notation as in (3.3)). We say that a property (P) holds for general deformations of f over g if there is an open neighborhood $[f] \in U \subset \operatorname{Hom}(C, X, g)$ and a dense open set $V \subset U$ such that (P) holds for every morphism $[f'] \in V$.

3.7 Proposition. *Let $f : C \to X$ be free over g. Let $Z \subset X$ be a subscheme of codimension at least two. Then*

$$f'(C - B) \cap Z = \emptyset$$

for a general deformation f' of f over g.

Proof. Let $[f] \in U \subset \operatorname{Hom}(C, X, g)$ be a connected open neighborhood such that

$$H^1(C, f_t^*T_X \otimes I_B) = 0$$

for every $[f_t] \in U$. By (1.9) U is smooth of dimension

$$h^0(C, f_t^*T_X \otimes I_B).$$

Let $p \in C - B$ be a closed point and $x = f_t(p)$ for some $[f_t] \in U$. Set $B_p = B \cup \{p\}$ and let g_p be given by $g_p|B = g$ and $g_p(p) = x$. By assumption

$$H^1(C, f_t^* T_X \otimes I_{B_p}) = 0,$$

thus $\operatorname{Hom}(C, X, g_p) \cap U$ is smooth of dimension

$$h^0(C, f_t^* T_X \otimes I_{B_p}) = h^0(C, f_t^* T_X \otimes I_B) - \dim X.$$

Since

$$\{[f_t] \in U | f_t(C - B) \cap Z \neq \emptyset\} = \bigcup_{p \in C - B} \bigcup_{x \in Z} \{[f_t] \in U | f_t(p) = x\},$$

we conclude that

$$\operatorname{codim}(\{[f_t] \in U | f_t(C - B) \cap Z \neq \emptyset\}, U) \geq \dim X - 1 - \dim Z \geq 1. \quad \square$$

The notion of free morphisms is especially useful for morphisms of $\mathbb{P}^1$. It is convenient to start with a little vector bundle theory on $\mathbb{P}^1$.

3.8 Definition – Proposition. *Let E be a vector bundle on $\mathbb{P}^1$. E is called semi positive (resp. ample) if any of the following equivalent conditions is satisfied:*

(3.8.1) $H^1(\mathbb{P}^1, E(-1)) = 0$ *(resp.* $H^1(\mathbb{P}^1, E(-2)) = 0$*);*
(3.8.2) E *(resp.* $E(-1)$*) is generated by global sections;*
(3.8.3) $E \cong \sum \mathcal{O}(a_i)$ *and* $a_i \geq 0$ *(resp.* ≥ 1*) for every i.*
(3.8.4) $H^0(\mathbb{P}^1, E) \to E \otimes k(p)$ *(resp.* $H^0(\mathbb{P}^1, E(-1)) \to E(-1) \otimes k(p)$*) is surjective for some $p \in \mathbb{P}^1$.*

Proof. By [Hartshorne77, V.Exercise 2.6] every vector bundle on $\mathbb{P}^1$ is a direct sum of line bundles. The rest is easy. $\square$

3.8.5 Exercise. Show that E is ample (resp. semi positive) iff $\mathcal{O}_{\mathbb{P}(E)}(1)$ is ample (resp. nef) on $\mathbb{P}(E)$.

Some simple properties are summarized in the next lemmas:

3.9.1 Lemma. (3.9.1.1) *Let $h : \mathbb{P}^1 \to \mathbb{P}^1$ be a nonconstant morphism. Then E is semi positive (resp. ample) iff $h^* E$ is.*

(3.9.1.2) *Let $p : E_1 \to E_2$ be a morphism of locally free sheaves on $\mathbb{P}^1$ which is surjective at the generic point. If E_1 is semi positive (resp. ample), then E_2 is also semi positive (resp. ample).* $\square$

3.9.2 Lemma. *Let E be a vector bundle on $Z \times \mathbb{P}^1$. For every geometric point $z \in Z$ write*

$$E|\{z\} \times \mathbb{P}^1 \cong \sum \mathcal{O}(a_i(z)).$$

(3.9.2.1) *For every j, the functions*

$$\sum_{a_i(z)\geq j} a_i(z); \quad \max a_i(z) \quad \text{and} \quad -\min a_i(z)$$

are upper semi continuous on Z. Informally speaking, the sequence $\{a_i\}$ is more evenly distributed for general z.

(3.9.2.2) *If $E|\{z\} \times \mathbb{P}^1$ is semi positive (resp. ample) for some $z \in Z$, then there is an open neighborhood $z \in U \subset X$ such that $E|\{z'\} \times \mathbb{P}^1$ is semi positive (resp. ample) for every $z' \in U$.*

Proof. The first claim follows from the upper semi continuity of $h^0(\{z\} \times \mathbb{P}^1, (E|\{z\} \times \mathbb{P}^1) \otimes \mathcal{O}(-j-1))$. This also implies the upper semi continuity of $\max a_i$. By looking at E^* we obtain the result for $-\min a_i$. (3.9.2.2) is a special case of the first part. □

3.9.3 Exercise (Notation as above). Find examples where the $a_i(z)$ are not constant.

3.10 Proposition (Notation as in (3.3)). *Assume that $C \cong \mathbb{P}^1$ and $|B| \leq 2$ and write $f^*T_X \otimes I_B = \sum \mathcal{O}(a_i)$. Then*

$$\#\{i|a_i \geq 0\} = \operatorname{rank} dF(p, [f]) \ \forall p \in \mathbb{P}^1 - B.$$

In particular, if

$$\operatorname{red} F : \mathbb{P}^1 \times \operatorname{red} \operatorname{Hom}(\mathbb{P}^1, X, g) \to X$$

is smooth at $(p, [f])$ for some $p \in \mathbb{P}^1$, then f is free over g.

Proof. Since $|B| \leq 2$, $H^0(\mathbb{P}^1, T_{\mathbb{P}^1} \otimes I_B) \to T_{\mathbb{P}^1} \otimes k(p)$ is surjective. Therefore

$$\operatorname{rank} \phi(p, f) = \operatorname{rank} dF(p, [f]).$$

This implies the first claim.

If $d(\operatorname{red} F)(p, [f])$ is surjective, then as in the proof of (3.5.1) we see that

$$\phi(p, f) : H^0(\mathbb{P}^1, f^*T_X \otimes I_B) \to f^*T_X \otimes k(p)$$

is also surjective. By (3.8) $f^*T_X \otimes I_B$ is semi positive and $H^1(\mathbb{P}^1, f^*T_X \otimes I_B) = 0$. Thus $\operatorname{Hom}(\mathbb{P}^1, X, g)$ is smooth at $(p, [f])$, hence by (3.5.1) f is free. □

3.10.1 Corollary. *Let X be a smooth, proper variety over a field of characteristic zero, $x \in X$ a closed point. Assume that there is an irreducible component*

$$V \subset \operatorname{Hom}(\mathbb{P}^1, X, 0 \mapsto x) \quad \text{such that} \quad u : \mathbb{P}^1 \times V \to X \quad \text{is dominant.}$$

Then f^*T_X ample for general $[f] \in V$.

Proof. By (3.10) f is free over $0 \mapsto x$ for general $[f] \in V$. Thus by (3.10) $f^*T_X(-1)$ is semi positive and f^*T_X is ample. □

The following result is crucial in many applications.

3.11 Theorem [KoMiMo92c, 1.1]. *Let X be a smooth projective variety over a field of* char 0. *Let $B \subset \mathbb{P}^1$ be a subscheme, $|B| \leq 2$ and $g : B \to X$ a morphism. There are countably many subvarieties $V_i = V_i(B, g) \subsetneq X$ such that if $f : \mathbb{P}^1 \to X$ is a nonconstant morphism such that $f|B = g$ and* $\operatorname{im} f \not\subset \cup V_i$, *then f is free over B.*

3.11.1 Comments. (3.11.1.1) This result is frequently (and somewhat imprecisely) evoked by saying that a morphism $f : \mathbb{P}^1 \to X$ passing through a very general point is free. Similarly, a morphism $f : \mathbb{P}^1 \to X$ passing through one or two preassigned points and a very general point is free. In the latter case one has to be careful to keep in mind that the notion of "very general point" depends on the preassigned points.

(3.11.1.2) The example given in (4.15.3) shows that it is sometimes necessary to leave out countably many subvarieties V_i.

(3.11.1.3) See (III.2.5) for some examples which show that the char $= 0$ assumption is necessary.

Proof. Let Z_i be the irreducible components of $\operatorname{Hom}(\mathbb{P}^1, X, g)$ with universal morphisms $F_i : \mathbb{P}^1 \times Z_i \to X$. Let

$$V_i = \begin{cases} \overline{\operatorname{im} F_i} & \text{if } F_i \text{ is not dominant, and} \\ X - U_{F_i} & \text{if } F_i \text{ is dominant,} \end{cases}$$

where $U_{F_i} \subset X$ is an open and dense subset such that $\operatorname{red} F_i : \mathbb{P}^1 \times \operatorname{red} Z_i \to X$ is smooth over U_{F_i}. (This is where we use the char $= 0$ assumption.)

Let $f : \mathbb{P}^1 \to X$ be a morphism such that $f|B = g$ and assume that $\operatorname{im} f \not\subset \cup V_i$. Let $Z_j \subset \operatorname{Hom}(\mathbb{P}^1, X, g)$ be an irreducible component containing $[f]$. If $F_j : \mathbb{P}^1 \times Z_j \to X$ is not dominant, then $\operatorname{im} f \subset \operatorname{im} F_j \subset V_j$, a contradiction. Thus F_j is dominant hence smooth above U_{F_j}. By assumption $\operatorname{im} f$ intersects U_{F_j}, hence by (3.10) f is free over B. □

3.11.2 Exercises. (3.11.2.1) Let $X \subset \mathbb{P}^n$ be a smooth hypersurface of degree $3 \leq d \leq 2n - 3$. Show that X contains a nonfree line.

(3.11.2.2) Let $X \subset \mathbb{P}^4$ be a smooth cubic hypersurface. Show that every smooth conic in X is free.

(3.11.2.3) Let $X \subset \mathbb{P}^n$ be a smooth quadric hypersurface. Show that every nonconstant morphism $f : \mathbb{P}^1 \to X$ is free.

(3.11.2.4) Notation as in (3.11) and let C be a smooth, proper curve. Show that there are $V_i \subsetneq X$ such that if $f : C \to X$ is a morphism, then f^*T_X/T_C is generated by global sections at p if $f(p) \notin \cup V_i$.

In using the above result the following observation is frequently useful:

3.11.3 Lemma. *Let X be a scheme and $V_i \subset X$ constructible subsets. Let $p : U \to S$ and $u : U \to X$ be morphisms. Assume that p has irreducible fibers. Let $S_i = \{s \in S | U_s \subset u^{-1}(V_i)\}$. Then $S_i \subset S$ is constructible and*

$$u(U_s) \not\subset \cup V_i \quad \Leftrightarrow \quad s \notin \cup S_i. \qquad \square$$

3.11.4 Comment. We sometimes refer to this lemma by saying that if a morphism $f : U_0 \to X$ passes through a very general point, then a very general deformation of f also passes through a very general point.

3.11.5 Corollary. *Let X be a smooth projective variety over a field of* char 0. *Then* $\operatorname{RatCurves}^n(x, X)$ *is smooth for $x \in X$ very general.*

Proof. Pick $x \in X - \cup V_i$. Then $\operatorname{Hom}(\mathbb{P}^1, X, 0 \mapsto x)$ is smooth by (1.7) and therefore $\operatorname{RatCurves}^n(x, X)$ is smooth by (2.16). $\square$

3.12 Notation. (3.12.1) Let $f : \mathbb{P}^1 \to X$ be a morphism over g. By [Hartshorne77, V.Exercise 2.6] we can write

$$f^*T_X \otimes I_B \cong \mathcal{O}(a_1) + \ldots + \mathcal{O}(a_n) \quad \text{where} \quad a_1 \geq \ldots \geq a_n.$$

The numbers $a_i = a_i(f, B)$ are important invariants of f.

$\deg_{(-K)}(f) = \deg f^*T_X = -\deg f^*K_X$ is called the *anticanonical degree* of f. In the above notation

$$\deg_{(-K)}(f) = \sum a_i + \dim X \cdot \deg B.$$

(3.12.2) If $V \subset \operatorname{Hom}(\mathbb{P}^1, X)$ is an irreducible subvariety, then $\deg_{(-K)}(f)$ is independent of $[f] \in V$ and $a_1(f), \ldots, a_n(f)$ are independent of $[f] \in V$ for general f. We denote these common values by

$$\deg_{(-K)}(\operatorname{Fam} V) \quad \text{resp.} \quad a_1(\operatorname{Fam} V), \ldots, a_n(\operatorname{Fam} V).$$

3.13 Lemma. *Let $f : \mathbb{P}^1 \to X$ be a nonconstant morphism. Then $a_1(f) \geq 2$. If f is not separable, then $a_1(f) \geq 2 \operatorname{char} k$.*

Proof. We can write

$$f : \mathbb{P}^1 \xrightarrow{h} \mathbb{P}^1 \xrightarrow{g} X$$

where g is an immersion at its generic point. Thus there is an injection

$$\mathcal{O}(2) \cong T_{\mathbb{P}^1} \to g^*T_X.$$

Pulling back via h we obtain an injection

$$\mathcal{O}(2\deg h) \to f^*T_X.$$

Thus $a_1(f) \geq 2\deg h$. □

3.13.1 Corollary (Notation as above). *If $f : \mathbb{P}^1 \to X$ is free, then* $\deg_{(-K)} f \geq 2$. □

(1.8) takes an especially simple form for morphisms of $\mathbb{P}^1$:

3.14 Theorem. *Let $f : \mathbb{P}^1 \to X$ be a free morphism over $g : B \to X$.*

(3.14.1) *A general deformation of f is separable.*

(3.14.2) *If $a_2(f, B) \geq 1$, then a general deformation of f over g is an immersion on $\mathbb{P}^1 - B$. Furthermore, if g is an immersion, then a general deformation of f over g is an immersion on $\mathbb{P}^1$.*

(3.14.3) *If $a_3(f, B) \geq 1$, then a general deformation of f over g is an embedding on $\mathbb{P}^1 - B$. Furthermore, if g is an embedding, then a general deformation of f over g is an embedding on $\mathbb{P}^1$.*

Proof. Let $[f] \in U \subset \operatorname{Hom}(\mathbb{P}^1, X, g)$ be an open set such that $h^1(\mathbb{P}^1, f_t^*T_X(-B)(-2)) \leq h^1(\mathbb{P}^1, f^*T_X(-B)(-2))$ for every $[f_t] \in U$. Fix $p \in \mathbb{P}^1$. If $[f_t] \in U$ is not an immersion at p, then set $g_p|2p = f_t|2p$. By (1.9)

$$\begin{aligned}\dim \operatorname{Hom}(\mathbb{P}^1, X, g_p) &\leq h^0(\mathbb{P}^1, f_t^*T_X(-B-2p))\\ &\leq h^0(\mathbb{P}^1, f_t^*T_X(-B)) - 2\dim X + h^1(\mathbb{P}^1, f_t^*T_X(-B-2p))\\ &\leq h^0(\mathbb{P}^1, f_t^*T_X(-B)) - \dim X - 1.\end{aligned}$$

Here the last inequality holds since

$$h^1(\mathbb{P}^1, f_t^*T_X(-B-2p)) = \#\{i : a_i = 0\} \leq \dim X - 1.$$

Since

$$\{[f_t] \in U | f_t \text{ is not an immersion at } p\} = \bigcup_{x\in X} \operatorname{Hom}(\mathbb{P}^1, X, g_p) \cap U,$$

we conclude that

$$\operatorname{codim}(\{[f_t] \in U | f_t \text{ is not an immersion at } p\}, U) \geq \dim X + 1 - \dim X \geq 1.$$

This shows the first claim. The other two follow from (1.18). □

3.14.4 Complement. *Let $f : \mathbb{P}^1 \to X$ be a free morphism. If* $\operatorname{char} = 0$, *then a general deformation f_t of f can be written as*

$$f_t : \mathbb{P}^1 \to \mathbb{P}^1 \xrightarrow{h_t} X$$

where h_t is an immersion.

Proof. Let f_t be a general deformation of f. Write $f_t = h_t \circ p_t$ where $h_t : \mathbb{P}^1 \to X$ is birational onto its image. Any deformation of h_t corresponds to a deformation of f_t, thus it is sufficient to prove the statement for h_t.

If $a_1(h_t) \leq 2$, then the natural map $T_{\mathbb{P}^1} \to f_t^* T_X$ is an isomorphism onto the first summand. Thus h_t is an immersion.

Thus assume that $a_1(h_t) \geq 3$. Fix $p \in \mathbb{P}^1$. $\dim_{[h_t]} \operatorname{Hom}(\mathbb{P}^1, X, h_t|p) \geq 3$ by (1.7). Let $Y \subset \operatorname{Hom}(\mathbb{P}^1, X, h_t|p)$ be an irreducible component containg $[h_t]$ and $Z \subset X$ the image of $F : \mathbb{P}^1 \times Y \to X$. If $\dim Z = 1$, then $\dim Y \geq 3$ shows that $h_t : \mathbb{P}^1 \to Z_t$ is not birational, a contradiction. Thus $\dim Z \geq 2$, hence $\operatorname{rank} dF(q, [h_s]) \geq 2$ where h_s is a general deformation of h_t fixing p. Thus by (3.10) $a_2(h_s) \geq 1$ and we are done by (3.14.2). □

3.14.5 Exercise. Analyze the example of (IV.2.12.1) for cubic surfaces of the form $x_0^3 = f(x_1.x_2, x_3)$ in characteristic 3 to show that (3.14.4) fails in positive characteristic. See also (V.5.19).

II.4 Cone of Curves

The aim of this section is to define the cone of curves and to illustrate its behaviour in some examples. One of the basic results is the theorem of the base of Néron-Severi (4.5) which says that the vectorspace of 1-cycles modulo numerical equivalence is finite dimensional. In characteristic zero this can be proved by topological methods. The general case is unfortunately more complicated and we do not prove it. (4.19–21) contain some results which imply finite dimensionality in certain special cases.

The cone of curves is computed in some examples in (4.13–18). These give ample illustration that the structure of this cone is rather complicated in general.

Most of the basic definitions are the same for cycles of arbitrary dimension. See [Fulton84] for a thorough treatment.

4.1 Definition. (4.1.1) Let $f : X \to S$ be a proper morphism of schemes. Let $Z_k(X/S)$ be the free abelian group generated by irreducible and reduced k-dimensional subvarieties $Y \subset X$ such that $f(Y) = \text{point}$. If $S = \operatorname{Spec} F$ for a field F, then $Z_k(X/\operatorname{Spec} F) = Z_k(X)$ as defined in (I.3.1). The effective cycles form a subsemigroup whch is denoted by $ZE_k(X/S)$. By our convention the empty k-cycle is effective.

(4.1.2) *Effective algebraic equivalence* of two effective cycles $D_1, D_2 \in ZE_k(X/S)$ is defined as follows:

(4.1.2.1) If W is a normal (possibly reducible) two dimensional scheme and $p : W \to C$ is a proper and flat morphism onto an irreducible one dimensional regular scheme, then the fundamental cycles (I.3.2.3) of any two fibers are effectively algebraically equivalent.

(4.1.2.2) If $g : W \to X$ is a proper morphism which sits in a commutative diagram

$$\begin{array}{ccc} W & \xrightarrow{g} & X \\ {\scriptstyle p}\downarrow & & \downarrow{\scriptstyle f} \\ C & \longrightarrow & S \end{array}$$

and $D_1, D_2 \in ZE_k(W)$ are effectively algebraically equivalent, then $g_*[D_1]$, $g_*[D_2] \in ZE_k(X)$ are also effectively algebraically equivalent.

(4.1.2.3) Finally, take the transitive hull of the relation given by (4.1.2.1) and (4.1.2.2). This relationship is denoted by

$$D_1 \overset{e}{\approx} D_2.$$

Equivalently, two effective cycles $D_1, D_2 \in ZE_k(X/S)$ are called effectively algebraically equivalent if $[D_1], [D_2] \in \mathrm{Chow}(X/S)$ are in the same connected component of $\mathrm{Chow}(X/S)$.

(4.1.3) Two cycles $D_1, D_2 \in Z_k(X/S)$ are called *algebraically equivalent* if there is a third cycle $E \in Z_k(X/S)$ such that $D_1 + E$ and $D_2 + E$ are effectively algebraically equivalent. This relationship is denoted by

$$D_1 \overset{a}{\approx} D_2.$$

All cycles algebraically equivalent to the zero cycle form a subgroup of $Z_k(X/S)$. There does not seem to be any standard notation for the quotient. I denote it by $B_k(X/S)$ following [Fulton84, 10.3].

(4.1.4) Assume that X is a scheme over a field. If in (4.1.2.1) we allow only rational curves for C, we obtain the notion of *effective rational equivalence* resp. *rational equivalence* (denoted by $\overset{r}{\approx}$). All cycles rationally equivalent to the zero cycle form a subgroup of $Z_k(X/S)$. We denote the quotient by $A_k(X)$. (The notation $Ch_k(X)$ is also customary.)

(4.1.5) Two one cycles $D_1, D_2 \in Z_1(X/S)$ are called *numerically equivalent* if for every Cartier divisor M on X the intersection numbers $M \cdot D_1$ and $M \cdot D_2$ agree. (See [Fulton84, 19.1] for a definition for k-cycles.) This relationship is denoted by

$$D_1 \equiv D_2.$$

All cycles numerically equivalent to the zero cycle form a subgroup of $Z_1(X/S)$. The quotient is denoted by $N_1(X/S)_{\mathbb{Z}}$.

(4.1.6) If B is a $\mathbb{Z}$-module, then we use the notation

$$Z_k(X/S)_B := Z_k(X/S) \otimes_{\mathbb{Z}} B,$$

and similarly for the other groups. We use this notation only for $B = \mathbb{Q}, \mathbb{R}, \mathbb{C}$.

It has been standard to write $N_1(X/S)$ instead of $N_1(X/S)_{\mathbb{R}}$.

(4.1.7) The effective B-cycles generate a subsemigroup $BE_k(X/S)_B \subset B_k(X/S)_B$ resp. $NE_1(X/S)_B \subset N_1(X/S)_B$. This subsemigroup is called the *cone of effective cycles.*

(4.1.8) We can define partial orders on the effective cycles as follows. Let $D_1, D_2 \in ZE_k(X/S)$ be two effective cycles. We say that D_1 is *effectively* (resp. *algebraically*) *less than* D_2 if there is an effective cycle E such that $D_1 + E \overset{e}{\approx} D_2$ (resp. $D_1 + E \overset{a}{\approx} D_2$). We use the notation

$$D_1 \overset{e}{\prec} D_2 \quad \text{resp.} \quad D_1 \overset{a}{\prec} D_2.$$

(4.1.9) Let $(f_i : U_i \to V_i)$ (i=1,2) be two algebraic families of 1-cycles of X/S. If the V_i are connected, then the notation

$$(f_1 : U_1 \to V_1) \overset{e}{\approx} \quad (\text{resp. } \overset{a}{\approx}, \overset{e}{\prec}, \overset{a}{\prec}) \quad (f_2 : U_2 \to V_2)$$

makes sense. If the choice of f_i and U_i is clear, we write $\mathrm{Fam}(V_1) \overset{e}{\approx} \mathrm{Fam}(V_2)$ etc.

4.1.10 Exercises. (4.1.10.1) Let $f : X \to Y$ be a proper morphism between schemes (or algebraic spaces). Let $\sum a_i[V_i]$ and $\sum b_i[U_i]$ be two effective k-cycles which are effectively algebraically equivalent. Assume that every V_i is contained in a fiber of f. Show that every U_i is also contained in a fiber of f.

(4.1.10.2) Let $Z \subset \mathbb{P}^4$ be a cubic hypersurface with a single ordinary node (that is, locally analytically given by the equation $xy - uv = 0$). Let $p : X \to Z$ be the blow-up of the node. The exceptional divisor of p is $E \cong \mathbb{P}^1 \times \mathbb{P}^1$. Let $L_1, L_2 \subset E$ be two intersecting lines. Show that mL_1 and mL_2 are not effectively algebraically equivalent for $m \geq 1$.

(4.1.10.3) With the above notation, show that L_1 and L_2 are algebraically equivalent.

(4.1.10.4) Let E be an elliptic curve and L a nontorsion degree zero line bundle. Let $X = \mathbb{P}(\mathcal{O} + L)$. X has two sections C_1 and C_2 with selfintersection zero. mC_1 and mC_2 are algebraically equivalent but not effectively algebraically equivalent for $m \geq 1$.

4.2 Definition. (4.2.1) Let X be a scheme. $\mathrm{CDiv}(X)$ denotes the group of Cartier divisors (I.1.12). (The notation $Z^1(X)$ would be more symmetrical but less standard.)

(4.2.2) If X is normal, then one can define $\mathrm{WDiv}(X)$ to be the free abelian group generated by irreducible Weil divisors. The Cartier divisors form a subgroup $\mathrm{CDiv}(X) \subset \mathrm{WDiv}(X)$.

(4.2.3) Let $f : X \to S$ be a proper morphism. Two Cartier divisors $M_1, M_2 \in \mathrm{CDiv}(X)$ are called *f-numerically equivalent* if for every 1-cycle $D \in Z_1(X/S)$ the intersection numbers $M_1 \cdot D$ and $M_2 \cdot D$ agree. Numerical equivalence is denoted by $\equiv$. All Cartier divisors f-numerically equivalent to the zero divisor form a subgroup of $\mathrm{CDiv}(X)$. The quotient is denoted by $N^1(X/S)_{\mathbb{Z}}$.

(4.2.4) Let D_i be Cartier divisors on X. Let B be a $\mathbb{Z}$-module (we use only $B = \mathbb{Q}, \mathbb{R}$) and $b_i \in B$. A finite formal linear combination $\sum b_i D_i$ is called a *B-Cartier B-divisor.* If X is not normal, we do not define the equality of two B-Cartier divisors. If X is normal, then every B-Cartier B-divisor can be identified with an element of $\mathrm{CDiv}(X) \otimes B \subset \mathrm{WDiv}(X) \otimes B$.

We can always define

$$N^1(X/S)_B := N^1(X/S) \otimes B.$$

It has been standard to write $N^1(X/S)$ instead of $N^1(X/S)_{\mathbb{R}}$.

(4.2.5)If $f : X \to S$ is proper, then the intersection number of curves and Cartier divisors gives a nondegenerate bilinear pairing

$$N_1(X/S)_{\mathbb{Q}} \times N^1(X/S)_{\mathbb{Q}} \to \mathbb{Q}.$$

(4.2.6) Let $M \in \mathrm{CDiv}(X/S)$. We write $M^{\cap}$ instead of M if we want to emphasize that we are considering M as a linear function on $Z_1(X/S)$. Similarly for $M \in N^1(X/S)$ etc.

(4.2.7) A Cartier divisor (or a $\mathbb{Q}$ or $\mathbb{R}$-Cartier divisor) M is called *f-nef* if $M \cdot C \geq 0$ for every $C \in ZE_1(X/S)$. If $g : Y/S \to X/S$ is a morphism and M is f-nef, then g^*M is $g \circ f$-nef.

4.3 Proposition. *Let X_F be a scheme over a field F and $E \supset F$ a normal extension with Galois group G. There is a natural isomorphism*

$$g^* : Z_k(X_F)_{\mathbb{Q}} \to Z_k(X_E)^G_{\mathbb{Q}},$$

which induces isomorphisms

$$A_k(X_F)_{\mathbb{Q}} \cong A_k(X_E)^G_{\mathbb{Q}}, \quad B_k(X_F)_{\mathbb{Q}} \cong B_k(X_E)^G_{\mathbb{Q}}, \quad N_1(X_F)_{\mathbb{Q}} \cong N_1(X_E)^G_{\mathbb{Q}}.$$

4.3.1 Remark. The $\mathbb{Q}$ coefficients are needed even in the simplest situations. Let X_F be a smooth conic whithout any points in F. Then $A_0(X_F) \cong \mathbb{Z}$ and it is generated by a degree 2 cycle. If X_E has a point in E, then $A_0(X_E) \cong \mathbb{Z}$ and it is generated by a degree 1 cycle. Thus $A_0(X_F) \to A_0(X_E)$ is multiplication by 2.

Proof. Any k-cycle over F gives a k-cycle over E. Conversely, if a k-cycle over E is G-invariant, then it is defined over a finite inseparable extension of F, hence by (I.4.5.4) a suitable multiple is defined over F.

Let $D_F \in Z_k(X_F)$ and assume that it is rationally (resp. algebraically) equivalent to zero. This equivalence is exhibited by a family over a connected curve C_F. Since C_F has a point in F, it stays connected after any field extension, hence D_E is rationally (resp. algebraically) equivalent to zero.

Conversely assume that D_E is rationally (resp. algebraically) equivalent to zero. We may assume that the equivalence is exhibited by a curve defined over a finite extension $F' \supset F$. The same curve viewed as a curve over F shows that (a multiple of) D_F is rationally (resp. algebraically) equivalent to zero.

Assume that D_F is numerically zero and let A_E be a Cartier divisor. We need to show that $D_E \cdot A_E = 0$. A_E is algebraically equivalent to a Cartier divisor A'_E defined over a finite normal extension of $F' \supset F$ with Galois group G' and degree of inseparability p^i. $p^i \sum_{g \in G'} gA'_E$ is a Cartier divisor which is defined over F and

$$0 = D_F \cdot p^i \sum_{g \in G'} gA'_E = \deg(F'/F)(D_E \cdot A_E).$$

Conversely, if D_E is numerically zero, then D_F is also numerically zero since any Cartier divisor on X_F gives a Cartier divisor on X_E. □

4.4 Proposition. *Let X/S and Y/S be proper schemes and $g : X/S \to Y/S$ a morphism.*

(4.4.1) *The push forward of cycles*

$$g_* : Z_k(X/S) \to Z_k(Y/S)$$

defined in (I.3.1) *is compatible with any of the above equivalence relations. In particular we obtain maps*

$$g_* : B_k(X/S) \to B_k(Y/S) \quad \text{and} \quad g_* : N_1(X/S) \to N_1(Y/S).$$

(4.4.2) *If in addition g is dominant, then*

$$g_* \otimes \mathbb{Q} : Z_k(X/S)_{\mathbb{Q}} \to Z_k(Y/S)_{\mathbb{Q}} \quad \text{and}$$
$$g_* \otimes \mathbb{Q} : ZE_k(X/S)_{\mathbb{Q}} \to ZE_k(Y/S)_{\mathbb{Q}}$$

are surjective. Similarly for A_k, B_k and N_1.

(4.4.3) *We have a complex*

$$Z_k(X/Y) \to Z_k(X/S) \to Z_k(Y/S)$$

which is usually not exact. Similarly for A_k, B_k and N_1.

Proof. The first part is clear from the definitions. To see the second claim let $C \subset Y$ be an irreducible curve. Then $g^{-1}(C) \to C$ is dominant. Let $D \subset g^{-1}(C)$ be any irreducible curve which is not mapped to a point by g. Then $g_*[D] = \deg(D/C)[C]$. This also explains why we need $\mathbb{Q}$ coefficients. Finally the last part is again clear. □

4.5 Theorem (Theorem of the Base of Néron-Severi). *If S is of finite type and $f : X \to S$ is proper, then $N_1(X/S)_{\mathbb{Q}}$ and $N^1(X/S)_{\mathbb{Q}}$ are finite dimensional vector spaces of the same dimension.*

Proof. We prove (4.5) in case X is proper over $\mathbb{C}$. The first Chern class defines a morphism

$$c_1 : \mathrm{CDiv}(X)_{\mathbb{R}} \to H^2(X, \mathbb{R}).$$

Clearly every divisor in $\ker c_1$ is numerically trivial. Thus $N^1(X)_{\mathbb{R}}$ is the quotient of a subvectorspace of $H^2(X,\mathbb{R})$, hence finite dimensional. In fact one can prove (cf. [Griffiths-Harris78, p.462]) that c_1 descends to an embedding

$$c_1 : N^1(X)_{\mathbb{R}} \to H^2(X,\mathbb{R}).$$

Unfortunately I do not know any simple argument in the general case. See [Kleiman66, p.334] for a proof. □

4.5.1 Remark. In some cases we are able to prove by different methods that $N_1(X/S)_{\mathbb{Q}}$ is finite dimensional. See for example (4.19.1).

4.6 Definition. The dimension of $N_1(X/S)_{\mathbb{Q}}$ is called the *Picard number* of X/S. It is denoted by $\rho(X/S)$. If $S = \operatorname{Spec} k$ for a field k, then we use $\rho(X)$ (or $\rho_k(X)$) instead of $\rho(X/\operatorname{Spec} k)$.

4.7 Notation. (4.7.1) We usually write $N_1(X)$ instead of $N_1(X)_{\mathbb{R}}$ and $NE(X)$ instead of $NE_1(X)_{\mathbb{R}}$. $NE(X)$ is called the *cone of curves* of X.

(4.7.2) Let $\overline{NE}(X)$ be the closure of $NE(X)$ in the Euclidean topology of $N_1(X)$.

4.8 Proposition. *Assume that X is projective and H is an ample Cartier divisor. Then $H^\cap$ is strictly positive on $\overline{NE}(X) - \{0\}$. Thus $\overline{NE}(X)$ does not contain a line and*

$$\overline{NE}(X) \cap \{z \in N_1(X) | z \cdot H \le const\} \tag{4.8.1}$$

is compact.

Proof. $H^\cap$ is a positive linear function on $NE(X)$, thus it is nonnegative on $\overline{NE}(X)$. Assume that $H \cdot z = 0$ for some $z \in \overline{NE}(X) - \{0\}$. Let $M \in \operatorname{CDiv}(X)$ be such that $M \cdot z < 0$. $M + kH$ is ample for $k \gg 1$, thus

$$0 \le (M + kH) \cdot z = M \cdot z < 0,$$

a contradiction.

Assume that $\overline{NE}(X)$ contains a line. If a closed cone contains a line, then it contains a line L through the origin. $H^\cap | L$ is a linear function, positive outside the origin, which is impossible.

Finally let $[M_i]$ be a basis of $N^1(X)$. $k_i H \pm M_i$ are ample for $k_i \gg 1$, thus if $z \in \overline{NE}(X)$, then

$$|z \cdot M_i| \le k_i (z \cdot H).$$

Thus the set defined by (4.8.1) is contained in a large cube. It is also closed, hence compact. □

4.8.2 Exercise. Let X_k be a proper variety and M a Cartier divisor on X_k. Let $K \supset k$ be any field extension. Show that $M^\cap$ is positive on $NE(X_k)$ (resp. on $\overline{NE}(X_k)$) iff $M^\cap$ is positive on $NE(X_K)$ (resp. on $\overline{NE}(X_K)$).

4.9 Definition. (4.9.1) Let $V \subset \mathbb{R}^n$ be a closed convex cone. A subcone $W \subset V$ is called *extremal* if

$$u, v \in V,\ u + v \in W \Rightarrow u, v \in W.$$

(4.9.2) A one dimensional subcone is called a *ray*.

(4.9.3) A nonzero linear function $L : \mathbb{R}^n \to \mathbb{R}$ is called a *supporting function* of V if L is nonnegative on V and

$$V_L := V \cap (L = 0)$$

is positive dimensional.

4.9.4 Note on Terminology. Let X be a proper variety. An extremal ray $R \subset \overline{NE}(X)$ is called an *extremal ray of* X. If L is a Cartier divisor on X and $R \cdot L < 0$, then R is called an *L-negative extremal ray*. Sometimes the expression K_X-extremal ray is used to refer to a K_X-negative extremal ray.

In [Mori82] the term "extremal ray" is reserved for K_X-negative extremal rays, but the terminology has been changing recently.

We use some easy properties of closed cones:

4.10 Lemma. *Notation as above. Let ∂V denote the boundary of V in $\mathbb{R}^n$. (If $\dim V < n$, then $\partial V = V$.)*

(4.10.1) *For every supporting function L, $V_L \subset \partial V$.*

(4.10.2) *If $z \in \partial V$, then there is a supporting function $L(z)$ such that $z \in V_{L(z)}$.*

(4.10.3) *A subcone $W \subset V$ is extremal iff there is a supporting function L such that $W \subset V_L$ and W is extremal in V_L.*

(4.10.4) *If V is a closed convex cone which does not contain a line, then V is the convex hull of its extremal rays. More precisely, any $v \in V$ can be written as a linear combination of at most n extremal rays.*

(4.10.5) *If $W \subsetneq V$ are closed cones, then there is a linear function L such that L is semi positive on V, vanishes on some ray of V and is strictly positive on $W - \{0\}$.* □

Next we discuss the cone of curves of smooth surfaces through some examples.

4.11 Definition. (4.11.1) Let S be a smooth projective surface. By the Hodge index theorem the intersection product on S has one positive eigenvalue. Thus in a suitable basis of $N_1(S)$ the intersection form is given by

$$x_1^2 - \sum_{i=2}^{m} x_i^2.$$

(4.11.2) Let $Q \subset NE(S)$ be the open subset given by $Z \cdot Z > 0$. Q has two connected components:

$$Q^+ := \left\{ \mathbf{x} \in N_1(S) | x_1 > \left(\sum_{i=2}^{m} x_i^2\right)^{1/2} \right\}$$

and

$$Q^- := \left\{ \mathbf{x} \in N_1(S) | x_1 < -\left(\sum_{i=2}^{m} x_i^2\right)^{1/2} \right\}$$

If H is ample, then $[H] \in Q^+$ and by the index theorem $H \cdot Z > 0$ for every $Z \in Q^+$.

The following lemma describes basic properties of the cone of curves of surfaces:

4.12 Lemma (Notation as in (4.11)).

(4.12.1) $Q^+ \subset NE(S)$. *Thus every vector in Q^+ is contained in the interior of $NE(S)$.*

(4.12.2) *Let $C \subset S$ be an irreducible curve. If $C \cdot C \leq 0$, then $[C] \in \partial \overline{NE}(S)$. If $(C^2) < 0$, then $\mathbb{R}^+[C]$ is an extremal ray of $\overline{NE}(S)$.*

(4.12.3) *If $Z \in \overline{NE}(S)$ generates an extremal ray and $(Z^2) < 0$, then there is an irreducible curve $C \subset S$ such that $[C] \in \mathbb{R}^+ Z$.*

Proof. (4.12.1) is a reformulation of [Hartshorne77, V.1.8].

If $C \cdot C = 0$, then $C^\cap$ is a supporting function of $[C]$, thus by (4.10.1) $[C] \in \partial \overline{NE}(S)$. If $C \cdot C < 0$, then $V := \{z \in \overline{NE}(S) | z \cdot C \geq 0\}$ is a closed subcone of $\overline{NE}(S)$ which does not contain C. If $C \neq D \subset S$ is an irreducible curve, then $[D] \in V$. Thus $\overline{NE}(S) = \langle V, \mathbb{R}^+[C] \rangle$, and so $[C]$ generates an extremal ray. This shows (4.12.2).

Finally let $Z \in \overline{NE}(S)$ be an extremal ray such that $Z \cdot Z < 0$. Let $Z_i = \sum_j a_{ij} C_j$ be a sequence of effective cycles whose limit is Z; $C_j \subset S$ distinct and irreducible curves. Since $Z \cdot Z = \lim Z \cdot Z_i$, there is at least one curve $C = C_k$ such that $Z \cdot C < 0$. Write $Z_i = c_i C + \sum_{j \neq k} a_{ij} C_j$. Then

$$0 > C \cdot Z = \lim C \cdot Z_i \geq (\lim c_i) \cdot (C^2).$$

Thus $(C^2) < 0$ and $\lim c_i > 0$. Pick $0 < \eta < \lim c_i$. Then $Z_i - \eta C$ is effective for $i \gg 1$ and $Z = \eta C + \lim(Z_i - \eta C)$. Since Z is extremal, this implies that $[C] \in \mathbb{R}^+ Z$. □

The following theorem gives a seemingly complete description of the cone of curves of surfaces. Unfortunately, as we see in some examples, it is sometimes very difficult to determine all curves with negative selfintersection on a given surface S.

4.13 Theorem. *Let S be a smooth projective surface. Then*

$$\overline{NE}(S) = \overline{Q^+} + \sum_D \mathbb{R}^+[D],$$

where the sum is over all irreducible curves $D \subset S$ such that $(D^2) < 0$.

Proof. By (4.10.4) it is sufficient to show that every extremal ray of $\overline{NE}(S)$ is contained in the right hand side of (4.13). Let $Z \in \overline{NE}(S)$ be an extremal ray. If $(Z^2) \geq 0$, then $Z \in \overline{Q^+}$. If $(Z^2) < 0$, then by (4.12.3) $\mathbb{R}^+ Z$ is generated by a curve with negative selfintersection. □

4.14 Corollary. *Assume that $-K_S$ is nef. Then*

$$\overline{NE}(S) = \overline{Q^+} + \sum_D \mathbb{R}^+[D],$$

where the sum is over all curves $D \subset S$ such that $D \cong \mathbb{P}^1$ and $(D^2) \in \{-1, -2\}$.

Proof. By adjunction

$$2g(D) - 2 = D \cdot D + D \cdot K_S \leq (D^2).$$

Thus if $(D^2) < 0$, then $D \cong \mathbb{P}^1$ and $(D^2) \in \{-1, -2\}$. □

4.15 Examples. (4.15.1) Let S be an Abelian surface. K_S is trivial and S does not contain rational curves. Thus $\overline{NE}(S) = \overline{Q^+}$ is a cone whose boundary is given by the equation

$$x_1^2 - \sum_{i=2}^m x_i^2 = 0.$$

If $m \geq 3$, then $\overline{NE}(S)$ has no "edges". This happens for instance if E is an elliptic curve and $S = E \times E$. $\{e\} \times E, E \times \{e\}$ and the diagonal are independent in $N_1(S)$, thus $m = \dim N_1(S) \geq 3$.

(4.15.2) Let S be a Del Pezzo surface. By (V.1.6) $\overline{NE}(S)$ is generated by K-negative extremal rays. By (III.2.1.7) if $\rho(S) \geq 3$, then every ray is a (-1)-curve which is a line on S. For instance, let $S \subset \mathbb{P}^3$ be a smooth cubic surface. S contains 27 lines L_i and

$$\overline{NE}(S) = \sum_i \mathbb{R}^+[L_i] \subset \mathbb{R}^7.$$

(4.15.3) Let $C_1, C_2 \subset \mathbb{P}^2$ be two smooth cubics with nine intersection points P_i. Assume for simplicity that every member of the pencil $|C_1, C_2|$ is irreducible. Let S be the surface obtained from $\mathbb{P}^2$ by blowing up the nine points P_i. The pencil $|C_1, C_2|$ becomes base point free on S, thus we have a morphism $f : S \to \mathbb{P}^1$ whose fibers are irreducible plane cubics. Let $D \subset S$ be the finite set of singular points of the fibers of f. The nine exceptional curves E_i over the points P_i become sections of f. Fixing one section E_9 makes $S - D$ into a family of algebraic groups over $\mathbb{P}^1$. For sufficiently general choice of C_1, C_2 the sections $E_1, \ldots, E_8$ generate a free abelian subgroup of rank 8 of $\mathrm{Aut}(S - D)$. Since D is a finite set of points, $\mathrm{Aut}(S - D) \subset \mathrm{Aut}(S)$. For $m =$

$(m_1, \ldots, m_8) \in \mathbb{Z}^8$ let $\rho_m \in \mathrm{Aut}(S)$ denote the automorphism corresponding to $\sum m_i E_i$. $\rho_m(E_9) \subset S$ is a smooth rational curve with selfintersection (-1). Thus for every $m \in \mathbb{Z}^8$ we obtain an extremal ray of $\overline{NE}(S)$.

(4.15.4) Let X be a surface such that there is a dominant morphism $X \to S$ where S is as in (4.15.3). Show that on X there are infinitely many curves with negative selfintersection.

4.16 Exercise. The aim of this exercise is to compute the cone of curves of surfaces in two interesting cases.

(4.16.1) Let C_1 and C_2 be smooth curves. Pulling back from the factors gives an exact sequence

$$0 \to \mathrm{Pic}(C_1) \times \mathrm{Pic}(C_2) \to \mathrm{Pic}(C_1 \times C_2) \to K \to 0.$$

Show that

$$K \cong \mathrm{Hom}(\mathrm{Pic}^\circ(C_1), \mathrm{Pic}^\circ(C_2)),$$

which is a finitely generated free Abelian group.

(4.16.2) Let E be a very general elliptic curve over $\mathbb{C}$. Show that $N_1(E \times E)$ is generated by the classes $E_{10} = E \times \{0\}$, $E_{01} = \{0\} \times E$ and the diagonal E_{11}. We always use this basis of $N_1(E \times E)$.

(4.16.3) $SL(2, \mathbb{Z})$ acts on $E \times E$ by

$$(x, y) \begin{pmatrix} a & b \\ c & d \end{pmatrix} = (ax + cy, bx + dy).$$

(4.16.4) For any pair of coprime integers n, m let E_{nm} be the image of $E \to E \times E$ given by $x \mapsto (nx, mx)$. Show that $(E_{nm}^2) = 0$ and every irreducible curve C with $(C^2) = 0$ is a translate of some E_{nm}. The $SL(2, \mathbb{Z})$ action is transitive on the set of curves $\{E_{nm}\}$.

(4.16.5) $\overline{NE}(E \times E) \subset N_1(E \times E)$ is one half of the closed cone $uv + vw + wu \geq 0$. The only rational points on the boundary are the multiples of $[E_{nm}]$.

(4.16.6) $SL(2, \mathbb{Z})$ acts on $N_1(E \times E)$ and in the above basis the action is given by

$$\begin{pmatrix} a & b \\ c & d \end{pmatrix} \mapsto \begin{pmatrix} a^2 - ab & c^2 - cd & (a+c)^2 - (a+c)(b+d) \\ b^2 - ab & d^2 - cd & (b+d)^2 - (a+c)(b+d) \\ ab & cd & (a+c)(b+d) \end{pmatrix}$$

(4.16.7) If $(u, v, w) \in N_1(E \times E)$ represents an effective curve, then $u + v, v + w, w + u$ are all nonnegative.

(4.16.8) Let C be an effective curve on $E \times E$. Show that under the $SL(2, \mathbb{Z})$-action it is equivalent to a curve C' all of whose coordinates are nonnegative.

(4.16.9) Let $p : S = B_{(0,0)}(E \times E) \to E \times E$ be the blow up of the origin and $F \subset S$ the exceptional curve of p. Let $E'_{nm} \subset S$ be the birational transform of E_{nm}. Show that $({E'_{nm}}^2) = -1$.

(4.16.10) Let $D \subset S$ be an effective curve. Assume that the coordinates of $p(D)$ are all nonnegative. Show that $[D]$ is a nonnegative linear combination of the classes $[F], [E'_{10}], [E'_{01}], [E'_{11}]$.

(4.16.11) Show that the classes $[F]$ and $[E'_{nm}]$ generate $NE(S)$.

(4.16.12) Describe all extremal rays in $\overline{NE}(S)$. Show that every ray not listed in (4.16.11) is a limit of rays $\mathbb{R}^+[E'_{nm}]$.

4.17 Exercise. Let X be a proper variety and $Y = X \times \mathbb{P}^1$. The zero section $X \cong X \times \{0\} \subset Y$ gives an map $i : A_1(X) \to A_1(Y)$. Show that it is an injection. Let $[\mathbb{P}^1] \in B_1(Y)$ denote the class of a fiber. Show that

$$B_1(Y) = B_1(X) + \mathbb{R}[\mathbb{P}^1] \quad \text{and} \quad BE_1(Y) = \langle BE_1(X), \mathbb{R}^+[\mathbb{P}^1] \rangle.$$

In general one expects that in (4.13) both $\overline{Q^+}$ and $\sum \mathbb{R}^+[D]$ are needed to generate the cone of curves. The following surprising result of [Kovács94] asserts that for K3 surfaces this is not the case. For the proof see the original article.

4.18 Theorem [Kovács94]. *Let S be a smooth K3 surface. Assume that $\dim N_1(S) \geq 3$. Then one of the following conditions is satisfied:*

(4.18.1) *S does not contain any curves with negative selfintersection, thus $\overline{NE}(S) = \overline{Q^+}$; or*

(4.18.2) *$\overline{NE}(S) = \overline{\sum \mathbb{R}^+[D]}$ where the sum runs over all smooth rational curves with selfintersection -2.* □

The following result can be used to prove the finite dimensionality of $N_1(X)$ in some situations:

4.19 Proposition. *Let $f : X \to Y$ be a morphism between proper schemes. Assume that every fiber of f is a connected curve with geometrically rational components. Let $i : Z \hookrightarrow X$ be a closed subscheme such that $f \circ i : Z \to Y$ is surjective. Then*

$$A_1(X)_{\mathbb{Q}} = \langle [\textit{irred. components of fibers}], i_* A_1(Z)_{\mathbb{Q}} \rangle.$$

Therefore also

$$N_1(X) = \langle [\textit{irred. components of fibers}], i_* N_1(Z) \rangle.$$

Proof. Let $C \subset X$ be an irreducible curve. Let $Y_C = f(C)$, $X_C = f^{-1}(Y_C)$ and $Z_C = Z \cap X_C$. It is sufficient to show that

$$[C] \in \langle [\text{irred. components of fibers of } f_C], i_* A_1(Z_C)_{\mathbb{Q}} \rangle \subset A_1(X_C)_{\mathbb{Q}}.$$

Thus assume that Y is an irreducible curve. We may assume that Z is also irreducible.

Next we reduce to the case when X is also irreducible. Let $X_1 \subset X$ be an irreducible component and $X^1 \subset X$ the union of the other components. By suitable choice of X_1 we may assume that $Z \subset X^1$ and that the fibers of $X^1 \to Y$ are connected. Let $Z' = X^1 \cap X_1$ and $j_1 : Z' \to X_1$ (resp. $j^1 : Z' \to X^1$) the inclusions. By induction on the number of irreducible components of X

$$A_1(X^1)_{\mathbb{Q}} = \langle [\text{irred. components of fibers}], i_* A_1(Z)_{\mathbb{Q}} \rangle, \quad \text{and}$$
$$A_1(X_1)_{\mathbb{Q}} = \langle [\text{irred. components of fibers}], (j_1)_* A_1(Z')_{\mathbb{Q}} \rangle.$$

Also

$$\begin{aligned} &\operatorname{im}[A_1(Z')_{\mathbb{Q}} \xrightarrow{(j_1)_*} A_1(X_1)_{\mathbb{Q}} \to A_1(X)_{\mathbb{Q}}] \\ &\quad = \operatorname{im}[A_1(Z')_{\mathbb{Q}} \xrightarrow{j^1_*} A_1(X^1)_{\mathbb{Q}} \to A_1(X)_{\mathbb{Q}}] \subset \operatorname{im}[A_1(X^1)_{\mathbb{Q}} \to A_1(X)_{\mathbb{Q}}]. \end{aligned}$$

Thus

$$\begin{aligned} A_1(X)_{\mathbb{Q}} &= \operatorname{im}[A_1(X^1)_{\mathbb{Q}} \to A_1(X)_{\mathbb{Q}}] + \operatorname{im}[A_1(X_1)_{\mathbb{Q}} \to A_1(X)_{\mathbb{Q}}] \\ &\subset \langle [\text{irred. components of fibers}], i_* A_1(Z)_{\mathbb{Q}} \rangle. \end{aligned}$$

Finally assume that X is an irreducible surface and Y is a curve. Let

$$\begin{array}{ccc} X' & \xrightarrow{f'} & Y' \\ \downarrow & & \downarrow \\ X & \xrightarrow{f} & Y \end{array}$$

be a commutative diagram where the vertical arrows are dominant. Let $Z' \subset X'$ be the preimage of Z. If the proposition is true for $f' : X' \to Y'$, then by (4.4.2) it also holds for $f : X \to Y$. Thus we may assume that Y and X are normal. Furthermore, we may also assume that the generic fiber is isomorphic to $\mathbb{P}^1$. Thus there is an open set $Y^0 \subset Y$ such that $U = f^{-1}(Y^0) \cong \mathbb{P}^1 \times Y^0$.

Let $D \subset X$ be an irreducible curve. By [Hartshorne77, II.6.6] $D \cap U$ is linearly equivalent to a divisor of the form

$$m[\{0\} \times Y^0] + \sum_i n_i [\mathbb{P}^1 \times \{y_i\}]$$

for some integers m, n_i and $y_i \in Y^0$. Since $Z \to Y$ is surjective,

$$Z \cap U \sim m(Z)[\{0\} \times Y^0] + \sum_i n_i(Z)[\mathbb{P}^1 \times \{y_i\}],$$

where $m(Z) \neq 0$. If $C \subset X$ is an arbitrary curve, then by suitable choice of $a, b \in Z$, $(a \neq 0)$ we can achieve that

$$(aC - bZ) \cap U \sim \sum_i n_i [\mathbb{P}^1 \times \{y_i\}].$$

Thus

$$a[C] \quad \text{and} \quad b[Z] + \sum_i n_i[\mathbb{P}^1 \times \{y_i\}] + \sum_j m_j[F_j]$$

are rationally equivalent where the F_j are irreducible components of $f^{-1}(Y - Y^0)$. □

4.19.1 Corollary. *Let $f : X \to C$ be a morphism from a proper surface to an irreducible curve. Assume that every fiber of f is a connected curve with rational components. Then*

$$\dim B_1(X)_{\mathbb{Q}} < \infty, \quad \textit{hence } \rho(X) < \infty.$$

If every fiber of f is irreducible, then

$$\dim B_1(X)_{\mathbb{Q}} = \rho(X) = 2.$$

Proof. For $Z \subset X$ choose any irreducible curve which is not contained in a fiber of f. Then $A_1(Z) \cong \mathbb{Z}$. Up to algebraic equivalence the components of the fibers form finitely many families. If all fibers are irreducible, then they are algebraically equivalent to each other. A fiber is not numerically equivalent to Z, hence the Picard number can not be one. □

4.20 Corollary. *Let $f : X \to Y$ and $i : Z \to X$ be as in* (4.19). *Let $p : X \to W$ be a proper and dominant morphism. Assume that $p \circ i(Z) =$ point. Then*

$$A_1(W)_{\mathbb{Q}} = \langle p_*[\textit{irred. components of fibers}]\rangle.$$

Therefore also

$$N_1(W) = \langle p_*[\textit{irred. components of fibers}]\rangle.$$

Proof. By (4.4.2) and (4.19)

$$\begin{aligned} A_1(W)_{\mathbb{Q}} &= p_* A_1(X)_{\mathbb{Q}} \\ &= p_*\langle[\text{irred. components of fibers}]\rangle + (p \circ i)_* A_1(Z)_{\mathbb{Q}} \\ &= p_*\langle[\text{irred. components of fibers}]\rangle. \end{aligned}$$

□

4.21 Corollary. *Let X be a projective variety. Let $V \subset \mathrm{RatCurves}(x, X)$ be an irreducible and proper subvariety. Then $\dim B_1(\mathrm{Locus}(V))_{\mathbb{Q}} = 1$.*

Proof. Let $f : U \to V$ be the universal rational curve and $p : U \to X$ the cycle map. The point x gives a (possibly multiple) section $i : V \to U$. (4.20) applied to $W = \mathrm{Locus}(V)$ gives that

$$B_1(\mathrm{Locus}(V))_{\mathbb{Q}} = \langle p_*[\text{ fibers of } f]\rangle.$$

Since V is connected, all the fibers of f are algebraically equivalent. □

II.5 Bend and Break

The aim of this section is to present a method of finding rational curves which is of fundamental importance in higher dimensional geometry. The basic idea, due to [Mori79], is very simple. Let X be a smooth projective variety and $f: C \to X$ a morphism from a smooth projective curve to X. If $\deg_C f^*K_X$ is sufficiently negative, then from (1.2) we conclude that f has nontrivial deformations $f_t : C \to X$. If we let f_t degenerate, we are likely to find a rational curve on X.

The simplest version of bend–and–break is explained in (5.1). In the applications we need to have more control over the rational curve obtained. The precise version of bend–and–break is in (5.4–6), with an easy application in (5.7). The main application is (5.8). In (III.1) this result is used to prove the Cone Theorem. The proof of (5.8) is of considerable interest in itself. Even for the characteristic zero version, it is essential to go through positive characteristic.

In the proof of (5.8) it is crucial that we are dealing with projective varieties but the result is probably true for algebraic spaces as well (5.16–17).

5.1 Bend-and-Break Technique. Let X be a smooth projective variety, $f : C \to X$ a morphism from a smooth projective curve to X, $p \in C$ a point and $g = f|p$. Assume that $\mathrm{Hom}(C, X, g)$ is positive dimensional. Let $B^0 \to \mathrm{Hom}(C, X, g)$ be a smooth curve through $[f]$ with compactification B. The natural morphism $F^0 : C \times B^0 \to X$ extends to a rational map $F : C \times B \dashrightarrow X$. F has the property that $F(\{p\} \times B) = f(p)$, i.e. $\{p\} \times B \subset C \times B$ is contracted by F.

By (5.3.1) F is not a morphism. There is a succession of blowing ups $q : Y \to C \times B$ such that $F \circ q$ is a morphism. The exceptional curves of these blow ups are rational and at least one of them is mapped to a curve in X. Thus X contains a rational curve.

5.1.1 Exercise. Let $C \subset \mathbb{P}^3$ be a space curve and $L \subset \mathbb{P}^3$ a line intersecting C at a point p. Rotate C around L. Where is the rational curve promised by (5.1)?

In applications it is important to control the resulting rational curve. This can be done by studying deformations that keep many points fixed. The precise version is given in (5.4–6). We start with some preliminary results.

5.2 Definition. Let S be a proper surface and $B \subset S$ a proper curve. We say that B is *contractible* in S if there is a surface S' and a dominant morphism $g : S \to S'$ such that $g(B)$ is zero dimensional.

Fibers of morphisms are rarely contractible:

5.3 Proposition (Rigidity lemma). *Let $f : X \to Y$ be a proper morphism such that $f_*\mathcal{O}_X = \mathcal{O}_Y$. Let $g : X \to Z$ be a morphism. Assume that for some $y \in Y$ there is a factorization*

$$g : f^{-1}(y) \xrightarrow{f} y \xrightarrow{h_y} Z.$$

Then there is an open neighborhood $y \in U \subset Y$ and a factorization

$$g : f^{-1}(U) \xrightarrow{f} U \xrightarrow{h_U} Z.$$

Proof. Let $\Gamma \subset Y \times Z$ be the image of (f,g). Then $p : \Gamma \to Y$ is proper and $p^{-1}(y) = (y, h_y(y))$ is finite over y. Thus there is an open neighborhood $y \in U \subset Y$ such that $p^{-1}(U) \to U$ is finite. Since

$$f_*\mathcal{O}_{f^{-1}(U)} \supset p_*\mathcal{O}_{p^{-1}(U)} \supset \mathcal{O}_U = f_*\mathcal{O}_{f^{-1}(U)},$$

$p^{-1}(U) \to U$ is an isomorphism. □

5.3.1 Exercise. Let B be an irreducible and proper curve and C an arbitrary curve. Let $B_c = B \times \{c\} \subset B \times C$ where $c \in C$ is arbitrary. Then B_c is not contractible in $B \times C$.

Another way of obtaining (5.3.1) is the following:

5.3.2 Proposition. *Let S be a proper surface and $B \subset S$ a contractible curve. Then $B \cdot B < 0$.*

Proof. We prove this only for S smooth and S' projective. The general case works the same once the definition of intersection numbers is established in general.

Let $S' \to \mathbb{P}^2$ be a finite morphism. We can pull back $\mathcal{O}_{\mathbb{P}^2}(1)$ to S to get a line bundle H such that $H \cdot H > 0$ and $H \cdot B = 0$. By the index theorem, then $B \cdot B < 0$. □

5.4 Theorem (Bend and Break) [Mori79; Miyaoka-Mori86]. *Let B be a smooth proper and irreducible curve over a field and S an irreducible, proper and normal surface. Let $p : S \to B$ be a morphism. Assume that there is an open subset $B^0 \subset B$, a smooth projective curve C and an isomorphism*

$$f : [C \times B^0 \xrightarrow{\pi} B^0] \cong [p^{-1}(B^0) \xrightarrow{p} B^0],$$

where π is the second projection.

A section $s : B \to S$ is called flat if $s(B^0) = \{c\} \times B^0$ under the above isomorphism.

(5.4.1) *If there is a contractible flat section* $s_1 : B \to S$, *then for some* $b \in B - B^0$ *the fiber* $p^{-1}(b)$ *contains a rational curve intersecting* $s_1(B)$.

(5.4.2) *If* $g(C) = 0$ *and there are two contractible sections* $s_1, s_2 : B \to S$, *then for some* $b \in B - B^0$ *the fiber* $p^{-1}(b)$ *is either reducible or nonreduced.*

(5.4.3) *Let* L *be a nef* $\mathbb{R}$*-Cartier divisor on* S. *If there are* $k \geq 1$ *contractible flat sections* $s_i : B \to S$ *such that* $L \cdot s_i(B) = 0$ *for every* i, *then for some* $b \in B - B^0$ *the fiber* $p^{-1}(b)$ *contains a rational curve* D *intersecting a section* $s_i(B)$ *such that*

$$L \cdot D \leq \frac{2}{k} L \cdot C,$$

where, by a slight abuse of notation, C *also denotes a general fiber of* p.

(5.4.4) *Let* L *be a nef* $\mathbb{R}$*-Cartier divisor on* S *such that* $(L^2) > 0$. *If there are* k *flat sections* $s_i : B \to S$ *such that* $L \cdot s_i(B) = 0$ *for every* i, *then for some* $b \in B - B^0$ *the fiber* $p^{-1}(b)$ *contains a rational curve* D *intersecting a section* $s_i(B)$ *such that*

$$0 < L \cdot D < \frac{2}{k} L \cdot C.$$

Proof. Let $X = C \times B$ and $\Gamma \subset X \times_B S$ the closure of the graph of the isomorphism f. Let $p_X : \Gamma \to X$ and $p_S : \Gamma \to S$ be the projections. Every flat section $s_i : B \to S$ induces a flat section $s_i^X : B \to X$. By (5.3.1) $f : X \dashrightarrow S$ is not defined somewhere along $s_i^X(B)$.

Let $D' \subset \Gamma$ be an exceptional curve of p_X intersecting the birational transform of $s_i^X(B)$. By (VI.1.2) D' is rational. Taking $D = p_S(D')$ proves (5.4.1).

In order to see (5.4.2) assume the contrary. Then every fiber of p is irreducible and reduced. Thus $h^1(\mathcal{O}_{p^{-1}(b)}) = 1 - \chi(\mathcal{O}_{p^{-1}(b)})$ is independent of $b \in B$ and every fiber of p is isomorphic to $\mathbb{P}^1$. Therefore S is a minimal ruled surface over B. By (4.19.1) $[s_1(B)]$ and $[C \times \{b\}]$ generate $N_1(S)$. The matrix of the intersection form of $[s_1(B)]$, $[s_2(B)]$ and $[C \times \{b\}]$ is

$$\begin{pmatrix} -a_1 & c & 1 \\ c & -a_2 & 1 \\ 1 & 1 & 0 \end{pmatrix}$$

where $-a_i = s_i(B) \cdot s_i(B) < 0$ by (5.3.2) and $c = s_1(B) \cdot s_2(B) \geq 0$. The determinant is $2c + a_1 + a_2 > 0$ which is impossible since $\dim N_1(S) = 2$.

The last two cases are considered simultaneously. We resolve the indeterminacies of $f : X \dashrightarrow S$ by repeatedly blowing up closed points if necessary. We obtain

$$C \times B = X \xleftarrow{\rho_X} Y \xrightarrow{\rho_S} S. \tag{5.4.5}$$

In case (5.4.3) let $Y' = Y, \pi_X = \rho_X$ and $L' = \rho_S^* L$.

In case (5.4.4) we construct Y' as follows. Let $F \subset Y$ be a (-1)-curve which is ρ_X -exceptional and $F \cdot \rho_S^* L = 0$. Let $\rho_1 : Y \to Y_1$ be the contraction of F. There is an $\mathbb{R}$-Cartier divisor L_1 such that $\rho_S^* L = \rho_1^* L_1$. We continue in this manner until we finally obtain

$$Y \to Y' \xrightarrow{\pi_X} X \quad \text{and a nef } \mathbb{R}\text{-Cartier divisor } L' \text{ on } Y',$$

such that L' has positive intersection with every (-1)-curve which is also π_X-exceptional. In general the resulting map $\pi_S : Y' \dashrightarrow S$ is not a morphism. By construction $L' \cdot L' = L \cdot L > 0$.

Let $B_i = \operatorname{im} s_i^X$, $B \subset X$ a general flat section and $C \subset X$ a general fiber of $X \to B$. Let $B_i' \subset Y'$ be the birational transform of B_i on Y'. Write

$$\pi_X^* B \overset{a}{\approx} \pi_X^* B_i = B_i' + E_i,$$

where $E_i \subset Y'$ is π_X-exceptional and effective. We have the following table of intersection numbers

	$\pi_X^* B$	$\pi_X^* C$	E_j
$\pi_X^* B$	0	1	0
$\pi_X^* C$	1	0	0
E_i	0	0	$-c_i\delta_{ij}$

where

$$c_i = -E_i \cdot E_i = -(E_i - \pi_X^* B) \cdot (E_i - \pi_X^* B) = -B_i' \cdot B_i'.$$

We claim that $c_i > 0$. If $s_i(B)$ is contractible, then by (5.3.1) f is not defined somewhere along B_i. Thus we performed at least one blow up along B_i, hence $c_i > 0$. In case (5.4.4), $L' \cdot B_i' = 0$ and the Hodge index theorem implies that $c_i > 0$. Let

$$L' \overset{a}{\approx} \beta \pi_X^* B + \gamma \pi_X^* C - \sum_i e_i E_i + F,$$

where F is orthogonal to $\langle \pi_X^* B, \pi_X^* C, E_1, E_2, \ldots \rangle$. Observe that $\beta = \pi_X^* C \cdot L' = C \cdot L$. Since L is nef,

$$0 \le L' \cdot E_i = e_i c_i, \quad \text{hence} \quad e_i \ge 0.$$

By assumption $0 = s_i(B) \cdot L = B_i' \cdot L' = \pi_X^* B \cdot L' - E_i \cdot L' = \gamma - e_i c_i$, hence $\gamma = e_i c_i$ for every i. Since L' is nef, $L' \cdot L' \ge 0$ (VI.2.17) and $F \cdot F \le 0$ by the Hodge index theorem. Thus

$$0 \le L' \cdot L' = 2\beta\gamma - \sum e_i^2 c_i + F \cdot F \le 2\beta\gamma - \sum e_i^2 c_i,$$

and the first inequality is strict in case (5.4.4) by assumption. Substituting $e_i c_i = \gamma$ we obtain that

$$\text{(5.4.6)} \qquad \begin{aligned} \sum e_i \gamma &\le 2\beta\gamma \quad \text{in case (5.4.3), and} \\ \sum e_i \gamma &< 2\beta\gamma \quad \text{in case (5.4.4).} \end{aligned}$$

If $\gamma = 0$ (this can happen only for (5.4.3)), then $\pi_X^* B \cdot L' = 0$ and $E_i \cdot L' = e_i c_i = \gamma = 0$. Hence we can take D to be any curve in $\pi_S(E_i)$. Otherwise

$$\begin{aligned} &\sum e_i \le 2\beta \quad \text{in case (5.4.3), and} \\ &\sum e_i < 2\beta \quad \text{in case (5.4.4).} \end{aligned} \tag{5.4.7}$$

We sum over k indices, thus

$$e_j \le \frac{2}{k}\beta \quad \text{resp.} \quad e_j < \frac{2}{k}\beta \quad \text{for some } j. \tag{5.4.8}$$

5.4.9 Sublemma. *Let X be a smooth projective surface and $B \subset X$ a smooth curve. Assume that $p : Y \to X$ is a succession of smooth blow ups. Let $B' \subset Y$ be the birational transform of B and $p^* B = B' + \sum_i f_i F_i$ where the F_i are p-exceptional. Then*

$$B \cdot B - B' \cdot B' \le \sum_{i:(F_i^2)=-1} f_i. \tag{5.4.9.1}$$

Proof (By induction on the number of blow-ups). Let $Y \to Y_1$ be a single blow-up with center y_1. Let $F_1 \subset Y$ be the exceptional curve over y_1.

If y_1 is not contained in the birational transform of B or in a (-1)-curve, then the left side of (5.4.9.1) is unchanged and the right side increases by one.

If y_1 is contained in the birational transform of B but not in any (-1)-curve, then both sides of (5.4.9.1) increase by one.

If y_1 is not contained in the birational transform of B but is contained in a (-1)-curve F_j, then the left side of (5.4.9.1) is unchanged and in the right side f_1 replaces f_j and $f_1 \ge f_j$.

If y_1 is contained in the birational transform of B and in a (-1)-curve F_j, then the left side of (5.4.9.1) increase by one and in the right side f_1 replaces f_j and $f_1 \ge f_j + 1$. □

Let $F_1, \dots, F_t$ be the (-1)-curves in $\operatorname{Supp} E_j$ and $f_1, \dots, f_t$ their coefficients in E_j. By (5.4.8)

$$\begin{aligned} &\frac{2}{k}\beta c_j \ge e_j c_j = L' \cdot E_j \ge \sum f_i L' \cdot F_i \quad \text{in case (5.4.3), and} \\ &\frac{2}{k}\beta c_j > e_j c_j = L' \cdot E_j \ge \sum f_i L' \cdot F_i \quad \text{in case (5.4.4).} \end{aligned} \tag{5.4.10}$$

$\sum f_i \ge c_j$ by (5.4.9), thus there is an index, say $i = 1$ such that

$$\frac{2}{k}\beta \ge L' \cdot F_1 \quad \text{resp.} \quad \frac{2}{k}\beta > L' \cdot F_1.$$

In case (5.4.3) we performed the minimal necessary number of blowing-ups to get Y, thus $D = \pi_S(F_1)$ is not a point and

$$L \cdot D = L' \cdot F_1 \le \frac{2}{k}\beta = \frac{2}{k} L \cdot C.$$

In case (5.4.4) observe that F_1 is a π_X-exceptional (-1)-curve, thus $F_1 \cdot L' > 0$ by construction. Let D be the birational transform of F_1 on S. Then

$$0 < L \cdot D = L' \cdot F_1 < \frac{2}{k}\beta = \frac{2}{k} L \cdot C. \qquad \square$$

5.5 Corollary. *Let C be an irreducible, proper and smooth curve and X a proper variety. Let $p_1, \dots, p_k \in C$ be k distinct points and $g : \{p_1, \dots, p_k\} \to X$ a morphism (i.e. a choice of k not necessarily distinct points of X).*

Assume that there is a smooth, irreducible, proper curve B, an open set $B^0 \subset B$ and a morphism

$$[h^0 : C \times B^0 \to X \times B^0] \in \operatorname{Hom}(C, X, g)(B^0)$$

such that $h^0(C \times b)$ and $p_X \circ h^0(c \times B^0)$ are one dimensional for some $b \in B^0$ and $c \in C$ where $p_X : X \times B \to X$ is the natural projection.

There is a unique normal compactification $S \supset C \times B^0$ such that h^0 extends to a finite morphism $h : S \to X \times B$. Let $p : S \to B$ denote the natural projection.

(5.5.1) *If $k \ge 1$, then for some $b \in B - B^0$ the 1-cycle $h_*(p^{-1}(b))$ contains a rational curve D which passes through $g(p_1)$.*

(5.5.2) *If $C \cong \mathbb{P}^1$, $\dim \operatorname{im}(p_X \circ h^0) = 2$ and $k \ge 2$, then for some $b \in B - B^0$ the 1-cycle $h_*(p^{-1}(b))$ is either reducible or nonreduced.*

(5.5.3) *Let L be a nef $\mathbb{R}$-Cartier divisor on X and $k \ge 1$. Then for some $b \in B - B^0$ the 1-cycle $h_*(p^{-1}(b))$ contains a rational curve D such that*

$$0 \le L \cdot D \le \frac{2}{k} L \cdot h_* C \quad \text{and} \quad \{g(p_1), \dots, g(p_k)\} \cap D \ne \emptyset.$$

(5.5.4) *Let L be a nef $\mathbb{R}$-Cartier divisor on X such that $(h^* L^2) > 0$ and $k \ge 1$. Then for some $b \in B - B^0$ the 1-cycle $h_*(p^{-1}(b))$ contains a rational curve D such that*

$$0 < L \cdot D < \frac{2}{k} L \cdot h_* C \quad \text{and} \quad \{g(p_1), \dots, g(p_k)\} \cap D \ne \emptyset.$$

Proof. If $h^0(C \times b)$ is a point for some $b \in B^0$, then by (5.3) $h^0(C \times b)$ is a point for every $b \in B^0$, a contradiction. Thus h^0 is finite on every fiber of $C \times B^0 \to B^0$, hence the natural morphism $C \times B^0 \to X \times B^0$ is quasifinite. Thus there is a unique normal compactification $S \supset C \times B^0$ such that h^0 extends to a finite morphism $h : S \to X \times B$.

Let us deal first with the possibility that $E := \operatorname{im}(p_X \circ h^0)$ is one dimensional. This could happen only in the cases (5.5.1) and (5.5.3). If E is not rational, then by (1.9.2) $\operatorname{Hom}(C, E, g)$ is zero dimensional for $k \ge 1$, which is impossible since by assumption $p_X \circ h^0(c \times B^0)$ is one dimensional for some $c \in C$. Thus E is rational. Set $D = E$. We are done in case (5.5.1). In case (5.5.3) note that again by (1.9.2) $2\deg(C/E) \ge k$, which shows (5.5.3).

From now on assume that $\operatorname{im}(p_X \circ h^0)$ is two dimensional.

Every p_i determines a flat section of S given by $s_i : B^0 \to \{p_i\} \times B^0$. By assumption $h(s_i(B)) = g(p_i)$, thus all these sections are contractible since $\dim \operatorname{im}(p_X \circ h^0) = 2$. By construction $p_X \circ h$ is finite on every fiber of p, thus every curve $D \subset p^{-1}(b)$ is mapped to a nonzero curve by h_*. Therefore (5.4) implies (5.5). □

5.5.5 Exercise. It is interesting to note that the coefficient $2/k$ in (5.4.3–4) and (5.5.3–4) is optimal.

Let C be any smooth projective curve and $f : C \to \mathbb{P}^1$ a morphism of degree $d \geq 1 + g(C)$. Then $\dim \operatorname{Hom}_{[f]}(C, \mathbb{P}^1) = 2d + 1 - g(C)$. Choose $k = 2d - g(C)$ points $\{p_1, \dots, p_k\} \subset C$ and let $g = f|\{p_1, \dots, p_k\}$. Then $\dim \operatorname{Hom}_{[f]}(C, \mathbb{P}^1, g) \geq 1$. Choose a smooth curve $B^0 \subset \operatorname{Hom}(C, \mathbb{P}^1, g)$ and let $B \supset B^0$ be a smooth compactification.

Resolving the indeterminacies of $C \times B \dashrightarrow \mathbb{P}^1$ we obtain a surface $p : S \to B$ with k flat sections $B_i \subset S$ corresponding to the points p_i. Show that every B_i has negative selfintersection. Let $L = h^*\mathcal{O}_{\mathbb{P}^1}(1)$. $L \cdot C = d$ and if D is an irreducible component of $p^{-1}(b)$ for some $b \in B$, then D dominates $\mathbb{P}^1$ hence $L \cdot D \geq 1$.

$$1 \leq L \cdot D \leq \frac{2}{2d - g(C)} L \cdot C = \frac{2d}{2d - g(C)}$$

indeed holds but the factor 2 cannot be replaced by $2 - \epsilon$.

Let H be any ample divisor on S. For r sufficiently divisible,

$$H' := r\Big(H + \sum \frac{H \cdot B_i}{-B_i \cdot B_i} B_i\Big)$$

is a Cartier divisor which is nef, big and has zero intersection number with every B_i. Taking $L = h^*\mathcal{O}_{\mathbb{P}^1}(m) \otimes \mathcal{O}(H')$ for $m \gg 1$ shows that 2 is optimal in (5.4.4) and (5.5.4) as well. (In these examples there is a 2-dimensional algebraic space S' and a morphism $S \to S'$ which contracts the curves B_i but it is not clear that one can choose S' to be a scheme.)

5.6 Corollary. *Let C be an irreducible, proper and smooth curve and X a proper variety. Let $f : C \to X$ be a nonconstant morphism.*

(5.6.1) *If*

$$\dim_{[f]} \operatorname{Hom}(C, X) \geq \dim X + 1,$$

then for every $x \in f(C)$ there is a morphism $f_x : C \to X$ and a 1-cycle $\sum a_i D_i$ whose irreducible components are rational curves such that

$$f_*[C] \stackrel{e}{\approx} (f_x)_*[C] + \sum a_i [D_i] \quad \text{and} \quad x \in \operatorname{Supp} \sum a_i D_i.$$

(5.6.2) *If $g(C) = 0$ and*

$$\dim_{[f]} \operatorname{Hom}(C, X) \geq 2 \dim X + 2,$$

then for every $x_1, x_2 \in f(C)$ there is a 1-cycle $\sum a_i D_i$ whose irreducible components are rational curves such that

$$f_*[C] \overset{e}{\approx} \sum a_i[D_i]; \quad \sum a_i \geq 2 \quad \textit{and} \quad x_1, x_2 \in \operatorname{Supp} \sum a_i D_i.$$

(5.6.3) *Let L be a nef $\mathbb{R}$-Cartier divisor on X and $k \geq 1$. If*

$$\dim_{[f]} \operatorname{Hom}(C, X) \geq k \dim X + 1,$$

then for every $x \in f(C)$ there is a morphism $f_x : C \to X$ and a 1-cycle $\sum a_i D_i$ ($a_1 > 0$) whose irreducible components are rational curves such that

$$f_*[C] \overset{e}{\approx} (f_x)_*[C] + \sum a_i[D_i]; \quad x \in D_1 \quad \textit{and} \quad L \cdot D_1 \leq \frac{2}{k} L \cdot f_* C.$$

Proof. Choose $\{p_1, \dots, p_k\} \subset C$ and let $g = f|\{p_1, \dots, p_k\}$. By (1.6)

$$\dim_{[f]} \operatorname{Hom}(C, X, g) \geq \dim_{[f]} \operatorname{Hom}(C, X) - k \dim X. \tag{5.6.4}$$

Assume (5.6.1) and pick $p_1 \in C$ such that $f(p_1) = x$. Then $\dim_{[f]} \operatorname{Hom}(C, X, g) \geq 1$.

Let B^0 be the normalization of an irreducible curve in $\operatorname{Hom}(C, X, g)$ containing $[f]$ and $h^0 : C \times B^0 \to X \times B^0$ the natural morphism. The assumptions of (5.5) are satisfied; let B and S be as constructed there. As in (5.4.5) we have a diagram

$$\begin{array}{ccccc} C \times B = X & \xleftarrow{\rho_X} & Y & \xrightarrow{\rho_S} & S \\ {\scriptstyle q}\downarrow & & & & \downarrow{\scriptstyle p} \\ B & = & B & = & B \end{array}$$

Pick $b \in B - B^0$ as in (5.5.1).Then

$$[(p \circ \rho_S)^{-1}(b)] = [(q \circ \rho_X)^{-1}(b)] = [C_0] + \sum e_j [E_j],$$

where $C_0 \cong C$, and the E_j are rational since they are exceptional curves of ρ_X. Set $f_x = (h \circ \rho_S)|C_0$ and $\sum a_i D_i = (h \circ \rho_S)_*[\sum e_j E_j]$. This shows (5.6.1).

The other cases are very similar. If in (5.6.2) the morphism $C \to f(C)$ is not birational, then let $f_*[C] = \sum a_i D_i$. If $C \to f(C)$ is birational, then fix $p_1, p_2 \in C$ such that $f(p_i) = x_i$. By (5.6.4)

$$\dim_{[f]} \operatorname{Hom}(C, X, g) \geq 2.$$

Since the group of automorphisms of $\mathbb{P}^1$ fixing two points is one dimensional, we can choose $B^0 \to \operatorname{Hom}(C, X, g)$ such that the natural morphism $C \times B^0 \to X$ satisfies all the conditions of (5.5.2). Thus $\sum a_i D_i = h_*[p^{-1}(b)]$ has the required properties.

Finally consider (5.6.3). As before we obtain $D = D_1$ which satisfies all the requirements except that we only know that $D \cap \{f(p_1), \dots, f(p_k)\} \neq \emptyset$. By letting the points p_i vary, we conclude that (5.6.3) holds except possibly for $k - 1$ points of $f(C)$.

Let $W \subset \mathrm{Chow}_1(X)$ be the connected component of $f_*[C]$. Let $V \subset W$ be the set of those points such that the corresponding cycle Z has the form

$$Z = (f_x)_*[C] + \sum a_i[D_i] \quad \text{where the } D_i \text{ are rational.}$$

By (2.2) V is closed in W and hence proper. By (2.4) $\mathrm{RatLocus}(V) \subset X$ is closed. Thus $\mathrm{RatLocus}(V) \cap C$ is a closed subset whose complement has at most $k-1$ points. Therefore $C \subset \mathrm{RatLocus}(V)$ and this completes the proof. □

5.6.6 Exercise. Let X be a projective variety and assume that X does not contain any rational curves. Let C be any irreducible curve. Show that $\dim \mathrm{Hom}(C, X) \le \dim X$. Find examples where equality holds.

5.7 Theorem [Mori79; Mori82, 1.6]. *Let X be a smooth projective variety.*

(5.7.1) *Let $f : \mathbb{P}^1 \to X$ be a nonconstant morphism. Then for every $x \in f(\mathbb{P}^1)$ there is a 1-cycle $\sum a_i D_i$ whose irreducible components are rational curves such that*

$$f_*[\mathbb{P}^1] \overset{e}{\approx} \sum a_i[D_i]; -K_X \cdot D_i \le \dim X + 1 \ \forall i \quad \text{and} \quad x \in \mathrm{Supp} \sum a_i D_i.$$

(5.7.2) *Let C be a smooth, projective and irreducible curve and $f : C \to X$ a morphism. Assume that*

$$\deg_C f^*(-K_X) > \dim X \cdot g(C).$$

Then for every $x \in f(C)$ there is a morphism $f_x : C \to X$ and a 1-cycle $\sum a_i D_i$ whose irreducible components are rational curves such that

$$f_*[C] \overset{e}{\approx} (f_x)_*[C] + \sum a_i[D_i]; \quad x \in \mathrm{Supp} \sum a_i D_i;$$
$$\deg_C f_x^*(-K_X) \le \dim X \cdot g(C) \quad \text{and} \quad -K_X \cdot D_i \le \dim X + 1 \ \forall i.$$

Proof. Fix an ample divisor H on X. We prove (5.7) by induction on $\deg f^*H$.

Let us start with (5.7.1). If $\deg f^*(-K_X) \le \dim X + 1$, then let $\sum a_i D_i = f_*[\mathbb{P}^1]$. Otherwise by (1.2)

$$\dim_{[f]} \mathrm{Hom}(\mathbb{P}^1, X) \ge 2 \dim X + 2,$$

hence by (5.6.2) there is a reducible 1-cycle $\sum a_i D_i$ satisfying all the required properties except possibly the one about $-K_X \cdot D_i$. Note, however, that $H \cdot D_i < \deg f^*H \ \forall i$. Let $f_i : \bar{D}_i \to X$ be the normalization of D_i. By induction, for every i there is a 1-cycle $\sum_j a_{ij} D_{ij}$ such that

$$(f_i)_*[\bar{D}_i] \overset{e}{\approx} \sum a_{ij}[D_{ij}]; \quad x \in \mathrm{Supp} \sum_{i,j} a_{ij} D_{ij} \quad \text{and}$$
$$-K_X \cdot D_{ij} \le \dim X + 1 \ \forall i, j.$$

Thus $\sum_{i,j} a_{ij} D_{ij}$ satisfies the required properties.

The proof of (5.7.2) is similar. By (1.2) $\dim_{[f]} \operatorname{Hom}(C,X) \geq \dim X + 1$, thus we can apply (5.6.1) to obtain $f_x : C \to X$ and a 1-cycle $\sum a_i D_i$. We can apply (5.7.1) to $\sum a_i D_i$ and induction on $\deg f_x^* H < \deg f^* H$ to conclude. □

The main application is the following result which was gradually developped in the articles [Mori79; Mori82, 1.7; Miyaoka-Mori86; Kollár91a, 3.3].

5.8 Theorem. *Let X be a projective variety over an algebraically closed field, C a smooth, projective and irreducible curve, $f : C \to X$ a morphism and M any nef $\mathbb{R}$-divisor. Assume that X is smooth along $f(C)$ and $-K_X \cdot C > 0$.*

Then for every $x \in f(C)$ there is a rational curve $L_x \subset X$ containing x such that

$$M \cdot L_x \leq 2 \dim X \frac{M \cdot C}{-K_X \cdot C}.$$

Proof. First we prove the following slightly weaker version:

5.8.1 Proposition (Notation as above). *Let M be an ample $\mathbb{R}$-divisor. Then for every $\epsilon > 0$ we can find $L_{x,\epsilon}$ such that*

$$M \cdot L_{x,\epsilon} \leq (2 \dim X + \epsilon) \frac{M \cdot C}{-K_X \cdot C}.$$

Everything is over a field k of characteristic $p \geq 0$. There are several steps. The first four prove (5.8.1) and in the last one we derive (5.8) from (5.8.1).

5.9 Step 1: Proof of (5.8.1) for M ample and char $= p > 0$.

Let $F^m : C^m \to C$ be the Frobenius of degree p^m. Let $f^m : C^m \to C \to X$ be the composition of F^m and f. Then

$$-K_X \cdot C^m = p^m(-K_X \cdot C).$$

By (1.2)

$$\dim_{[f^m]} \operatorname{Hom}(C^m, X) \geq p^m(-K_X \cdot C) + \dim X \cdot \chi(\mathcal{O}_C).$$

If $\epsilon > 0$ is fixed, then for m sufficiently large,

$$\dim_{[f^m]} \operatorname{Hom}(C^m, X) \geq p^m \frac{-K_X \cdot C}{\dim X + \epsilon/2} \dim X + 2.$$

Thus by (5.6.3) for every $x \in f(C)$ there is a rational curve $L_{x,\epsilon}$ such that

$$M \cdot L_{x,\epsilon} \leq 2p^{-m} \frac{\dim X + \epsilon/2}{-K_X \cdot C} M \cdot C^m = (2 \dim X + \epsilon) \frac{M \cdot C}{-K_X \cdot C}. \qquad \square$$

5.9.1 Remark. The magic property of $C^m \to C$ that we need is that it is a high degree morphism which does not increase the genus. In characteristic zero such morphisms exist only if C is rational or elliptic. It is instructive to check that in the other cases the genus increase cancels out the increase of the intersection number in (1.2).

There is no known proof of (5.8) which does not use positive characteristic techniques.

5.10 Step 2: General theory of reduction $\bmod p$.

Assume that we are given finitely many schemes of finite type X_i, coherent sheaves F_i and maps g_i defined over a field k. All of these can be described by a finite number of equations. The schemes are given by affine charts and patching functions, the sheaves by finitely presented modules over the affine charts and patchings and the maps are described by their graphs which are schemes themselves. All these equations involve only finitely many elements a_j of the field k. Let $\mathbb{F} \subset k$ denote the prime field of p elements if $\operatorname{char} k = p$ and the ring of integers if $\operatorname{char} k = 0$. Let $R = \mathbb{F}[a_j]$. R is a finitely generated ring over $\mathbb{F}$. Any such ring has the following useful properties:

5.10.1 Lemma. *Let R be a finitely generated ring over $\mathbb{F}$. Then*

(5.10.1.1) *The residue field R/m of any maximal ideal $m \subset R$ is finite.*

(5.10.1.2) *The maximal ideals are dense in* $\operatorname{Spec} R$. □

By the choice of a_j the equations that we used to describe X_i etc. over k define schemes, sheaves or maps over $\operatorname{Spec} R$. These are denoted by X_i^R, F_i^R and g_i^R. One should keep in mind that these objects depend not only on R but also on the choice of the defining equations of X_i etc.

By construction these objects have the following property:

$$X_i^R \times_{\operatorname{Spec} R} \operatorname{Spec} k \cong X_i; \ F_i^R \times_{\operatorname{Spec} R} \operatorname{Spec} k \cong F_i; \ g_i^R \times_{\operatorname{Spec} R} \operatorname{Spec} k \cong g_i.$$

I.e. we constructed data $\{X_i^R, F_i^R, g_i^R\}$ over $\operatorname{Spec} R$ such that the fibers over $\operatorname{Spec} k$ are the original data $\{X_i, F_i, g_i\}$.

For every maximal ideal $m \subset R$ we can consider the fiber over $\operatorname{Spec} R/m$. This is denoted by $\{X_i^m, F_i^m, g_i^m\}$. Let $\operatorname{char} R/m = p$. The data $\{X_i^m, F_i^m, g_i^m\}$ are sometimes refered to as a $\bmod m$ *(or* $\bmod p$*) reduction* of the data $\{X_i, F_i, g_i\}$.

For a given m we should not expect any nice relationship between $\{X_i^m, F_i^m, g_i^m\}$ and $\{X_i, F_i, g_i\}$. There is, however, a dense open subset $U \subset \operatorname{Spec} R$ such that for $m \in U$ the data $\{X_i^m, F_i^m, g_i^m\}$ and $\{X_i, F_i, g_i\}$ have many similarities.

One can formulate this remark as a precise principle:

5.10.2 Definition. Let (P) be a property of schemes (morphisms etc.) in algebraic geometry. We say that (P) is of *finite type* if the following condition is satisfied:

Let $K \supset k$ be a field extension and X_k a k-scheme. Then (P) holds for X_K iff there is a finitely generated subextension $K \supset F \supset k$ such that (P) holds for X_L for every $L \supset F$.

Existence statements are usually of finite type. A typical property that is not of finite type is: "X_K has only finitely many K-points".

5.10.3 Meta Theorem. *Let (Property 1) $\Rightarrow$ (Property 2) be a statement in algebraic geometry that we want to prove. Assume the following four conditions:*

(5.10.3.1) *(Property 1) and (Property 2) are of finite type.*

(5.10.3.2) *If (Property 1) holds for the generic fiber of a morphism $X \to Y$, then it holds for every fiber over a nonempty open set.*

(5.10.3.3) *If (Property 2) holds for every fiber of a morphism $X \to Y$ over a (not necessarily open) dense set, then it holds for the generic fiber.*

(5.10.3.4) *(Property 1) $\Rightarrow$ (Property 2) holds in positive characteristic.*

Then (Property 1) $\Rightarrow$ (Property 2) always holds. □

In practice (5.10.3.1) is usually fairly easy. (5.10.3.2) is harder and (5.10.3.3) is the hardest. The most difficult aspect of (5.10.3) is to find situations where it is applicable. One of the nicest applications, discovered by [Mori79], is the next step.

5.11 Step 3: Proof of (5.8.1) for M an ample line bundle and char $= 0$.

Let $f : C \to X$ be given over a field of characteristic zero. Let $p \in C$ be a point such that $f(p) = x$. As in Step 2 we construct a ring R such that

$$x \to C \xrightarrow{f} X \quad \text{and the locally free sheaf } M$$

are defined over R. Now consider

$$x^R \to C^R \xrightarrow{f^R} X^R \quad \text{and } M^R.$$

By shrinking $\operatorname{Spec} R$ we may assume that C^R is smooth over $\operatorname{Spec} R$, $X^R \to \operatorname{Spec} R$ is smooth along $f^R(C^R)$ and M^R is locally free and relatively ample.

Let $W_\epsilon \subset \operatorname{Chow}_1(X^R/\operatorname{Spec} R)$ be the subvariety parametrizing those 1-cycles $Z = \sum a_i D_i$ which satisfy the following properties:

(5.11.1.1) every D_i is rational;

(5.11.1.2) $Z \cdot M \le (2\dim X + \epsilon)\frac{M \cdot C}{-K_X \cdot C}$;

(5.11.1.3) $\operatorname{Supp} Z \cap f^R(x^R) \ne \emptyset$.

By (I.4.13) and (2.2) $\pi : W_\epsilon \to \operatorname{Spec} R$ is proper.

5.11.2 Claim (Notation as above). *Then $\pi : W_\epsilon \to \operatorname{Spec} R$ is surjective.*

Proof. It is sufficient to show that $\pi(W_\epsilon)$ contains all closed points of Spec R. Let $m \subset R$ be a maximal ideal,

$$x^m \to C^m \xrightarrow{f^m} X^m \quad \text{and} \quad M^m$$

the corresponding objects over Spec R/m. By assumption C^m is smooth and X^m is smooth along $f^m(C^m)$. Furthermore

$$-K_{X^m} \cdot C^m = -K_X \cdot C \quad \text{and} \quad M^m \cdot C^m = M \cdot C.$$

Thus by Step 1 there is a rational curve $L_{x^m,\epsilon}$ such that $[L_{x^m,\epsilon}] \in W_\epsilon^m$. □

By (5.11.2) $W_\epsilon \times_{\operatorname{Spec} R} \operatorname{Spec} k$ is not empty, thus it has a point $[Z]$ in some extension of k. Let $L_{x,\epsilon}$ be the irreducible component of Z which contains x.

5.12 Step 4: Proof of (5.8.1) for M an ample $\mathbb{R}$-divisor.

We already proved (5.8.1) in the case when M is an ample line bundle. Since the statement does not change if we replace M by a multiple of it, the result also holds for ample $\mathbb{Q}$-divisors. An ample $\mathbb{R}$-divisor M can be approximated by ample $\mathbb{Q}$-divisors M_j. This implies (5.8.1) for an arbitrary ample $\mathbb{R}$-divisor M.

5.13 Step 5: Proof of (5.8).

Let M be any nef $\mathbb{R}$-divisor and H ample. $kM + H$ is ample for every $k > 0$ (VI.2.18.7.1). Thus for every $\epsilon > 0$ there is a rational curve $L_{x,k,\epsilon}$ such that

$$(kM+H)\cdot L_{x,k,\epsilon} \le (2\dim X + \epsilon)k\frac{M\cdot C}{-K_X\cdot C} + (2\dim X+\epsilon)\frac{H\cdot C}{-K_X\cdot C}.$$

We can rearrange this to

$$\begin{aligned} &k\left(M\cdot L_{x,k,\epsilon} - 2\dim X\frac{M\cdot C}{-K_X\cdot C}\right) + H\cdot L_{x,k,\epsilon} \\ &\qquad \le (2\dim X+\epsilon)\frac{H\cdot C}{-K_X\cdot C} + k\epsilon\frac{M\cdot C}{-K_X\cdot C}. \end{aligned} \tag{5.13.1}$$

If

$$M\cdot L_{x,k_0,\epsilon} - 2\dim X\frac{M\cdot C}{-K_X\cdot C} \le 0$$

for some k_0 and ϵ, then take $L_x = L_{x,k_0,\epsilon}$. Otherwise we obtain that

$$H\cdot L_{x,k,\epsilon} \le (2\dim X+\epsilon)\frac{H\cdot C}{-K_X\cdot C} + k\epsilon\frac{M\cdot C}{-K_X\cdot C} \tag{5.13.2}$$

for every k and ϵ.

Set $\epsilon = 1/k$ and let k go to infinity. We obtain a sequence of curves $L_{x,k} = L_{x,k,1/k}$. By (5.13.2) $H\cdot L_{x,k}$ is uniformly bounded, thus the $L_{x,k}$ form a bounded family. By (I.4.13) $\operatorname{Chow}_1(X)$ has only finitely many components parametrizing 1-cycles of bounded degree. In particular there is a subsequence $k_i \to \infty$ such that

$$P = P(i) := M \cdot L_{x,k_i} - 2\dim X \frac{M \cdot C}{-K_X \cdot C}$$

is independent of i.

On the other hand, by (5.13.1)

$$k_i P \leq (2\dim X + 1)\frac{H \cdot C}{-K_X \cdot C} + \frac{M \cdot C}{-K_X \cdot C} \quad \text{and} \quad k_i \to \infty.$$

Thus $P \leq 0$ and we can take $L_x = L_{x,k_i}$. □

If X is smooth, then (5.8) can be strengthened. It is not known if the assumption of smoothness is really necessary.

5.14 Theorem. *Let X be a smooth projective variety, C a smooth, projective and irreducible curve and $f : C \to X$ a morphism. Let M be any nef $\mathbb{R}$-divisor. Assume that*

$$-K_X \cdot C > 0.$$

Then for every $x \in f(C)$ there is a rational curve $D_x \subset X$ containing x such that

$$M \cdot D_x \leq 2\dim X \frac{M \cdot C}{-K_X \cdot C} \quad \textit{and} \quad -K_X \cdot D_x \leq \dim X + 1.$$

Proof. By (5.8) we can find a rational curve L_x containing x which satisfies the first inequality. By (5.7)

$$[L_x] \overset{e}{\approx} \sum a_i [D_i] \quad \text{where} \quad -K_X \cdot D_i \leq \dim X + 1$$

and $x \in \operatorname{Supp} \sum D_i$. Let D_x be one of the D_i containing x. □

5.15 Remark. In (5.8) and in (5.14) the smoothness assumption on X is used only to obtain the lower bound for $\operatorname{Hom}(C, X)$ given by (1.2).

(5.8) remains true if X has local complete intersection singularities along $f(C)$ and $f(C)$ intersects the smooth locus.

(5.14) remains true if X has local complete intersection singularities and $f(C)$ intersects the smooth locus.

More generally, a variant of this method works if the singularities of X are only quotients of complete intersection singularities [Kollár91c; Kollár92]. This addition is especially interesting in dimension 3 since there terminal singularities are quotients of hypersurface singularities.

5.16 Conjecture. *Let X be a smooth and proper algebraic space which contains a curve $C \subset X$ such that $C \cdot K_X < 0$. Then X contains a rational curve.*

As a related problem one can ask even more:

5.17 Conjecture. *Let X be a smooth and proper algebraic space which is not projective. Then X contains a rational curve.*

5.17.1 Remark. During the proof of (5.8) the assumption that X be projective is used in the lifting of rational curves (5.11). (5.9) proves that (5.16) holds for char > 0.

II.6 Application: A Vanishing Theorem for H^1

The aim of this section is to prove a vanishing theorem for the first cohomology group of certain line bundles on a scheme X over a field of *positive* characteristic. The heart of the proof is a construction due to [Ekedahl88] which asserts that if vanishing fails, then X has a purely inseparable cover with rather unusual properties. In many cases these properties are impossible to satisfy, implying the vanishing result. In applications (see [Ekedahl88, Shepherd–Barron97]) it is usually necessary to investigate several properties of the covering in order to get all required vanishing results. Here I concentrate on one particular approach which relies on the bend-and-break techniques of II.5. This strategy was discovered in a conversation with Shepherd–Barron.

First we consider the construction of the cover in full generality.

6.1 Construction. Let X be a scheme and E a vector bundle of rank r over X. Let $P = \mathbb{P}_X(E)$ with tautological line bundle $\mathcal{O}_P(1)$ and projection $p : P \to X$. Let M be a line bundle on X, $s \in H^0(P, \pi^*M(k))$ a section and $Y = (s = 0) \subset P$ the corresponding Cartier divisor. Assume that ω_X is locally free (for instance, X is smooth). Then ω_P and ω_Y are also locally free and there is a natural exact sequence

$$(6.1.1) \qquad 0 \to \omega_P \to \omega_P(Y) \to \omega_Y \to 0.$$

Pushing it forward to X we obtain

$$0 \to \pi_*\omega_P(Y) \to \pi_*\omega_Y \to R^1\pi_*\omega_P \to 0.$$

This sequence is exact since

$$\pi_*\omega_P = 0 \quad \text{and} \quad R^1\pi_*\omega_P(Y) = 0 \quad \text{for } k > 0.$$

We know that

$$\omega_P(Y) = \pi^*(\omega_X \otimes \det E \otimes M)(k - r),$$

and it is easy to see that

$$R^1\pi_*\omega_P = \begin{cases} \omega_X & \text{if } \operatorname{rank} E = 2, \text{ and} \\ 0 & \text{otherwise.} \end{cases}$$

Thus we get exact sequences

$$(6.1.2)\qquad \begin{array}{ll} 0 \to \det E \otimes M \otimes S^{k-r}E \to \pi_*\omega_{Y/X} \to \mathcal{O}_X \to 0 & \text{if } r = 2, \\ 0 \to \det E \otimes M \otimes S^{k-r}E \to \pi_*\omega_{Y/X} \to 0 & \text{if } r > 2. \end{array}$$

Assume now that $r = 2$ and E is obtained as an extension

$$(6.1.3)\qquad 0 \to L \to E \to \mathcal{O}_X \to 0.$$

Let $\sigma : X \to P$ be the section corresponding to the above quotient, $D = \sigma(X)$. $\pi^*L^{-1}(1)$ has a natural section t corresponding to

$$1 \in H^0(X, \mathcal{O}_X) = H^0(X, L^{-1} \otimes L) \subset H^0(X, L^{-1} \otimes E) = H^0(P, \pi^*L^{-1}(1)).$$

By construction $(t = 0) = D$. Assume furthermore that $M = \mathcal{O}_X$ and $s \in H^0(P, \mathcal{O}(k))$ is a section such that its image under the natural maps

$$H^0(P, \mathcal{O}(k)) = H^0(X, S^kE) \to H^0(X, \mathcal{O}_X^k) \cong H^0(D, \mathcal{O}_D^k)$$

is nowhere zero. This means that $Y = (s = 0)$ and $D = (t = 0)$ are disjoint. In particular,

$$\pi^*L^{-1}(1)|Y \cong \mathcal{O}_P(D)|Y \cong \mathcal{O}_Y.$$

Therefore we obtain that

$$(6.1.4)\qquad \omega_Y = \pi^*(\omega_X \otimes \det E)(k-2)|Y = \pi^*(\omega_X \otimes L^{k-1}).$$

There are two special cases when these assumptions are satisfied

6.1.5 Classical Case. Let L be a line bundle and $E = L + \mathcal{O}_X$. Thus

$$S^kE = L^k + \cdots + L + \mathcal{O}_X.$$

Pick sections $s_i \in H^0(X, L^i)$ and let

$$s : \mathcal{O}_X \to S^kE \quad \text{be given by} \quad 1 \mapsto (s_k, \ldots, s_1, 1).$$

If $s_{k-1} = \ldots = s_1 = 0$, then we say that Y is obtained by *taking a kth root of s.*

Assume that the characteristic does not divide k and let ϵ be a primitive k^{th} root of unity. Let τ be the action of $\mathbb{Z}_k$ on E given as multiplication by ϵ on L and the identity on $\mathcal{O}_X$. If $s_{k-1} = \ldots = s_1 = 0$, then the above $s : \mathcal{O}_X \to S^kE$ is τ-invariant, and we obtain a $\mathbb{Z}_k$ action on Y.

One can easily see that $\pi : Y \to X$ is a Galois cover with Galois group $\mathbb{Z}_k$. It is called the *cyclic cover of X associated to $s \in H^0(X, L^k)$.*

6.1.6 Weird Case [Ekedahl88]. Let X be an integral scheme over a field of characteristic $p > 0$. Assume that X is normal (or at least weakly normal). Let L be a line bundle on X such that

(6.1.6.1) $$H^1(X,L) \neq 0 \quad \text{and} \quad H^1(X,L^p)=0.$$

The first condition implies that there is a nonsplit extension

(6.1.6.2) $$0 \to L \to E \to \mathcal{O}_X \to 0.$$

In characteristic p the pth symmetric power S^pE has a rank two subbundle spanned by p^{th} powers of local sections. We denote it by $E^{(p)}$. By construction there is an exact sequence

(6.1.6.3) $$0 \to L^p \to E^{(p)} \to \mathcal{O}_X^p = \mathcal{O}_X \to 0.$$

This sequence splits since $H^1(X,L^p)=0$. Thus $\mathcal{O}_X$ is a subbundle of $S^pE = \pi_*\mathcal{O}_P(p)$. Therefore we obtain a section $s \in H^0(P, \mathcal{O}_P(p))$. Let $Z \subset P$ be its zero set. (The notation is changed from Y to Z to emphasize the difference.)

6.1.6.4 Remark. One can see from the above construction that instead of $H^1(X,L^p)=0$ it is sufficient to assume that (6.1.6.3) is split.

6.1.7 Proposition (Notation as in (6.1.6)).

(6.1.7.1) *Z is an integral scheme;*

(6.1.7.2) *$K_Z \cong \pi^*(K_X \otimes L^{p-1})$; and*

(6.1.7.3) *there is an exact sequence*

$$0 \to L^{-1} \otimes S^{p-2}E \to \pi_*\omega_{Z/X} \to \mathcal{O}_X \to 0.$$

Proof. The second claim follows from (6.1.4), the third from (6.1.2); the main part is the first one.

Let $U \subset X$ be an affine open set and $P_U = \pi^{-1}(U)$. Over U the sequence (6.1.6.2) splits $\rho : \mathcal{O}_U \to E|U$. We can choose an isomorphism $P_U \cong U \times \mathbb{P}^1$ and coordinates $(u:v)$ on $\mathbb{P}^1$ such that u (resp. v) viewed as sections of $E|U$ are generators of $L|U \subset F|U$ (resp. $\rho(\mathcal{O}_X) \subset E|U$).

$Z \cap P_U$ is defined by an equation $f_u u^p - f_v v^p = 0$ where f_u, f_v are functions on U. Such an equation is either irreducible or a pth power. From this we see that Z is either integral or $Z = pZ'$ for a reduced divisor $Z' \subset Y$. In the latter case the projection $Z' \to X$ is one-to-one on closed points and an isomorphism generically. Since X is weakly normal, it is an isomorphism. Therefore Z' is a section of $\pi : P \to X$ which is disjoint from D. Thus

$$E = \pi_*(\mathcal{O}_P(1)) \to \pi_*(\mathcal{O}_P(1)|Z') + \pi_*(\mathcal{O}_P(1)|D) \cong \mathcal{O}_X + L$$

is an isomorphism. This means that the extension (6.1.6.2) is split, a contradiction. Thus Z is an integral scheme. □

6.1.7.3 Remark. If X is not weakly normal, it can happen that $Z = pZ'$ but $Z' \to X$ is not an isomorphism. See (6.6.2).

Kodaira vanishing can fail in positive characteristic [Raynaud78]. The following theorem shows that failure of vanishing has strong geometric consequences. The corollaries state some special cases which are easy to formulate.

6.2 Theorem. *Let X be a normal and projective variety over a field of* char p. *Let L be an ample line bundle on X such that $H^1(X, L^{-1}) \neq 0$. Assume that X is covered by a family of curves $\{D_t\}$ such that X is smooth along the general D_t and*

$$(p-1)(L \cdot D_t) - (K_X \cdot D_t) > 0. \tag{6.2.1}$$

Then through every point $x \in X$ there is a rational curve $C_x \subset X$ such that

$$(L \cdot C_x) \leq 2 \dim X \frac{(L \cdot D_t)}{(p-1)(L \cdot D_t) - (K_X \cdot D_t)}. \tag{6.2.2}$$

If X is smooth, then the following additional inequality is also satisfied

$$(p-1)(L \cdot C_x) - (K_X \cdot C_x) \leq 1 + \dim X. \tag{6.2.3}$$

6.2.4 Remark. It would be desirable to allow L nef and big instead of just L ample. This is possible if we know that $H^1(X, L^{-m}) = 0$ for $m \gg 1$. This is known for surfaces [Szpiro81, p.37] but fails in higher dimensions. Thus if $\dim X = 2$, then we can replace L ample by L nef and big in the statement of (6.2).

6.3 Corollary. *Let X be a smooth projective variety over a field of* char p. *Assume that X is not uniruled. Let L be an ample divisor on X such that $(p-1)L - K_X$ is ample. Then $H^1(X, L^{-1}) = 0$.*

Proof. Let $\{D_t\}$ be any covering family of curves. (6.2.1) is satisfied by assumption. Since X is not uniruled, it cannot be covered by rational curves. Thus by (6.2) $H^1(X, L^{-1}) = 0$. □

6.4 Corollary. *Let X be a smooth projective variety over a field of* char p. *Assume that $-K_X$ is ample (that is, X is Fano) and $p > 1 + \dim X$. Then $H^1(X, K_X^m) = 0$ for $m > 0$.*

Proof. Apply (6.2) with $L = -K_X$. (6.2.1) is automatic. If $H^1(X, K_X^m) \neq 0$, then by (6.2) we obtain a family of curves C_x such that

$$m(p-1)(L \cdot C_x) - (K_X \cdot C_x) = (mp - m + 1)(L \cdot C_x) \leq 1 + \dim X.$$

Thus $(L \cdot C_x) < 1$, which is impossible. □

6.5 Corollary. *Let X be a smooth projective variety over a field of* char p. *Assume that K_X is ample and $p > 3 + \dim X$. Then $H^1(X, K_X^{-m}) = 0$ for $m > 0$.*

Proof. Apply (6.2) with $L = K_X$. As in (6.4) we obtain a family of curves C_x such that

$$m(p-1)(L \cdot C_x) - (K_X \cdot C_x) = (mp - m - 1)(L \cdot C_x) \leq 1 + \dim X.$$

Again $(L \cdot C_x) < 1$, a contradiction. □

6.5.1 Remarks. (6.5.1.1) In (6.4–5) it is sufficient to assume that X is weakly normal and K_X is Cartier.

(6.5.1.2) [Ekedahl88, I.2.14] gives examples of smooth surfaces X in characteristic 2 such that K_X is ample and $h^1(X, K_X^{-1}) = 1$. I do not know higher dimensional examples in higher characteristics.

It is possible that for $p \geq 3$ there are only finitely many families of varieties X of fixed dimension such that K_X is ample and $H^1(X, K_X^{-1}) \neq 0$.

(6.5.1.3) [Reid94] gives examples of nonnormal Del Pezzo surfaces in positive characteristic with $H^1(X, \mathcal{O}_X) \neq 0$.

6.6 Corollary. *Let X be a normal projective variety over a field of characteristic* 0 *and L an ample line bundle on X. Then $H^1(X, L^{-1}) = 0$.*

Proof. As in (5.10) there is a finitely generated ring R and a scheme X_R flat and proper over $\operatorname{Spec} R$ such that the generic fiber of X_R is isomorphic to X (after suitable base field extension). We may also assume that there is an ample line bundle L_R extending L. By semicontinuity of cohomology groups, it is sufficient to prove that $H^1(X_s, L_s^{-1}) = 0$ where $s \in \operatorname{Spec} R$ with sufficiently large residue characteristic p.

Pick any covering family of curves $\{D_t\}$ of X_R whose general member is in the smooth locus. $(p-1)(L \cdot D_t) - (K_X \cdot D_t) > 0$ for $p \gg 1$. If $H^1(X_s, L_s^{-1}) \neq 0$, then by (6.2) X is covered by rational curves $C_x \subset X$ such that

$$(L \cdot C_x) \leq 2 \dim X \frac{(L \cdot D_t)}{(p-1)(L \cdot D_t) - (K_X \cdot D_t)}.$$

For $p \gg 1$ the right hand side is less than 1, which is impossible. □

6.6.1 Remark. In char 0 the above vanishing is essentially a surface question, since the higher dimensional case can be reduced to surfaces using the exact sequence

$$0 \to L^{-1}(-H) \to L^{-1} \to L^{-1}|H \to 0,$$

where $H \subset X$ is sufficiently ample.

In positive characteristic one can try to use a similar sequence. The problem is that if H is much bigger than L, then $L|H$ is much smaller than K_H and (6.2.1) fails on H.

6.6.2 Exercise. Give an example of an integral and projective scheme over a field of characteristic zero and an ample line bundle L on X such that

$H^1(X, L^{-1}) > 0$. Using the method of (6.6) this gives similar examples in positive characteristic.

6.7 Proof of (6.2). Let $m \geq 1$ be the largest integer such that $H^1(X, L^{-m}) \neq 0$. Since L is ample, $H^1(X, L^{-k}) = 0$ for $k \gg 1$, thus m exists.

Let $X_0 \subset X$ be the smooth locus and $L_0 = L^{-k}|X_0$. I claim that

$$H^1(X, L^{-k}) \hookrightarrow H^1(X_0, L_0) \tag{6.7.1}$$

Let E be the extension corresponding to an element $0 \neq \eta \in H^1(X, L^{-k})$. We need to prove that $E|X_0$ is also nonsplit, which is implied by

$$\operatorname{Hom}_X(\mathcal{O}_X, E) = \operatorname{Hom}_{X_0}(\mathcal{O}_{X_0}, E|X_0).$$

The latter is just a reformulation of the assertion that if a rational function is regular outside a codimension two set, then it is regular.

$H^1(X, L^{-pm}) = 0$, thus (cf. (6.1.6.4)) we can apply (6.1.6) to X_0 and L_0 and construct $\pi_0 : Z_0 \to X_0$ such that $K_{Z_0} = \pi_0^*(K_{X_0} \otimes L_0^{p-1})$. We can extend π_0 to a finite morphism $\pi : Z \to X$. The above formula for the canonical class gives $K_Z = \pi^*(K_X \otimes L^{m(1-p)})$.

Pick $x \in X$ such that Z is smooth at $x' = \pi^{-1}(x)$ and a curve D_t which contains x such that X is smooth along D_t. Let $D_t' = \operatorname{red} \pi^{-1}(D_t)$. Z is a Cartier divisor in a $\mathbb{P}^1$-bundle over X, thus Z has hypersurface singularities along D_t'. (6.2.1) and (6.1.7.2) guarantee that $D_t' \cdot K_Z < 0$. Thus by (5.8; 5.15) there is a rational curve C_x' containing x' such that

$$(\pi^* L^m \cdot C_x') \leq 2 \dim X \frac{(\pi^* L^m \cdot D_t')}{(p-1)(\pi^* L^m \cdot D_t') - (\pi^* K_X \cdot D_t')}.$$

Set $C_x = \pi(C_x')$ (with reduced scheme structure). Then

$$\begin{aligned} m(L \cdot C_x) &\leq (\pi^* L^m \cdot C_x') \\ &\leq 2 \dim X \frac{(\pi^* L^m \cdot D_t')}{(p-1)(\pi^* L^m \cdot D_t') - (\pi^* K_X \cdot D_t')} \\ &= 2 \dim X \frac{(L^m \cdot D_t)}{(p-1)(L^m \cdot D_t) - (K_X \cdot D_t)} \\ &\leq 2 \dim X \frac{m(L \cdot D_t)}{(p-1)(L \cdot D_t) - (K_X \cdot D_t)}. \end{aligned}$$

Dividing by m we obtain (6.2.2). (6.2.3) is obtained similarly using (5.14–15). □

6.7.2 Remark. The above proof shows that (6.2) is valid more generally if X is weakly normal, K_X is Cartier outside a codimension two set, L is a rank one torsion free sheaf such that L is locally free outside a codimension two set and the double dual of L^t is locally free for some $t > 0$.

6.8 Exercise. Let X be a smooth projective surface and L an ample divisor. Assume that $L = -K_X$ and char $= 3$ or $L = K_X$ and char $= 5$. Show that $H^1(X, L^{-1}) = 0$.

Hints: Assume the contrary. Use the method of (6.7) to prove that for every $x \in X$ there is a 1-parameter family of rational curves $C_{x,t} \subset X$ such that $L \cdot C_{x,t} = 1$. Use the Hodge index theorem to conclude that $C_{x,t}^2 = 1$. Conclude that every $C_{x,t}$ is smooth. Get a contradiction using the adjunction formula.

II.7 Smoothing of Morphisms of Curves

The aim of this section is to consider the following problem. Let C be a reducible curve and $f : C \to X$ a morphism. Under what conditions can it be obtained as a limit of morphisms from smooth curves to X?

The results of this section are taken from [KoMiMo92a,b,c].

7.1 Definition. Let $f : C \to X$ be a morphism. A *smoothing* of $f : C \to X$ is a diagram

$$\begin{array}{ccc} Y & \xrightarrow{F} & X \\ {\scriptstyle q}\downarrow & & \\ 0 \in T & & \end{array}$$

where $0 \in T$ is a connected, smooth and pointed curve such that:

(7.1.1) q is flat and proper over T;
(7.1.2) q is smooth over $T - 0$;
(7.1.3) $[F|q^{-1}(0) : q^{-1}(0) \to X] \cong [f : C \to X]$.

$F|q^{-1}(t) : q^{-1}(t) \to X$ is called a *nearby smoothing* for $t \neq 0$.

7.2 Definition. Let $f : C \to X$ be a morphism and $x_1, \dots, x_n \in C$ distinct points. We say that the above smoothing fixes $f(x_1), \dots, f(x_n)$ if there are sections $t_i : T \to Y$ such that

(7.2.1) $t_i(0) = x_i \in C \subset Y$, and
(7.2.2) $F \circ t_i(T) = f(x_i)$ for every i.

We say that f is *smoothable fixing* $f(x_1), \dots, f(x_n)$ if there is a smoothing which fixes $f(x_1), \dots, f(x_n)$.

The existence of a smoothing follows from suitable properties of the Hom scheme:

7.3 Lemma. *Let $q : R \to (0 \in S)$ be a smoothing of the curve $q^{-1}(0)$ (1.10). Let $s_i : S \to R$ be sections and $x_i = s_i(0)$. Let $f : q^{-1}(0) \to X$ be a morphism and define G by $G \circ s_i(S) = f(x_i)$. Assume that* $\mathrm{Hom}(R, X \times S/S, G) \to S$ *is dominant at $[f]$. Then f is smoothable fixing the $f(x_i)$.*

Proof. By assumption there is a morphism from a smooth pointed curve $r : (0 \in T) \to \operatorname{Hom}(R, X \times S/S, G)$ such that $r(0) = [f]$ and $Y := R \times_S T \to T$ is a smoothing of $q^{-1}(0)$. $F : Y \to X$ exists by construction. □

7.4 Definition. (7.4.1) Let C be a connected curve and $D \subset C$ a subcurve, C_i the irreducible components of C not contained in D. We say that C is *obtained from D by attaching trees* if

(7.4.1.1) every C_i is isomorphic to $\mathbb{P}^1$, and

(7.4.1.2) C_i intersects $D + \sum_{j=1}^{i-1} C_j$ in a single point which is an ordinary node of C.

(7.4.2) If $D \cong \mathbb{P}^1$, then we say that C is a *tree of smooth rational curves.*

(7.4.3) A tree of smooth rational curves is called a *chain of smooth rational curves* (of length n) if C_{i+1} intersects $\sum_{j=1}^{i} C_j$ in a single point which is a point of C_i $(1 \le i \le n-1)$.

The next easy result is used to control obstructions to smoothing:

7.5 Lemma. *Let $C = D + \sum C_i$ be a curve which is obtained from D by attaching trees. Let E be a vector bundle on C such that $E|C_i$ is semi positive for every i. Then $H^1(C, E) = H^1(D, E|D)$.*

Proof. Set $C^i = D + \sum_{j=1}^{i} C_j$. We have an exact sequence

$$0 \to E \otimes \mathcal{O}_{C_i}(-p_i) \to E \otimes \mathcal{O}_{C^i} \to E \otimes \mathcal{O}_{C^{i-1}} \to 0,$$

where $p_i \in C_i$ is the intersection point of C_i and C^{i-1}. By (3.8) $H^1(C_i, E \otimes \mathcal{O}_{C_i}(-p_i)) = 0$. Thus

$$H^1(C^i, E \otimes \mathcal{O}_{C^i}) \cong H^1(C^{i-1}, E \otimes \mathcal{O}_{C^{i-1}}),$$

and we are done by induction. □

The following is the easy smoothing result: there are no obstructions to smoothing.

7.6 Theorem. *Let $C = \sum C_i$ be a tree of smooth rational curves, and $f : C \to X$ a morphism such that X is smooth along $f(C)$.*

(7.6.1) *Fix a smooth point $0 \in C$. If $f^*T_X|C_i$ is semi positive for every C_i, then $f : C \to X$ can be smoothed keeping $f(0)$ fixed.*

(7.6.2) *Fix smooth points $p_i \in C_i$. If $f^*T_X|C_i$ is ample for every C_i, then $f : C \to X$ can be smoothed keeping the $f(p_i)$ fixed.*

(7.6.3) *Let $f_t : \mathbb{P}^1 \to X$ be a nearby smoothing. Then $f_t^*T_X$ is semi positive in the first case and ample in the second case.*

Proof. Let $q : C_T \to T$ be a smoothing of C (this exists by (1.12)). Consider first (7.6.1). We may assume that there is a section $t : T \to C_T$ such that $t(0) = 0 \in C$. Let $G : t(T) \to X$ be the constant morphism to $f(0)$.

I claim that $H^1(C, f^*T_X(-0)) = 0$. Let $D = C_i$ be the curve containing 0. C is obtained from D by attaching trees, thus by (7.5)

$$H^1(C, f^*T_X(-0)) = H^1(D, f^*T_X \otimes \mathcal{O}_D(-0)) = 0,$$

where the last equality is by (3.8).

By (1.7) $\mathrm{Hom}(C_T/T, X \times T/T, G) \to T$ is smooth at $[f]$. Thus f is smoothable fixing $f(0)$ by (7.3).

The arguments are similar in case (7.6.2).

Finally $H^1(C, f^*T_X(-0)) = 0$ implies that $H^1(\mathbb{P}^1, f_t^*T_X(-0)) = 0$ for nearby smoothings, hence $f_t^*T_X$ is semi positive by (3.8). In the second case fix an extra point $p \in C$. As before we obtain that $H^1(C, f^*T_X(-0-p)) = 0$, which gives that $f_t^*T_X$ is ample. (In fact $f_t^*T_X = \sum \mathcal{O}(a_i)$ and $a_i \geq$ number of irreducible components of C.) □

7.6.4 Exercise. Let X be a smooth variety and $D = \sum a_i D_i$ a connected 1-cycle such that every D_i is rational.

(7.6.4.1) Show that there is a tree of smooth rational curves C and a morphism $f : C \to X$ such that $D = f_*[C]$.

(7.6.4.2) Let X be a homogeneous space under a linear algebraic group such that the stabilizer of a point is reduced (V.1.4). Show that there is a family of morphisms $f_t : \mathbb{P}^1 \to X$ such that $\lim_{t\to 0}(f_t)_*[\mathbb{P}^1] = D$.

7.7 Definition. A *comb with n teeth* is a connected and reduced curve C with irreducible components $D, C_1, \ldots, C_n$ with the following properties:

(7.7.1) D is smooth and every C_i is isomorphic to $\mathbb{P}^1$; (D is called the handle and C_i the teeth)

(7.7.2) the only singularities of C are ordinary nodes;

(7.7.3) every C_i intersects D in a single point x_i and $x_i \neq x_j$ for $i \neq j$. These are the only intersections between the irreducible components.

A *subcomb* of C is a subcurve which contains the handle.

7.8 Construction. Let $C = D + \sum C_i$ be a comb with m teeth intersecting D at the points x_i. Let $Z_i \subset \{x_i\} \times \mathbb{A}^m \subset D \times \mathbb{A}^m$ be defined by the equation $y_i = 0$ where the y_i are coordinates on $\mathbb{A}^m$. The Z_i are disjoint codimension two subvarieties of $D \times \mathbb{A}^m$. Let $p : Y \to D \times \mathbb{A}^m \to \mathbb{A}^m$ be obtained by blowing up every Z_i. The following properties are clear:

7.8.1 Claim. $p : Y \to \mathbb{A}^m$ *is flat. Its fibers are reduced curves which are combs. The fiber over a point* $(y_1, \ldots, y_m)$ *is a comb with handle* D *and with* $\#\{i | y_i = 0\}$ *teeth.*

In particular $p^{-1}(0)$ *is the only fiber isomorphic to* C. □

The following is the hard smoothing result: there are obstructions but some smoothing is still possible.

7.9 Theorem. *Let $C = D + \sum_{1\le i\le m} C_i$ be a comb, $p_1,\dots,p_k \in D$ points different from the x_i. Let $f: C \to X$ be a morphism such that X is smooth along $f(C)$ and $f^*T_X|C_i$ is semi positive for every i. Then there is a subcomb $C' = D + \sum' C_i$ with m' teeth such that $f' := f|C' : C' \to X$ is smoothable fixing $f'(p_1),\dots,f'(p_k)$ and m' satisfies both of the following inequalities:*

$$m' \ge m - h^1(D,(f|D)^*T_X(-\textstyle\sum p_i)), \quad \textit{and}$$
$$m' \ge m - (K_X \cdot D) - k\dim X + \dim X \cdot \chi(\mathcal{O}_D) - \dim_{[f|D]} \operatorname{Hom}(D,X,f|\{p_1,\dots,p_k\}).$$

Proof. Let $t_i : \mathbb{A}^m \to D \times \mathbb{A}^m \dashrightarrow Y$ be the constant p_i section. Set $G|\operatorname{im} t_i = f(p_i)$. Let c be the codimension of the image of a small neighborhood of

$$[f] \in \operatorname{Hom}(Y/\mathbb{A}^m, X \times \mathbb{A}^m/\mathbb{A}^m, G) \quad \text{in} \quad \mathbb{A}^m.$$

Then there is a curve

$$r : (0 \in T) \to ([f] \in \operatorname{Hom}(Y/\mathbb{A}^m, X \times \mathbb{A}^m/\mathbb{A}^m, G))$$

such that the image of $r(T)$ in $\mathbb{A}^m$ is contained in at most c of the coordinate hyperplanes $(y_i = 0)$. By base change we obtain a flat family $Y \to T$ whose general fiber is a comb with at most c teeth. Let $Y' \subset Y$ be the irreducible component containing $D \subset C$. $Y' \to T$ is flat, proper, its generic fiber is isomorphic to D and the central fiber $C' \subset C$ is a subcomb of C with at least $m - c$ teeth. By construction $f|C'$ extends to $F : Y' \to X$.

It remains to get an upper bound for c. By (I.2.10, I.2.17, 1.7) there is a smooth morphism $\operatorname{Spec} R \to \mathbb{A}^m$ such that (the completion of) $\operatorname{Hom}(Y/\mathbb{A}^m, X \times \mathbb{A}^m/\mathbb{A}^m, G)$ is defined in $\operatorname{Spec} R$ by at most $h^1(C, f^*T_X(-\sum p_i))$ equations. Thus

$$c \le h^1(C, f^*T_X(-\textstyle\sum p_i)) = h^1(D,(f|D)^*T_X(-\sum p_i)),$$

where the last equality is by (7.5). This gives the first estimate.

To get the second estimate note that by upper semi continuity of fiber dimension,

$$c \le m - \dim_{[f]} \operatorname{Hom}(Y/\mathbb{A}^m, X \times \mathbb{A}^m/\mathbb{A}^m, G) + \dim_{[f]} \operatorname{Hom}(C,X,f|\{p_1,\dots,p_k\}).$$

(1.7) gives that

$$\begin{aligned}&\dim_{[f]} \operatorname{Hom}(Y/\mathbb{A}^m, X \times \mathbb{A}^m/\mathbb{A}^m, G)\\ &\quad \ge -K_X \cdot C - k\dim X + \dim X \cdot \chi(\mathcal{O}_C) + m\\ &\quad = -K_X \cdot D - k\dim X + \dim X \cdot \chi(\mathcal{O}_D) + \textstyle\sum(-K_X \cdot C_i) + m.\end{aligned}$$

There is a forgetful morphism $\operatorname{Hom}(C, X) \to \operatorname{Hom}(D, X)$. The fiber of it over $[f]$ is the same as $\operatorname{Hom}(\sum C_i, X, f|\{x_1, \dots, x_m\})$. Since $f|C_i$ is free we obtain that

$$\begin{aligned} &\dim_{[f]} \operatorname{Hom}(C, X, f|\{p_1, \dots, p_k\}) \\ &\quad \leq \dim_{[f|D]} \operatorname{Hom}(D, X, f|\{p_1, \dots, p_k\}) + \sum \dim_{[f|C_i]} \operatorname{Hom}(C_i, X, f|\{x_i\}) \\ &\quad = \dim_{[f|D]} \operatorname{Hom}(D, X, f|\{p_1, \dots, p_k\}) + \sum(-K_X \cdot C_i). \end{aligned}$$

Combining the above two estimates gives

$$\begin{aligned} c \leq K_X \cdot D + k \dim X - \dim X \cdot \chi(\mathcal{O}_D) \\ + \dim_{[f|D]} \operatorname{Hom}(D, X, f|\{p_1, \dots, p_k\}). \end{aligned}$$ □

In some cases one can show that the smoothing of f' is free over nearby fibers:

7.10 Proposition (Notation as above). *Assume in addition that $f^*T_X|C_i$ are ample and $H^1(D, M \otimes f^*T_X(-\sum p_i)|D) = 0$ for every line bundle M of degree at least d. Let $C' \subset C$ be a subcomb with at least $m - c$ teeth and $F' : Y' \to X$ a smoothing of $f|C'$ fixing $f(p_1), \dots, f(p_k)$ constructed above. If $m - c > d$, then $F'|C'_t$ is free over $t_1(t), \dots, t_k(t) \in C_t$ for $t \neq 0$.*

Proof. In order to see that $F'|C'_t$ is free over $t_1(t), \dots, t_k(t) \in C_t$ we need to prove that $H^1(C'_t, F'^*T_X(-p - \sum t_i(t))) = 0$ for every $p \in C'_t$. This follows from the next lemma.

7.10.1 Lemma. *Let $q : Y \to T$ be a smoothing of a comb with n teeth. Let E be a vector bundle on Y such that the $E|C_i$ are ample and $H^1(D, M \otimes (E|D)) = 0$ for every line bundle M of degree $\geq n$.*

Then $H^1(C_t, E|C_t) = 0$ for general t.

Proof. Let us consider first the simpler case when Y is smooth. Let $L = \mathcal{O}_Y(\sum C_i)$. E and $L \otimes E$ are isomorphic outside the central fiber. Furthermore, $L \otimes E|C_i \cong (E|C_i)(-1)$ and $L \otimes E|D \cong (E|D)(\sum x_i)$. Thus by (7.5) $H^1(C, L \otimes E|C) = H^1((E|D)(\sum x_i)) = 0$. We obtain (10.1) by upper semi continuity.

In general one would like to prove that there are natural numbers n_i such that $n_i C_i$ is Cartier and $L = \mathcal{O}_Y(\sum n_i C_i)$ has the same properties as above. For instance this can be seen by showing that Y has only singularities of the form $xy = z^n$ and, then computing explicitly the minimal resolution. Details are left as an exercise. □

Chapter III. The Cone Theorem and Minimal Models

The most important application of the bend-and-break techniques of [Mori79] is his proof of the Cone Theorem and the applications to minimal models of smooth varieties. Later development of this theory strayed far from the methods considered in this book. The cohomological methods of Kawamata–Reid–Shokurov are described in detail in [CKM88]. Minimal models are not fully covered in any monograph,[1] since the proof of the crucial flip theorem [Mori88] is rather long and complicated. The ideas of the Cone Theorem and of Mori's Program (or Minimal Model Program) are very simple, powerful and influential. The aim of this short chapter is to illustrate these in rather special cases where the technical details do not overwhelm them.

Section 1 contains the proof of the Cone Theorem for smooth varieties, closely following the original arguments of [Mori82]. The main advantage of the cohomological approach is in proving that every extremal ray can be contracted (1.8). The geometric method does not seem to yield contractibility, but some important special cases are easy to get (1.6). In many applications these already give enough information.

The philosophy of Mori's Program is the following. We aim to study algebraic varieties modulo birational equivalence. It is rather hard to do birational geometry, thus we try to find a special variety where the birational questions reduce to biregular ones. In the case of surfaces this is possible in most cases. If S is not birationally ruled, then there is a unique surface S' birational to S such that $K_{S'}$ is nef. Thus any biregular property of S' is a birational property of S. The situation becomes more complicated for birationally ruled surfaces. The program selects a specially nice model for these, but there are infinitely many such models and it is rather hard to see which properties are the attributes of the special model and which are the properties of the birational equivalence class. It is expected that for these varieties an extension of Mori's program eventually will enable us to describe all different birational models, but at the moment only the beginnings of this theory are firmly established [Corti94].

The philosophy of Mori's Program is illustrated in Sect. 2. The theory of minimal models of surfaces is classical, but the unified view provided by Mori's Program is new. The treatment is very short except for the last step

[1] See [Kollár-Mori98] for a recent treatment.

(2.3.3) which requires a characterization of $\mathbb{P}^2$. In characteristic zero one can use topological methods to get a rather quick proof (3.7). The positive characteristic case was first treated in [Zariski58a]. Subsequently different arguments were found by many authors.

The original arguments of [Zariski58a] have the advantage that they are very geometric, use little machinery and can be combined with the bend-and-break technique. This approach is presented in Section 3. It relies on a complete classification of Del Pezzo surfaces of degree at most four (3.5). This result is of considerable interest in itself and is also used later in (IV.6). Other aspects of the theory of Del Pezzo surfaces are developed in a series of exercises (3.8–13).

III.1 The Cone Theorem

The aim of this section is to present a proof of the Cone Theorem of [Mori82]. This result is the starting point of Mori's program. The current popular approach (see e.g. [CKM88]) proceeds through different methods, but the ideas of Mori's original proof are still influential in similar problems.

1.1 Notation. Let X be a proper scheme over a field k, L a Cartier divisor on X and $S \subset N_1(X)$ a subset. We use the notation

$$S_{L\geq 0} := S \cap \{z \in N_1(X) | z \cdot L \geq 0\}.$$

$S_{L\leq 0}, S_{L<0}$ and $S_{L>0}$ are defined similarly.

1.2 Theorem [Mori82]. *Let X be a smooth projective variety over an algebraically closed field. There are countably many rational curves $L_i \subset X$ such that $0 < -K_X \cdot L_i \leq \dim X + 1$ and*

$$\overline{NE}(X) = \overline{NE}(X)_{K_X\geq 0} + \sum_i \mathbb{R}^+[L_i]. \tag{1.2.1}$$

The rays $\mathbb{R}^+[L_i] \subset N_1(X)$ are locally discrete in the open halfspace $N_1(X)_{K_X<0}$ and every extremal ray R of $\overline{NE}(X)$ lying in $N_1(X)_{K_X<0}$ is of the form $R = \mathbb{R}^+[L_i]$ for some i.

Proof. There are only countably many families of rational curves $\{L_t\}$ in X such that $0 < -K_X \cdot L_t \leq \dim X + 1$. Pick one curve L_i from every family.

Let W denote the closure of the right hand side of (1.2.1). As a first step we prove that $\overline{NE}(X) = W$. Assume the contrary. Then by (II.4.10.5) there is a linear function M which is nonnegative on $\overline{NE}(X)$, $M \cdot z = 0$ for some $z \in \overline{NE}(X) - \{0\}$ and M is strictly positive on $W - \{0\}$. By replacing M with a suitable multiple, we may assume that if $D \subset X$ is an irreducible curve such that $[D] \in W$, then $M \cdot D \geq 1$. $-K_X \cdot z > 0$ since M is strictly positive on $\overline{NE}(X)_{K_X\geq 0} - \{0\}$.

Let $C_i = \sum a_{ij} C_{ij}$ be effective 1-cycles (with $\mathbb{R}$-coefficients) on X such that $\lim[C_i] = z$.

1.2.2 Lemma (Notation as above). *For every i there is a j such that $-K_X \cdot C_{ij} > 0$ and*

$$\frac{M \cdot C_{ij}}{-K_X \cdot C_{ij}} \leq \frac{M \cdot C_i}{-K_X \cdot C_i}.$$

Proof. Let $d = (M \cdot C_i)/(-K_X \cdot C_i)$. Then

$$\sum a_{ij} \left[(M \cdot C_{ij}) - d(-K_X \cdot C_{ij}) \right] = 0.$$

Thus either $0 \leq (M \cdot C_{ij}) < d(-K_X \cdot C_{ij})$ for some j and we are done or $(M \cdot C_{ij}) = d(-K_X \cdot C_{ij})$ for every j. $-K_X \cdot C_{ij} > 0$ for some j, and we are done again. □

Apply (II.5.14) to C_{ij} as in (1.2.2) and to the divisor M. We obtain that there are rational curves $L_{k(i)} \subset X$ such that $0 < -K_X \cdot L_{k(i)} \leq \dim X + 1$ and

$$M \cdot L_{k(i)} \leq 2 \dim X \frac{M \cdot C_{ij}}{-K_X \cdot C_{ij}} \leq 2 \dim X \frac{M \cdot C_i}{-K_X \cdot C_i}. \tag{1.2.3}$$

The limit of the right hand side of (1.2.3) is $2 \dim X (M \cdot z)/(-K_X \cdot z) = 0$. Thus for $i \gg 1$ we obtain that $M \cdot L_{k(i)} < 1$, which implies that $L_{k(i)} \not\subset W$, a contradiction.

$\overline{NE}(X)_{K_X<0}$ is the union of the open subsets $\overline{NE}(X)_{K_X+\epsilon H<0}$ for $\epsilon > 0$, thus it is sufficient to prove that every $\overline{NE}(X)_{K_X+\epsilon H<0}$ contains only finitely many classes $[L_i]$. If $(K_X + \epsilon H) \cdot L_i < 0$, then

$$H \cdot L_i < \epsilon^{-1}(-K_X \cdot L_i) \leq \epsilon^{-1}(\dim X + 1).$$

By (I.3.21.3; I.4.13.4; II.2.2) there are only finitely many families of rational curves satisfying this inequality.

It remains to show that $\overline{NE}(X)_{K_X \geq 0} + \sum_i \mathbb{R}^+[L_i]$ is a closed cone. Let $z \in \overline{NE}(X)$ and set $m = \dim \overline{NE}(X)$. By (II.4.10.4) we can write

$$z = \lim_i \Big(z_i + \sum_{1 \leq j \leq m} a_{ij}[L_{ij}] \Big) \quad \text{where } z_i \in \overline{NE}(X)_{K_X \geq 0}.$$

If H is ample on X, then $H \cdot z_i$ and $a_{ij}(H \cdot L_i)$ are bounded by $(H \cdot z) + 1$ (at least for $i \gg 1$). By (II.4.8) there is a subsequence such that

$$\lim z_i \quad \text{and} \quad \lim_i a_{ij}[L_{ij}] \quad \text{all exist.}$$

Clearly $\lim z_i \in \overline{NE}(X)_{K_X \geq 0}$. The rays $\mathbb{R}^+[L_{ij}]$ are discrete in the open halfspace where $K_X^{\cap}$ is negative. Thus if $\lim_i a_{ij}[L_{ij}]$ exists, then either

$[L_i] := [L_{ij}]$ is independent of i for $i \gg 1$ and $\lim_i a_{ij}[L_{ij}] = (\lim_i a_{ij})[L_i]$, or $\lim_i a_{ij}[L_{ij}] \in \overline{NE}(X)_{K_X \geq 0}$. Thus

$$z = (\lim z_i) + \sum_j (\lim_i a_{ij}[L_{ij}]),$$

and the summands on the right hand side are in $\overline{NE}(X)_{K_X \geq 0} + \sum_i \mathbb{R}^+[L_i]$. □

1.2.5 Remarks. (1.2.5.1) The example given in (II.4.15.3) shows that there can be infinitely many extremal rays $\mathbb{R}^+[L_i]$ such that $K_X \cdot L_i < 0$. In that example X is uniruled. The conjecture [KaMaMa87, 4-2-5] asserts that if X is not uniruled, then there are only finitely many K_X-negative extremal rays.

(1.2.5.2) (1.2) is frequently formulated by saying that $\overline{NE}(X)_{K_X<0}$ is "locally polyhedral". One should be careful about the precise meaning of this statement. Let x_i be coordinates in $\mathbb{R}^4$. Consider the cone which is generated by

$$(x_1 - 1)^2 + x_2^2 + x_3^2 + x_4^2 = 1/4 \quad \text{and} \quad (0,0,0,1).$$

The extremal ray $\mathbb{R}^+(0,0,0,1)$ is not a limit of other extremal rays, but the cone is not the cone over a polyhedron near that ray. By (II.4.15.1) and (II.4.17) this cone is isomorphic to the closed cone of curves of $E \times E \times \mathbb{P}^1$ where E is any elliptic curve without complex multiplication.

(1.2.5.3) The part $\overline{NE}(X)_{K_X \geq 0}$ occuring in the cone theorem is sometimes rather mysterious. For instance it is not known if $NE(X)_{K_X \geq 0}$ is dense in $\overline{NE}(X)_{K_X \geq 0}$. Even in very special cases this seems quite hard. Assume for instance that $-K_X$ has nonnegative intersection number with every curve $C \subset X$. By (VI.2.19) $-K_X$ is ample iff it is strictly positive on $\overline{NE}(X) - \{0\}$. By assumption $-K_X \cdot L_i > 0$, thus $-K_X$ is ample iff it is strictly positive on $\overline{NE}(X)_{K_X \geq 0}$. In our case $\overline{NE}(X)_{K_X \geq 0} = \overline{NE}(X)_{K_X = 0}$. Thus we obtain:

1.2.5.4 Conjecture. *Let X be a smooth projective variety. Assume that $-K_X$ has positive intersection number with every curve $C \subset X$. Then $-K_X$ is ample.*

1.2.5.5 Remark. The surface case of the above conjecture is an exercise. For $\dim X = 3$ it is proved in [Matsuki87]. See also [Serrano95].

Over nonclosed fields the situation is very similar:

1.3 Theorem [Mori82]. *Let X be a smooth and projective variety over a field k. There are countably many k-irreducible 1-cycles $L_i \subset X$ such that if $L_i = \sum a_{ij} L_{ij}$ is the irreducible decomposition over $\bar{k}$, then $0 < -K_X \cdot L_{ij} \leq \dim X + 1$ for every i, j, and*

(1.3.1) $$\overline{NE}(X) = \overline{NE}(X)_{K_X \geq 0} + \sum_i \mathbb{R}^+[L_i].$$

Moreover, the rays $\mathbb{R}^+[L_i] \subset N_1(X)$ are locally discrete in the open halfspace $N_1(X)_{K_X<0}$.

Proof. It is possible to derive this from the corresponding cone theorem on $X_{\bar{k}}$ (cf. [Mori82, 2.3]). Here I would like to point out that a proof parallel to that of (1.2) is equally natural. Let us define first the cycles L_i.

Let $\bar{X}$ be an irreducible component of $X_{\bar{k}}$. There are countably many irreducible components $U_i \subset \mathrm{RatCurves}(\bar{X})$ which parametrize rational curves C_t such that $0 < -K_{\bar{X}} \cdot C_t \le \dim X + 1$. In every U_i pick a point $[C_i]$ which is defined over an algebraic extension $k_i \supset k$. We may assume that $k_i \supset k$ is normal with Galois group G_i.

$\sum_{h \in G_i} h(C_i)$ is a 1-cycle on X_{k_i} which is invariant under G_i. In char 0 it is therefore a 1-cycle on X. In char p we need to take $p^{e_i} \sum_{h \in G_i} h(C_i)$ for suitable $e_i \ge 0$ (I.4.5.4). Set

$$L_i = \frac{1}{|G_i| p^{e_i}} \left[p^{e_i} \sum_{h \in G_i} h(C_i)\right].$$

This is a 1-cycle in $N_1(X)_{\mathbb{Q}}$. Observe that if M is any Cartier divisor, then

(1.3.2) $$C_i \cdot M = L_i \cdot M.$$

The proof of (1.2) uses two properties of the curves L_i. The main one is (II.5.14) about the existence of rational curves. The rational curve D_x whose existence is given by (II.5.14) is algebraically equivalent to one of the above C_i, so we can replace D_x by one of the L_i and (II.5.14) stays valid. (We can not, however, guarantee that the 1-cycle L_i passes through a given point.)

The second property we used (after (1.2.4)) is that there are only finitely many L_i such that $H \cdot L_i \le$ const. This follows from (1.3.2) and from the corresponding fact about the C_i. □

1.4 Corollary. *Let X be a smooth and projective variety over a field k and $R \subset \overline{NE}(X)$ an extremal ray such that $R \cdot K_X < 0$. Then*

(1.4.1) *there is a Cartier divisor H_R on X such that $H_R^{\cap}$ is a supporting function of R (i.e. H_R is nef and if $z \in \overline{NE}(X)$, then $H_R \cdot z = 0$ iff $z \in R$), and*

(1.4.2) *$mH_R - K_X$ is ample for $m \gg 1$.*

Proof. Let

$$V = \overline{NE}(X)_{K_X \ge 0} + \sum_{[L_i] \not\in R} \mathbb{R}^+[L_i].$$

As in the proof of (1.2) we see that V is a closed cone. Also, $\overline{NE}(X) = V + R$. Since R is extremal $R \not\subset V$. By (II.4.10.5) there is a supporting function M of $\overline{NE}(X)$ such that M is strictly positive on $V - \{0\}$. Thus $M \cdot R = 0$.

By (1.2) there is a rational curve $L \subset X$ such that $R = \mathbb{R}^+[L]$. Thus $R^\perp = \{M \in N^1(X) | M \cdot R = 0\}$ is a hyperplane with a rational equation. If L is any linear function in $R^\perp$ which is close to M, then L is still positive on $V - \{0\}$ (use (II.4.8)). Thus it is also a supporting function of R. Therefore

R has a supporting function L in $N^1(X)_{\mathbb{Q}}$. A suitable multiple of L is in $N^1(X)_{\mathbb{Z}}$, hence corresponds to a Cartier divisor H_R.

In order to see the second part observe that H_R is positive on $V-\{0\}$, thus by (II.4.8) $mH_R - K_X$ is positive on $V - \{0\}$ for $m \gg 1$. $(mH_R - K_X)\cdot R = -K_X \cdot R > 0$, thus $mH_R - K_X$ is positive on $\overline{NE}(X) - \{0\}$. Therefore it is ample by (VI.2.19). □

The next result says that if an irreducible curve C is sufficiently close to a K_X-negative extremal ray R, then C is contained in the union of all rational curves which belong to R.

1.5 Corollary. *Let X be a smooth and projective variety over an algebraically closed field k. Let $R \subset \overline{NE}(X)$ be an extremal ray such that $R \cdot K_X < 0$ and H_R a Cartier divisor on X such that $H_R^{\frown}$ is a supporting function of R. Let $C \subset X$ be an irreducible curve such that*

$$H_R \cdot C < \frac{1}{2\dim X}(-K_X \cdot C).$$

Then $C \subset$ Locus R *(see (1.5.1) for the definition of* Locus*).*

Proof. By (II.5.8) for every point $x \in C$ there is a rational curve L_x containing x such that

$$H_R \cdot L_x \le 2\dim X \frac{H_R \cdot C}{-K_X \cdot C} < 1.$$

Since H_R is a Cartier divisor, this implies that $H_R \cdot L_x = 0$, hence $[L_x] \in R$. □

1.5.1 Definition. Let X be a proper variety and $V \subset N_1(X)$. Locus V denotes the set of closed points $x \in X$ such that there is a curve $x \in C \subset X$ such that $[C] \in V$. It is called the *locus* of V.

If $V \subset N_1(X)$ is an extremal subcone, then Locus V is closed under specialization.

1.5.2 Exercise. (1.5.2.1) Let R be a K_X-negative extremal ray. Show that Locus R is covered by rational curves.

(1.5.2.2) Assume that $\dim \operatorname{im}[N_1(\operatorname{Locus} R) \to N_1(X)] \ge 2$. Show that there are irreducible curves $C_i \subset X$ such that $[C_i] \notin R$ but the rays $\mathbb{R}^+[C_i]$ converge to R. Show that such curves do not exist if $\dim \operatorname{im}[N_1(\operatorname{Locus} R) \to N_1(X)] = 1$.

The following result can be viewed as a very weak version of the contraction theorem (1.8) which holds in all characteristics.

1.6 Theorem [Kollár91a]. *Let X be a smooth projective variety and $R \subset \overline{NE}(X)$ a K_X-negative extremal ray with supporting Cartier divisor H_R. Then:*

(1.6.1) $(H_R^{\dim X}) > 0$ *iff there is an irreducible divisor* $E \subset X$ *such that* $R \cdot E < 0$. *In particular* $\mathrm{Locus}(R) \subset E$.

(1.6.2) $(H_R^{\dim X}) = 0$ *iff* $\mathrm{Locus}(R) = X$.

Proof. Assume that $(H_R^{\dim X}) > 0$ and let L be an ample divisor on X. By (VI.2.16) $mH_R - L$ is linearly equivalent to an effective $\mathbb{Q}$-divisor D for $m \gg 1$. $(mH_R - L) \cdot R = -L \cdot R < 0$, thus there is an irreducible component E of D such that $E \cdot R < 0$.

Conversely, assume that $\mathrm{Locus}(R) \neq X$ and pick $x \notin \mathrm{Locus}(R)$. Choose m such that $mH_R - K_X$ is ample (1.4.2). Let $C \subset X$ be any curve through x. By (1.5) $2 \dim X(H_R \cdot C) \geq -K_X \cdot C$. Thus

$$(m + 2\dim X)(H_R \cdot C) \geq (mH_R - K_X) \cdot C,$$

which says that the Seshadri constant of H_R is positive at $x \in X$ (VI.2.18.6). By (VI.2.18.6.1) this implies that $(H_R^{\dim X}) > 0$. □

The following exercises give examples of birational contractions of K-negative extremal rays.

1.6.3 Exercise. (1.6.3.1) Let X be a smooth variety and $Z \subset X$ a smooth subvariety. Show that $B_Z X \to X$ is the contraction of a K-negative extremal ray.

(1.6.3.2) Let $X \subset \mathbb{P}^{n+1}$ be a hypersurface with a single singular point $p \in X$. Characterize those cases when $B_p X$ is smooth. Show that $B_p X \to X$ is the contraction of an extremal face. Characterize those cases when $B_p X \to X$ is the contraction of a K-negative extremal ray. In some of these cases the exceptional divisor is singular.

1.6.4 Exercises. Let X be a smooth projective variety and R a K_X-negative extremal ray.

(1.6.4.1) Assume that $\mathrm{Locus}\, R \neq X$. Show that if $\mathrm{Locus}\, R$ contains a divisor D, then $\mathrm{Locus}\, R = D$.

(1.6.4.2) If $\dim X = 3$, then either $\mathrm{Locus}\, R = X$ or $\mathrm{Locus}\, R$ is an irreducible divisor.

(1.6.4.3) Find examples for $\dim X \geq 4$ such that $\dim \mathrm{Locus}\, R \leq \dim X - 2$. (I do not know any examples where $\dim \mathrm{Locus}\, R < \dim X/2$.)

(1.6.4.4) Find examples for $\dim X \geq 4$ such that $\mathrm{Locus}\, R$ is not connected.

1.6.5 Exercise. Fix $n > r > 0$. Let

$$Y := \{M | M \in \mathrm{End}(k^n); \mathrm{rank}\, M \leq r\},$$
$$X := \{(V, M) | M \in \mathrm{End}(k^n); \mathrm{rank}\, M \leq r; V \subset \ker M; \dim V = n - r\}.$$

Let $f : X \to Y$ be the forgetful morphism. Using this construct examples of extremal rays on smooth projective varieties such that $\mathrm{Locus}\, R$ is a nonsmooth divisor.

1.6.6 Exercise. Let C be a smooth projective curve with a distinguished point $p \in C$. Let S^mC denote the mth symmetric product of C and $J(C)$ the Jacobian variety of C.

(1.6.6.1) The choice of p gives imbeddings $S^mC \subset S^{m+1}C$ and S^mC is an ample divisor in $S^{m+1}C$.

(1.6.6.2) Show that $\rho(S^mC) = \rho(J(C)) + 1$ for $m \geq 2g(C) - 1$.

(1.6.6.3) Show that $\rho(S^mC) = \rho(S^{m+1}C)$ for $m \geq 3$, thus $\rho(S^mC) = \rho(J(C)) + 1$ for $m \geq 3$. (This needs the Lefschetz theorem on hyperplane sections, see e.g. [Grothendieck68].)

(1.6.6.4) The choice of p gives morphisms $u_m : S^mC \to J(C)$ which are surjective for $m \geq g(C)$, birational for $m = g(C)$. Show that u_m is the contraction of a K-negative extremal ray on S^mC for $m \geq g(C)$.

This gives many examples of extremal contractions between smooth varieties where the structure of the exceptional set is complicated.

1.7 Remark. The above proof of the cone theorem relies on the bend-and-break method which requires information about the deformation space of morphisms of curves. The requisite dimension estimates for deformation spaces are known for smooth varieties (II.1.2), for varieties with isolated complete intersection singularities (II.1.3) or more generally for varieties which are are locally the quotients of isolated complete intersection singularities [Kollár92]. This is a reasonably large class in dimension 3, but not in higher dimensions.

A cohomological approach to the cone theorem was developped in a series of articles by Kawamata, Reid and Shokurov. This method is very powerful, but so far it works only in characteristic zero. See [KaMaMa87] for a technically complete review or [CKM88] for a more leisurely introduction. For smooth varieties the result implies one crucial piece of information which so far proved unattainable using the geometric method. It is not used in the rest of the book.

1.8 Theorem. *Let X be a smooth and projective variety over a field of* char 0. *Let $V \subset \overline{NE}(X)$ be an extremal subcone such that K_X is negative on $V - \{0\}$. Let H_V be a supporting Cartier divisor of V. Then $|mH_V|$ is base point free for $m \gg 1$. Thus it defines a morphism*

$$\operatorname{cont}_V : X \to X_V$$

such that if $C \subset X$ is an irreducible curve, then $\operatorname{cont}_V(C) = $ *point iff* $[C] \in V$.

We may assume that cont_V *has connected fibers and X_V is normal. With these assumptions* cont_V *is unique. It is called the contraction of V.*

In some cases the following generalization of Castelnuovo's contractibility criterion can be used to obtain the existence of a contraction:

1.8.1 Exercise. Let W be a proper scheme (or algebraic space) and $S \subset W$ a Cartier divisor. Let M be a line bundle on W generated by global sections,

$\phi : W \to W'$ the Stein factorization of the corresponding morphism. Assume that

(1.8.1.1) $H^1(W, M) = 0$,
(1.8.1.2) $H^1(S, \mathcal{O}_S \otimes M(jS)) = 0$ for $1 \le j \le k-1$, and
(1.8.1.3) $\mathcal{O}_S \otimes M(kS)$ is generated by global sections. Let $cont : S \to \bar{S}$ be the Stein factorization of the corresponding morphism.

Then $M(kS)$ is generated by global sections and the Stein factorization $Cont : W \to \bar{W}$ of the corresponding morphism has the following properties:

(1.8.1.4) $Cont|W - S = \phi|W - S$,
(1.8.1.5) $Cont|S = cont$, and
(1.8.1.6) $\bar{W}$ is projective if W is.

As an application show that if W is a smooth projective variety and $\mathbb{P}^{n-1} \cong S \subset W$ is a Cartier divisor such that $N_{S|W} \cong \mathcal{O}(m)$ for some $m < 0$, then there is a morphism onto a projective variety $f : W \to W'$ such that f contracts S to a point and is an isomorphism outside S.

1.9 Exercise [Wiśniewski91b]. Let X be an irreducible and proper variety of dimension n. On $N^1(X)$ let Φ denote the homogeneous polynomial $\Phi(D) = (D^n)$ (cf. (VI.2)). Let $f : X \to Y$ be a morphism to a variety of dimension m. This induces a map $f^* : N^1(Y) \to N^1(X)$. Show that Φ vanishes along $f^*(N^1(Y))$ with multiplicity $n - m$.

Let $f_i : X \to Y_i$ be different dominant morphisms with connected fibers such that $\dim N^1(X) = 1 + \dim N^1(Y_i)$ for every i. (This assumption is satisfied if each f_i is the contraction of a K-negative extremal ray, see e.g. [KaMaMa87, 3-2-5].) Show that $\dim X \ge \sum_i(\dim X - \dim Y_i)$.

Find examples where equality holds.

III.2 Minimal Models of Surfaces

The aim of this section is to present the classical theory of minimal models of surfaces from the point of view of extremal ray theory. It turns out that it is not much harder to consider surfaces over arbitrary fields. One could also consider surfaces with a group action; everything that we do in this section holds if we consider only equivariant morphisms.

The classical theory goes back to Castelnuovo. In characteristic p several problems arise, these were considered by [Zariski58b] over algebraically closed fields. Over arbitrary fields the theory is due to [Manin66, Iskovskikh80c]. This presentation follows [Mori82].

2.1 Theorem. *Let X be a smooth proper algebraic surface over a field k. If K_X is not nef, then there is a morphism $f : X \to Y$ onto a smooth k-variety Y such that $\rho_k(Y) = \rho_k(X) - 1$. Furthermore f is one of the following types:*

(2.1.1) *Y is a smooth surface and X is obtained from Y by blowing up a zero dimensional k-irreducible and geometrically reduced subscheme $P \subset Y$. (I.e. P is a collection of points of $Y_{\bar{k}}$, conjugate over k.)*

(2.1.2) *$\rho_k(X) = 2$, Y is a smooth curve and X is a conic bundle over Y.*

(2.1.3) *$\rho_k(X) = 1$, Y is a point and $-K_X$ is ample.*

Proof. We are done if $\rho_k(X) = 1$. Thus assume that $\rho_k(X) \geq 2$ and let $\mathbb{R}^+[L] \subset \overline{NE}(X)$ be an extremal ray (1.3). As we discussed in (1.3), one can write $L = p^e \sum_{i=1}^{s} C_i$ where $p = \operatorname{char} k$ and the C_i are irreducible and reduced curves on $X_{\bar{k}}$, which are conjugate under the Galois group of $\bar{k}/k$.

By (II.4.12) $(L^2) \leq 0$. This gives that

$$0 \geq \frac{1}{p^e s}(L^2) = C_1 \cdot (\sum C_i) = (C_1^2) + C_1 \cdot (\sum_{i \geq 2} C_i).$$

Thus $(C_1^2) \leq 0$. By assumption $C_1 \cdot K_X < 0$, hence by adjunction we have one of the following two possibilities:

(2.1.4) $$(L^2) < 0 \quad \text{and} \quad C_i^2 = -1, C_i \cdot C_j = 0 \ \forall i \neq j, \text{ or}$$

(2.1.5) $$(L^2) = 0, \quad C_i \cdot L = 0 \ \forall i, \quad \text{and}$$

(2.1.6) $$\text{either } (C_i^2) = -1, C_i \cdot K_X = -1 \text{ or } (C_i^2) = 0, C_i \cdot K_X = -2.$$

In the first case let H be a very ample divisor on X. Then

$$M = (H \cdot C_1)L + (-L \cdot C_1)H$$

is a line bundle on X which is very ample outside L by construction and has degree zero on L. The usual proof of the contractibility of -1-curves on a surface (see e.g. [Hartshorne77, V.5.7] or (1.8.1)) shows that $|M|$ is base point free and contracts the irreducible components of L to smooth points. This gives (2.1.1).

In the second case we prove that $|mL|$ gives the required morphism for $m \gg 1$. Since L is effective, $H^2(X, \mathcal{O}(mL)) = 0$ for $m \gg 1$. By (2.1.6) $K_X \cdot L < 0$, thus

$$h^0(X, \mathcal{O}(mL)) \geq \chi(X, \mathcal{O}(mL)) = \frac{-K_X \cdot L}{2} m + \chi(\mathcal{O}_X) \geq 2 \quad \text{for } m \gg 1.$$

Since $C_i \cdot L = 0$ for every i, $|mL|$ is base point free for $m \gg 1$. Let $f : X \to Y$ be the Stein factorization of the corresponding morphism and F the generic fiber. By (2.1.8) $F \cdot K_X = -2$ and F is a conic. Using (2.1.6) we see that every fiber is either a smooth conic or a pair of intersecting lines, thus $f : X \to Y$ is a conic bundle. (If k is closed, then a $\mathbb{P}^1$-bundle.)

Finally, in order to see that $\rho_k(X) = 2$ we prove that if M is a Cartier divisor on X such that $M \cdot L = 0$, then $M \cong f^* f_* M$. If $F \subset X$ is any fiber, then $M|F$ is a line bundle which has degree zero on every irreducible component. Thus $h^0(F, M|F) = 1$ and $h^1(F, M|F) = 0$. Therefore f_*M is a line bundle and the natural map $f^*f_*M \to M$ is an isomorphism. □

The proof of (2.1) shows the following equivalent formulation of (2.1):

2.1.7 Theorem. *Let X be a smooth proper algebraic surface over a field k and $R \subset \overline{NE}(X)$ a K_X-negative extremal ray. There is a morphism $f : X \to Y$ onto a smooth k-variety Y such that f contracts precisely those curves $C \subset X$ such that $[C] \in R$. Moreover, f is one of the three types listed in (2.1).* □

2.1.8 Exercise. Let F be a one dimensional, geometrically connected and regular scheme over a field k. Assume that ω_F^{-1} is ample. Show that F is isomorphic to a conic in $\mathbb{P}^2_k$.

The contraction (2.1.1) introduces a new surface Y. We can apply (2.1) to Y and continue if possible. This gives the following:

2.2 Theorem. *Let X be a smooth proper algebraic surface defined over a field k. There is a sequence of contractions $X \to X_1 \to \ldots \to X_n = X'$ such that X' satisfies exactly one of the following conditions:*

(2.2.1) *$K_{X'}$ is nef;*
(2.2.2) *$\rho_k(X') = 2$ and X' is a conic bundle over a curve C;*
(2.2.3) *$\rho_k(X') = 1$ and $-K_{X'}$ is ample.*

Proof. The only part which does not follow from (2.1) is the assertion that any sequence of contractions must terminate. Since $\rho_k(X_n) = \rho_k(X) - n$, this follows from the Theorem of the Base (II.4.5).

For the surface case it is easy to provide a different argument.

Let $X_0 \to X_1 \to \ldots \to X_n$ be a sequence of contractions. Pick an ample divisor H_0 on X_0. Assume that we already defined a divisor H_i on X_i. Let $E_i \subset X_i$ be the exceptional divisor and $F_i \subset (E_i)_{\bar{k}}$ an irreducible and reduced component over $\bar{k}$.

Let $c_i = H_i \cdot F_i$ and set $H_{i+1} = (f_i)_*(H_i)$. It is easy to see that H_{i+1} is an ample divisor on X_{i+1}. Furthermore

$$K_{X_{i+1}} \cdot H_{i+1} = K_{X_i} \cdot H_i + c_i(E_i^2) < K_{X_i} \cdot H_i.$$

Thus if $n > K_{X_0} \cdot H_0$, then $K_{X_n} \cdot H_n < 0$. Since H_n is ample, through every point of X_n there is a curve which is numerically equivalent to a multiple of H_n. Thus by (II.5.8) X_n and hence also X are uniruled.

By (II.4.19.1) $\rho(X) < \infty$ if X is uniruled, thus we can argue as in the beginning. □

A complete description of the surfaces occuring in case (2.2.3) is possible. A lot of the relevant work is done in Sect. 3. If k is algebraically closed, the answer is especially simple. Using (3.7) we can put (2.2) into its final form for algebraically closed fields:

2.3 Theorem. *Let X be a smooth proper algebraic surface defined over an algebraically closed field k. There is a sequence of contractions of (-1)-curves $X \to X_1 \to \dots \to X_n = X'$ such that X' satisfies exactly one of the following conditions:*

(2.3.1) *$K_{X'}$ is nef;*
(2.3.2) *X' is a $\mathbb{P}^1$-bundle over a curve C ;*
(2.3.3) *$X' \cong \mathbb{P}^2$.* □

2.3.4 Complement. *If $K_{X'}$ is nef, then the morphism $X \to X'$ is unique, thus X' is determined by X.*

Proof. Let $E \subset X$ be the exceptional curve of $p : X \to X'$. Since p is the composite of blow-ups of points, we see that $K_X = p^*K_{X'} + F$ where F is an effective divisor and $\operatorname{Supp} F = \operatorname{Supp} E$.

Let $f : X \to Y$ be a contraction as in (2.1) and $C \subset X$ a curve such that $f(C) =$ point. Then $C \cdot K_X < 0$. Since $K_{X'}$ is nef we obtain that

$$C \cdot F = C \cdot K_X - p_*C \cdot K_{X'} \leq C \cdot K_X < 0.$$

Therefore $C \subset F$ and f is a contraction as in (2.1.1). Moreover, there is a factorization

$$p : X \xrightarrow{f} Y \to X'.$$

The uniqueness of X' follows by induction on the number of exceptional curves of $X \to X'$. □

It is rather easy to derive the following rationality criterion as well:

2.4 Theorem (Castelnuovo – Zariski Rationality Criterion). *Let X be a smooth projective surface over an algebraically closed field k.*

(2.4.1) *Let $p : \mathbb{P}^2 \dashrightarrow X$ be separable and dominant. Then X is a rational surface.*

(2.4.2) *X is rational iff $h^1(\mathcal{O}_X) = h^0(\mathcal{O}_X(2K_X)) = 0$.*

Proof. If X is rational, then the conditions in (2.4.2) are obviously satisfied. All the conditions are birationally invariant, thus it is sufficient to consider the case when X satisfies one of the conditions of (2.3). If $X \cong \mathbb{P}^2$, then X is rational. If X is a ruled surface over a curve C and $C \cong \mathbb{P}^1$, then X is rational by (IV.6.6.2). In case (2.4.1) let $\ell \subset \mathbb{P}^2$ be a general line. The resulting map $p_\ell : \ell \dashrightarrow C$ is dominant, hence C is rational. In case (2.4.2) $h^1(\mathcal{O}_C) = h^1(\mathcal{O}_X) = 0$, hence again C is rational.

We show that K_X can not be nef. In case (2.4.1) $\deg p_\ell^* K_X \leq \ell \cdot K_{\mathbb{P}^2} = -3$. In the second case $h^0(\mathcal{O}_X(2K_X)) = 0$, thus also $h^0(\mathcal{O}_X(K_X)) = 0$. Therefore

$$\chi(\mathcal{O}_X) = 1 - h^1(\mathcal{O}_X) + h^0(\mathcal{O}_X(K_X)) = 1.$$

Also, $h^2(X, \mathcal{O}(-K_X)) = h^0(\mathcal{O}_X(2K_X)) = 0$. Since K_X is nef, $(K_X^2) \geq 0$ (VI.2.17) hence by Riemann-Roch

$$h^0(\mathcal{O}(-K_X)) \geq \chi(\mathcal{O}(-K_X)) = K_X^2 + 1 \geq 1.$$

Let $D \in |-K_X|$. $-D$ is nef, thus $D = \emptyset$. Therefore K_X is trivial and $h^0(\mathcal{O}_X(2K_X)) = 1$, a contradiction. □

2.5 Exercise. (2.5.1) Show that in characteristic $p > 0$, surfaces of the form $z^p = f(x, y)$ are unirational but usually not rational.

(2.5.2) [Shioda74] The Fermat-type hypersurface $(\sum x_i^n = 0) \subset \mathbb{P}^m$ is unirational if m is odd and $p^r \equiv -1 \bmod n$ for some $r \geq 1$ ($p = \text{char}$). Thus, for instance, the quartic surface $x_0^4 + x_1^4 + x_2^4 + x_3^4 = 0$ is unirational in $\operatorname{char} p$ if $p \equiv -1 \bmod 4$. The proof in [Shioda74] is elementary but tricky. See [Shioda-Katsura79] for similar results.

2.6 Exercise. Let X be a projective surface over an algebraically closed field which has only *Du Val singularities.* (These are the same as *rational double points,* see e.g. [Reid87] for a definition). Work out the analogs of (2.1–2) in this setting.

III.3 Del Pezzo Surfaces

In this section we study surfaces X such that $-K_X$ is ample. Theses surfaces have been studied extensively from geometric and arithmetic points of view. My aim is to develop the theory sufficiently to complete the proof of (2.3). The main result of the section is (3.5) which is also used in (IV.6). In positive characteristic it is necessary for the proof of (2.3). In characteristic zero it can be avoided using more analytic tools (3.7). For related developments see e.g. [Demazure80, Manin72].

3.1 Definition. A *Del Pezzo surface* (over a field k) is a smooth, geometrically irreducible and proper surface X such that $-K_X$ is ample.

The first main step toward their classification is the following:

3.2 Proposition. *Let X be a Del Pezzo surface over an algebraically closed field. Then the general member of $|-K_X|$ is irreducible and reduced.*

3.2.1 Lemma. *If X is a Del Pezzo surface, then $h^1(X, \mathcal{O}_X) = h^2(X, \mathcal{O}_X) = 0$ and $\chi(X, \mathcal{O}_X) = 1$.*

Proof. $h^2(X, \mathcal{O}_X) = h^0(X, K_X) = 0$ since $-K_X$ is ample. By Serre duality $h^1(X, \mathcal{O}_X) = h^1(X, K_X)$.

In characteristic zero the Kodaira vanishing theorem implies that $h^1(X, K_X) = 0$ since $-K_X$ is ample.

We saw in (II.6.4) that $h^1(X, K_X) = 0$, at least if char $k \neq 2, 3$. The char $= 3$ case is treated in (II.6.8.1) Unfortunately I do not see what to do with the last case, especially if $(K_X^2) = 1$.

Another argument, which works in all cases, uses the Albanese morphism (cf.[Grothendieck62, 236]). If $h^1(X, \mathcal{O}_X) > 0$ and $h^2(X, \mathcal{O}_X) = 0$, then there is a nonconstant morphism alb : $X \to \mathrm{Alb}(X)$ where $\mathrm{Alb}(X)$ is an Abelian variety (see e.g. [Mumford66, lecture 27] for a complete proof of the main part). Let H be ample on $\mathrm{Alb}(X)$. There is a curve $D \subset X$ such that $\mathrm{alb}^* H \cdot D > 0$, hence by (V.1.6) there is such a rational curve $C \subset X$ as well. This gives a nonconstant morphism $\mathbb{P}^1 \to C \to \mathrm{Alb}(X)$ which is impossible by (VI.1.9.4). □

3.2.2 Lemma. *Let X be a Del Pezzo surface and $m \geq 0$. Then*
(3.2.2.1) $h^2(X, \mathcal{O}(-mK_X)) = 0$; *and*
(3.2.2.2) $h^0(X, \mathcal{O}(-mK_X)) \geq \chi(\mathcal{O}(-mK_X)) = \frac{m(m+1)}{2}(K_X^2) + 1 \geq 2$.

Proof. $h^2(X, \mathcal{O}(-mK_X)) = h^0(X, \mathcal{O}((m+1)K_X)) = 0$ for $m \geq 0$ since $-K_X$ is ample. Thus

$$h^0(X, \mathcal{O}(-mK_X)) \geq \chi(\mathcal{O}(-mK_X)) = \frac{m(m+1)}{2}(K_X^2) + \chi(\mathcal{O}_X). \qquad \square$$

3.2.3 Lemma. *If X is a Del Pezzo surface and $\sum a_i C_i \in |-K_X|$ is not irreducible and reduced, then every C_i is a smooth rational curve.*

Proof. By adjunction,

$$\begin{aligned} 2p_a(C_i) - 2 = C_i \cdot (C_i + K_X) &= C_i \cdot (C_i + \frac{1}{a_i} K_X + \frac{a_i - 1}{a_i} K_X) \\ &= -C_i \cdot \left(\sum_{j \neq i} \frac{a_j}{a_i} C_j \right) - \frac{a_i - 1}{a_i} C_i \cdot (-K_X). \end{aligned}$$

If $\sum a_i C_i$ is reducible, then the first summand is strictly negative (since $\sum C_i$ is connected), if $a_i \geq 2$, then the second. □

3.2.4 Proof of (3.2). Assume the contrary and let $C_1, \ldots, C_k$ be the moving parts of $|-K_X|$. If D is any irreducible curve on a surface S and $(D^2) \geq -1$, then from the sequence

$$H^0(S, \mathcal{O}_S) \to H^0(S, \mathcal{O}_S(D)) \to H^0(D, \mathcal{O}_D(D))$$

we obtain that $h^0(S, \mathcal{O}_S(D)) \leq (D^2) + 2$. If D is smooth and rational, then from adjunction $(D^2) + 2 = -K_X \cdot D$. Applying this to our situation we obtain that

$$h^0(X, \mathcal{O}(-K_X)) \le \sum h^0(X, \mathcal{O}(C_i)) \le \sum (2 + C_i^2)$$
$$= -K_X \cdot \sum C_i \le (-K_X)^2.$$

This contradicts (3.2.2.2). □

3.2.5 Corollary. *Let X be a Del Pezzo surface. Then*
(3.2.5.1) $h^1(X, \mathcal{O}(-mK_X)) = 0$ *for* $m \ge 0$, *and*
(3.2.5.2) $h^0(X, \mathcal{O}(-mK_X)) = \frac{m(m+1)}{2}(K_X^2) + 1$ *for* $m \ge 0$.

Proof. Both statements are stable under field extension, thus we may assume that everything is over an algebraically closed field. Let $C \in |-K_X|$ be the general member. By (3.2) and adjunction C is an irreducible and reduced curve of genus one. From the sequence

$$H^1(X, \mathcal{O}_X(-mK_X)) \to H^1(X, \mathcal{O}_X(-(m+1)K_X))$$
$$\to H^1(C, \mathcal{O}_C(-(m+1)K_X)) = 0$$

we can inductively prove (3.2.5.1) starting with the $m = 0$ case (which is (3.2.1)). The second part follows by Riemann-Roch as in (3.2.2.2). □

3.3 Definition. Let Y be a scheme and L a line bundle. The graded ring $\sum_{j \ge 0} H^0(Y, L^j)$ is denoted by $R(Y, L)$. If this ring is used, then $H^0(Y, L^m)$ is also denoted by $R_m(Y, L)$ (or simply R_m if no confusion is likely).

$R(X, K_X)$ is frequently called the *canonical ring* and $R(X, -K_X)$ the *anticanonical ring.*

For ease of notation define a function by the rules $\alpha(1) = 3$, $\alpha(2) = 2$ and $\alpha(m) = 1$ for $m \ge 3$.

3.4 Proposition. *Let X be a Del Pezzo surface. Then*
(3.4.1) $|-mK_X|$ *is free if* $m(K_X^2) \ge 2$;
(3.4.2) R_j *for* $j \le \alpha(K_X^2)$ *generate* $R(X, -K_X)$;
(3.4.3) $|-K_X|$ *is very ample if* $(K_X^2) \ge 3$.

Proof. As before, we may assume that everything is over an algebraically closed field. Let $C \in |-K_X|$ be irreducible and reduced. By (3.2.5.1) we have exact sequences

$$0 \to H^0(X, \mathcal{O}_X(-mK_X)) \xrightarrow{i_m} H^0(X, \mathcal{O}_X(-(m+1)K_X))$$
$$\longrightarrow H^0(C, \mathcal{O}_C(-(m+1)K_X)) \to 0$$

Thus (3.4.1) follows from (3.4.4.2).

(3.4.2) is proved by induction on m. Let $Q = \sum_j R_j$ where $j \le \alpha(K_X^2)$. By assumption Q generates $H^0(X, \mathcal{O}(-mK_X))$, hence also its image under i_m. By (3.4.4.4) $Q|C$ generates $H^0(C, \mathcal{O}_C(-(m+1)K_X))$, which shows (3.4.2).

Finally $|-mK_X|$ is very ample for $m \gg 1$, thus (3.4.2) implies (3.4.3). □

We have used the following well known results. (3.4.4.4) can be proved with the above method (by reducing it to the zero dimensional case).

3.4.4 Exercise. Let C be an irreducible and reduced curve of genus one over an algebraically closed field, L a line bundle on C. Then

(3.4.4.1) $h^1(C,L) = 0$ if $\deg L \geq 1$;

(3.4.4.2) L is generated by global sections if $\deg L \geq 2$;

(3.4.4.3) L is very ample if $\deg L \geq 3$;

(3.4.4.4) R_j for $j \leq \alpha(\deg L)$ generate $R(C,L)$.

The main theorem of this section describes Del Pezzo surfaces with $(K_X^2) \leq 4$ over an arbitrary field. In some sense the others are not very interesting from the arithmetic point of view (cf. [Manin72]). (See (V.1.3) for basic properties of weighted projective spaces.)

3.5 Theorem. *Let X be a Del Pezzo surface over a field k. Assume that $(K_X^2) \leq 4$. Then $X \cong \operatorname{Proj}(\sum_{m\geq 0} H^0(X, \mathcal{O}(-mK_X)))$ can be described as follows:*

(3.5.1) *if $(K_X^2) = 1$, then $X \cong X_6 \subset \mathbb{P}(1,1,2,3)$;*

(3.5.2) *if $(K_X^2) = 2$, then $X \cong X_4 \subset \mathbb{P}(1,1,1,2)$;*

(3.5.3) *if $(K_X^2) = 3$, then $X \cong X_3 \subset \mathbb{P}^3$;*

(3.5.4) *if $(K_X^2) = 4$, then $X \cong X_{2,2} \subset \mathbb{P}^4$.*

Conversely, any smooth weighted complete intersection as above is a Del Pezzo surface with the expected (K_X^2).

Proof. If $(K_X^2) = 3$ (resp. 4), then $|-K_X|$ is very ample and gives an embedding $X \to \mathbb{P}^3$ (resp. $\mathbb{P}^4$) whose image has degree 3 (resp. 4). If $(K_X^2) = 3$, then the image must be a cubic surface. If $(K_X^2) = 4$, then from (3.2.5.2) we compute that the image is contained in a pencil of quadrics. Therefore it is a complete intersection of any two of them.

If $(K_X^2) = 2$, then $|-K_X|$ is base point free and gives a morphism $X \to \mathbb{P}^2$. Thus $H^0(\mathcal{O}(-K_X))$ generates a six dimensional subspace of $H^0(\mathcal{O}(-2K_X))$. $h^0(\mathcal{O}(-2K_X)) = 7$, hence $H^0(\mathcal{O}(-K_X))$ and one element of $H^0(\mathcal{O}(-2K_X))$ generate the anticanonical ring. Therefore X is isomorphic to a hypersurface X' in $\mathbb{P}(1,1,1,2)$. The selfintersection of $\mathcal{O}_{\mathbb{P}(1,1,1,2)}(1)$ is $1/2$, thus X' is given by a degree 4 polynomial.

The arguments for the case $(K_X^2) = 1$ are similar. By computing $h^0(\mathcal{O}(-jK_X))$ for $j = 1,2,3$ we see that the anticanonical ring is generated by $H^0(\mathcal{O}(-K_X))$ and one element each of $H^0(\mathcal{O}(-jK_X))$ for $j = 2,3$. Thus X is isomorphic to a hypersurface X' in $\mathbb{P}(1,1,2,3)$. The selfintersection of $\mathcal{O}_{\mathbb{P}(1,1,2,3)}(1)$ is $1/6$, thus X' is given by a degree 6 polynomial. □

For our purposes the most important consequence is the following:

3.6 Corollary. *Let X be a Del Pezzo surface over an algebraically closed field. Assume that $(K_X^2) \leq 4$. Then X contains a smooth rational curve with selfintersection -1.*

Proof. Let $\mathbb{P}_{\mathbb{Z}}$ stand for any of the weighted projective spaces over $\operatorname{Spec}\mathbb{Z}$ occurring in the statement of (3.5). Let $X \to W$ be the universal family of smooth weighted complete intersections with degree as specified in (3.5). W is irreducible and $\mathcal{O}_{\mathbb{P}}(1)$ gives an ample line bundle on X/W.

3.6.1 Lemma. *Let Y be a Del Pezzo surface and $C \subset Y$ a Cartier divisor such that $-K_X \cdot C = 1$ and $\chi(\mathcal{O}_C) = 1$. Then C is a smooth rational curve with selfintersection -1.*

Proof. Since $-K_X$ is ample, C is irreducible and reduced. (C^2) can be computed from adjunction. □

Let $Z \subset \operatorname{CDiv}(X/W)$ be the family parametrizing Cartier divisors $C \subset X_w$ such that $-K_{X_w} \cdot C = \mathcal{O}_{\mathbb{P}}(1) \cdot C = 1$ and $\chi(\mathcal{O}_C) = 1$. $Z \to W$ is proper by (I.1.13) and smooth of reative dimension zero by (I.2.10). Thus $Z \to W$ is surjective if Z is not empty. Thus all we need is to find for every value of (K_X^2) at least one Del Pezzo surface which contains a line.

If $(K_X^2) \leq 3$, then let $(x_0 : x_1 : x_2 : x_3)$ be the coordinates on $\mathbb{P}$ in the order that they appear in (3.5). For $(K_X^2) = 4$ we take $(x_0 : x_1 : x_2 : x_3 : x_4)$. Then $L = (x_2 = x_3 = 0)$ resp. $L = (x_2 = x_3 = x_4 = 0)$ is a line in $\mathbb{P}$. In characteristic zero we can use Bertini to conclude that the generic complete intersection containing L is smooth.

It is also easy to write down numerical examples. □

The following theorem was used in Sect. 2:

3.7 Theorem. *Let X be a Del Pezzo surface over an algebraically closed field k. Assume that $\rho(X) = 1$. Then $X \cong \mathbb{P}^2$.*

Proof. Let H be an ample generator of $N^1(X)_{\mathbb{Z}}$/torsion and write $-K_X \equiv rH$. By Riemann–Roch and (3.2.1)

$$h^0(X, \mathcal{O}(H)) \geq \chi(\mathcal{O}(H)) = \frac{1+r}{2}(H^2) + 1 \geq 2.$$

Since H is a generator, every $C \in |H|$ is irreducible and reduced.

Assume first that $r > 1$. By adjunction

$$2g(C) - 2 = C \cdot (C + K_X) = (1 - r)(H^2) < 0.$$

Thus $g(C) = 0$ and either $r = 2, (H^2) = 2$ or $r = 3, (H^2) = 1$. From the sequence

$$0 \to \mathcal{O}_X \to \mathcal{O}_X(C) \to \mathcal{O}_C(C) \to 0$$

we see that $|H|$ is free. Therefore $|H|$ gives a finite morphism $\phi : X \to \mathbb{P}^k$ ($k = (H^2) + 1$) such that $(H^2) = \deg \phi \cdot \deg \operatorname{im} \phi$.

If $(H^2) = 1$, then ϕ has degree one, hence $X \cong \mathbb{P}^2$. If $(H^2) = 2$, then $\deg \operatorname{im} \phi \geq 2$ since $\operatorname{im} \phi$ spans $\mathbb{P}^3$. Thus $\deg \phi = 1$ and X is isomorphic to a smooth quadric $Q \subset \mathbb{P}^3$. $\rho(Q) = 2$, so this does not occur.

Next consider the case $r = 1$. By (II.5.14) there is a morphism $f : \mathbb{P}^1 \to X$ such that $-K_X \cdot f(\mathbb{P}^1) \leq 3$. By assumption $f(\mathbb{P}^1) \equiv -mK_X$ for some $m \geq 1$, thus

$$(K_X^2) \leq m(K_X^2) = -K_X \cdot f(\mathbb{P}^1) \leq 3.$$

By (3.6) X contains a line, which contradicts the assumption $\rho(X) = 1$. Thus $r = 1$ is impossible.

Over $\mathbb{C}$ one can avoid (3.6) as follows. From $h^1(X, \mathcal{O}_X) = 0$ we see that $b_1(X) = 0$. From $h^2(X, \mathcal{O}_X) = 0$ and $\rho(X) = 1$ we obtain that $b_2(X) = 1$. Thus $c_2(X) = 3$. From the Noether formula $(K_X^2) = 12\chi(\mathcal{O}_X) - c_2(X) = 9$. Thus $r = 1$ is impossible. □

The following exercises develop the theory of Del Pezzo surfaces further. These results are used in (IV.6).

3.8 Exercise. Let X be a smooth projective surface over an algebraically closed field k. Assume that $-K_X$ is nef. Then we have one of the following cases:

(3.8.1) $K_X \equiv 0$;

(3.8.2) $X \cong \mathbb{P}_{\mathbb{P}^1}(\mathcal{O} + \mathcal{O}(2))$ or $X \cong \mathbb{P}_E(F)$ where E is an elliptic curve and F a rank 2 vector bundle on E. (Classify the occurring vector bundles.)

(3.8.3) $X \cong \mathbb{P}^2$ or $X \cong \mathbb{P}^1 \times \mathbb{P}^1$;

(3.8.4) X is obtained from $\mathbb{P}^2$ by successively blowing up at most nine points.

3.9 Exercise. Let X be a smooth projective surface over an algebraically closed field. Assume that $-K_X$ is ample. Then we have one of the following cases:

(3.9.1) $X \cong \mathbb{P}^2$ or $X \cong \mathbb{P}^1 \times \mathbb{P}^1$;

(3.9.4) X is obtained from $\mathbb{P}^2$ by blowing up $9 - (K_X^2)$ points of $\mathbb{P}^2$.

3.10 Exercise. (3.10.1) Let X_k be a smooth surface over a field k. Let $\bar{k}$ be an algebraic closure of k and $C \subset X_{\bar{k}}$ a curve. Assume that

(3.10.1.1) $\operatorname{Hilb}(X_{\bar{k}})$ is smooth and zero dimensional at $[C]$, and

(3.10.1.2) C is invariant under $\operatorname{Gal}(\bar{k}/k)$.

Show that C is defined over k.

(3.10.2) Let X_k be a Del Pezzo surface over a field k. Assume that $X_{\bar{k}}$ is obtainable by blowing up one or two points in $\mathbb{P}^2_{\bar{k}}$. Then X_k is obtainable by blowing up one or two points in $\mathbb{P}^2_k$. In particular, X_k is k-rational.

3.11 Exercise. (3.11.1) $\mathrm{Aut}(\mathbb{P}^2_k) = PGL(3,k)$.

(3.11.2) The connected component of $\mathrm{Aut}(\mathbb{P}^1_k \times \mathbb{P}^1_k)$ is $PGL(2,k) \times PGL(2,k)$.

(3.11.3) Let $x_1, x_2, x_3 \in \mathbb{P}^2_k$ be three points not on a line. Let X_k be obtained from $\mathbb{P}^2_k$ by blowing up these points. Show that the connected component of $\mathrm{Aut}(X_k)$ is $GL(1,k) \times GL(1,k)$. X_k contains six smooth rational curves with selfintersection -1. Their complement is a principal homogeneous space under the connected component of $\mathrm{Aut}(X_k)$.

(3.11.4) Let X_k be a proper surface. Assume that there is an open subset $U \subset X_{\bar{k}}$ which is a homogeneous space over $\mathrm{Aut}(X_{\bar{k}})$. Show that U is defined over k and it is a homogeneous space over $\mathrm{Aut}(X_k)$.

3.12 Exercise. Let X_k be a Del Pezzo surface over a field k. Assume that the Brauer group of k is trivial (see [Serre79], especially X.6–7 for the definition and relevant facts).

(3.12.1) Assume that $(K^2_{X_k}) = 6, 9$ or $(K^2_{X_k}) = 8$ and $X_{\bar{k}} \cong \mathbb{P}^1_{\bar{k}} \times \mathbb{P}^1_{\bar{k}}$. Show that X_k is k-rational.

(3.12.2) Assume that $(K^2_{X_k}) \geq 6$. Show that X_k is k-rational.

3.13 Exercise (Enriques). Let X_k be a Del Pezzo surface over a field k such that $(K^2_{X_k}) = 5$. Then X_k has a k-point and is rational over k.

The first complete proof by [Swinnerton-Dyer72] is elementary but rather tricky. It has two parts. The first step produces two points P_1, P_2 (which may coincide) such that $P_1 + P_2$ is defined over k. The second step starts with P_1, P_2 and shows that X_k has a k-point and is rational over k.

Below I outline two different approaches. One is due to [Shepherd-Barron92], the other uses some results about projective resolutions of ideals. The steps (3.13.1–3–4) are the same, the only difference is in the two parts of (3.13.2).

(3.13.1) $|-K_{X_k}|$ embedds X_k into $\mathbb{P}^5_k$; let $C(X_k) \subset \mathbb{A}^6_k$ be the cone over the image. Show that $C(X_k)$ is Gorenstein and it is defined by 5 quadratic equations.

(3.13.2.1) Show that the five quadrics define a morphism $B_X\mathbb{P}^5 \to \mathbb{P}^4$ whose general fiber is a secant line of X. Every fiber is either a secant line or a plane intersecting X in a conic. Use this to find two points $P_1, P_2 = X \cap L$ (which may coincide) such that $P_1 + P_2$ is defined over k.

(3.13.2.2) Use [Buchsbaum-Eisenbud74] to conclude that there is a 5×5 skew-symmetric matrix $M = (m_{ij})$ whose entries are linear polynomials such that the Pfaffians of the five 4×4 skew symmetric minors of M give the 5 quadratic equations defining $C(X)$.

Let $L \subset \mathbb{P}^5_k$ be the line given by $m_{1,2} = \ldots = m_{1,5} = 0$. L intersects X_k in two points P_1, P_2 (which may coincide) and $P_1 + P_2$ is defined over k. (Discuss the case when L is not a line.)

(3.13.3) Let $E_k \subset X_k$ be the intersection of X_k with a hyperplane containg L. We distinguish two cases:

(3.13.3.1) $E_{\bar{k}}$ is not irreducible and reduced. By (3.2.3) every irreducible component is a smooth rational curve. Use $P_1 + P_2$ to show that one of the irreducible components is defined over k and has odd degree in $\mathbb{P}^5_k$. Show that any smooth rational curve of odd degree in $\mathbb{P}^m_k$ is k-rational. This gives a k-rational point in X_k.

(3.13.3.2) $E_{\bar{k}}$ is irreducible and reduced. The restriction $\mathcal{O}_E(1)$ of $\mathcal{O}_{\mathbb{P}^5}(1)$ to E gives a degree 5 line bundle on E. $\mathcal{O}_E(1)(-2P_1 - 2P_2)$ is a line bundle of degree 1 on E thus it is isomorphic to $\mathcal{O}_E(Q)$ for some $Q \in E$. Q is a k-point on E and on X_k.

(3.13.4) Project X_k from L. We obtain a cubic surface $X'_k \subset \mathbb{P}^3$. If P_1, P_2 are in general position, then X'_k contains a pair of skew lines defined over k, hence it is rational over k. Study the degenerate cases.

3.14 Remark. Let X_k be a Del Pezzo surface with a k-point. [Manin72, IV.7.8] shows that X_k is k-rational if $(K^2_{X_k}) \geq 5$. The proof is selfcontained and geometric.

Chapter IV. Rationally Connected Varieties

This is the key chapter of the book. Its aim is to study the birational properties of those varieties which are covered by rational curves. It is reasonable to expect that many of their properties can be detected by studying the rational curves on them.

In surface theory the two relevant classes are ruled and rational surfaces. Both of these capture a very important feature of a surface. The usefulness of these properties in higher dimension is less clear. The main problem is that varieties very similar to a ruled or rational variety fail to be ruled or rational. This led to the discovery of the concepts unirational and uniruled. The latter seems to be a very good notion, but the former is probably still too restrictive. The class of rationally connected varieties was introduced to remedy the situation.

Uniruled varieties are of interest for several reasons. For us the most important part is that these are the varieties for which Mori's program does not produce a minimal model, only a Fano fiber space. The real trouble is that these fiber spaces are not unique and one variety can lead to infinitely may different Fano fiber spaces. Also, it is very hard to decide which Fano fiber spaces are birational to each other. Therefore, at least for now, it seems reasonable to study birational properties of these varieties directly.

In positive characteristic various complications arise, which I ignore for the moment.

Roughly speaking, we say that a smooth and proper variety is uniruled (resp. rationally connected) if through any point there is a rational curve (resp. through any pair of points there is a rational curve). (See (1.1) and (3.2) for the precise definitions.) Both of these properties can be detected by a single rational curve on X and, at least conjecturally, both of these properties can be characterized by holomorphic tensors (1.11–12, 3.8).

The following table summarizes the most important parallels:

	uniruled	rationally connected
(1)	$\forall x \in X, \exists$ a rational curve such that $x \in C \subset X$	$\forall x_1, x_2 \in X, \exists$ a rational curve such that $x_1, x_2 \in C \subset X$
(2)	$\exists\, f : \mathbb{P}^1 \to X$ such that f^*T_X is semi positive	$\exists\, f : \mathbb{P}^1 \to X$ such that f^*T_X is ample
(3)	$H^0(X, \omega_X^m) = 0\ \forall m > 0$	$H^0(X, \Omega_X^{\otimes m}) = 0\ \forall m > 0$

It turns out that these properties are invariant under smooth deformations of the variety, at least in characteristic zero (1.10, 3.11).

For singular varieties, and especially in positive characteristic, the situation is more complicated. Property (2) of the table implies the others, but they are not equivalent any more. Many of the interesting features are studied in detail in Sect. 1 and 3.

The aim of Sect. 2 is to replace a family of rational curves covering a subvariety by another family which is as small as possible. The results are especially nice for families that cover a smooth variety (2.10).

Quotients of varieties by algebraic equivalence relations are studied in Sect. 4. Let X be an algebraic variety and $R \subset X \times X$ an equivalence relation. Assume first that R is an algebraic subvariety of $X \times X$. The quotient X/R exists as a set, but in general there is no algebraic morphism $q : X \to Z$ whose fibers are exactly the equivalence classes of R. The situation is even worse for equivalence relations generated by algebraic relations. It is somewhat surprising that under mild conditions there is an open subset $X^0 \subset X$ and a morphism $X^0 \to Z^0$ whose fibers are the equivalence classes of R on X^0. The main results of the section are (4.13, 4.16). These assert the existence of the quotient for open (resp. proper) equivalence relations and also give a measure of how complicated the equivalence classes can be.

The first application of the quotient theorems is in Sect. 5. We declare two points of a variety to be equivalent if they can be connected by a chain of rational curves. The resulting quotient map is called the maximal rationally connected fibration or MRC-fibration. This provides a conceptually pleasing way of writing any variety as a fiber space where the fibers are rationally connected and most of the rational curves on the variety are in the fibers. Conjecturally, the target of the MRC-fibration is not even uniruled (5.6).

This conjecture can be formulated as a statement about families of rationally connected varieties over $\mathbb{P}^1$. Equivalently, it is about rationally connected varieties defined over the function field $\mathbb{C}(t)$. In this form it has been extensively studied and proved in dimensions 1 and 2. These questions are considered in Sect. 6. The main result is (6.8) which says that if S is a smooth surface over $\mathbb{C}(t)$ which is rational over the algebraic closure of $\mathbb{C}(t)$, then S has a $\mathbb{C}(t)$-valued point. (6.10) shows that S has lots of points in $\mathbb{C}(t)$. Both of these results apply more generally to function fields of curves.

IV.1 Ruled and Uniruled Varieties

If X is an arbitrary variety, then $X \times \mathbb{P}^1$ is a variety with lots of rational curves. If $X \times \mathbb{P}^1 \to Y$ is dominant and generically finite, then Y again has lots of rational curves. The aim of this section is to study the properties of such varieties.

The definitions and elementary properties are covered in (1.1–5). There seem to be at least three versions of the definition that are of importance. The behaviour of these notions in families of varieties is discussed in (1.6–8). All of these results go back to [Matsusaka68]. It turns out that a smooth and proper variety X is separably uniruled iff there exists a free morphism $\mathbb{P}^1 \to X$ (1.9). This implies that for smooth and proper varieties in characteristic zero, being uniruled is a deformation invariant property (1.10). Also, if X is separably uniruled, then $H^0(X, K_X^m) = 0$ for every $m > 0$ (1.11). It is conjectured that the converse also holds (at least in characteristic zero). This is one of the hardest problems in higher dimensional geometry. Further characterizations of uniruled varieties are considered in (1.13–16).

1.1 Definition. (1.1.1) Let $f : X \to S$ be an irreducible and reduced scheme of dimension n over S. We say that X is *ruled* (resp. *uniruled*, resp. *separably uniruled*) *over* S if there is a scheme Y of dimension $n-1$ over S and a map

$$\mathbb{P}^1 \times Y \dashrightarrow X/S$$

which is birational (resp. dominant, resp. dominant and separable). It is clear that X is ruled (resp. uniruled, resp. separably uniruled) over S iff the fiber of f over the generic point of $f(X)$ is ruled (resp. uniruled, resp. separably uniruled) over S.

If X/S is proper, then by shrinking Y we may assume that $\mathbb{P}^1 \times Y \dashrightarrow X/S$ is a morphism. This is frequently convenient.

(1.1.2) A scheme Z over S is said to have *ruled* (resp. *uniruled*, resp. *separably uniruled*) *components over* S if every irreducible component of red Z is ruled (resp. uniruled, resp. separably uniruled) over S.

(1.1.3) Let X be a variety. A map $g : \mathbb{P}^1 \times Y \dashrightarrow X$ is called a *degree d uniruling* of X if g is dominant, generically finite and $d = \deg g$. In this case we also say that X is *degree d uniruled*. (This notion should not be confused with the one to be introduced in (1.4).)

(1.1.4) Let X_k be a variety over a field k and $\bar{k}$ an algebraic closure of k. We say that X_k is *geometrically* ruled if $X_{\bar{k}}$ is ruled. Similarly for geometrically uniruled, etc.

1.1.5 Note on Terminology. In the classical literature a rational variety is called "birational": there is a correspondence $\mathbb{P}^n \leftrightarrow X$ which is a rational map in both directions. Thus "unirational" indicated that the correspondence is a rational map in one direction only. To me the term "uniruled" does not make any sense, but it is generally accepted.

1.1.6 Exercises. (1.1.6.1) In characteristic zero uniruled is equivalent to separably uniruled.

(1.1.6.2) If $\dim X \leq 2$ and $S = \operatorname{Spec} k$ for an algebraically closed field of characteristic zero, then uniruled is equivalent to ruled.

(1.1.6.3) Let $S = \operatorname{Spec} k[u, v]$ and $X = (x_0^2 + ux_1^2 + vx_2^2 = 0) \subset \mathbb{P}^2 \times S$. Then X is uniruled over S but it is not ruled over S.

(1.1.6.4) Let $C_{\mathbb{R}}$ be the conic $(\sum x_i^2 = -1) \subset \mathbb{P}^2_{\mathbb{R}}$. Show that $C_{\mathbb{R}}$ is uniruled but not ruled. Show that $C_{\mathbb{R}}$ is geometrically ruled.

(1.1.6.5) Let char $= p$. $(z^p - f(x, y) = 0) \subset \mathbb{A}^3$ is uniruled but usually not separably uniruled. (The last assertion follows easily from (1.11).)

1.2 Lemma. *Assume that X is irreducible and uniruled (resp. separably uniruled) over S. Let $f : X/S \dashrightarrow Y/S$ be a generically finite map which is dominant (resp. separable and dominant). Then Y is also uniruled (resp. separably uniruled) over S.* □

There are other ways of approaching uniruledness through families of rational curves. The next result shows that they are mostly equivalent.

1.3 Proposition. *Let X be a variety over a field k. The following are equivalent:*

(1.3.1) *X is uniruled over k;*

(1.3.2) *X is geometrically uniruled;*

(1.3.3) *There is a variety Z over k and a dominant map $\mathbb{P}^1 \times Z \dashrightarrow X$ such that for some $z \in Z$ the induced map $\mathbb{P}^1_z \dashrightarrow X$ is nonconstant.*

(1.3.4) *There are varieties Z, U over k, a k-morphism $g : U \to Z$ whose fibers are rational curves and a dominant map $p : U \dashrightarrow X$ such that for some $z \in Z$ the induced map $g^{-1}(z) \dashrightarrow X$ is nonconstant.*

(1.3.5) *There is an uncountable algebraically closed field $K \supset k$ and a dense open subset $V \subset X_K$ such that every K-point $v \in V$ is contained in a K-rational curve $v \in C_v \subset X_K$.*

(1.3.6) *There is a k-irreducible component $W' \subset \operatorname{RatCurves}^n(X)$ such that the cycle map $\operatorname{Univ}_{W'} \to X$ is dominant.*

(1.3.7) *There is a closed irreducible k-subvariety $W \subset \operatorname{Chow}(X)$ parametrizing 1-cycles with rational components such that the cycle map $\operatorname{Univ}_W \to X$ is dominant.*

Proof. Being uniruled is a birational property, thus we may assume that X is projective. Clearly (1.3.1) implies (1.3.2).

Assume (1.3.2). Let

$$p : \mathbb{P}^1_k \times \mathrm{Hom}(\mathbb{P}^1_k, X) \to X$$

be the universal morphism. By assumption there is a field $K \supset k$, a K-variety Y and a dominant, generically finite morphism $\mathbb{P}^1 \times Y \to X_K$. This corresponds to a morphism $b : Y \to \mathrm{Hom}(\mathbb{P}^1_k, X)_K$. Let $Z \subset \mathrm{Hom}(\mathbb{P}^1_k, X)$ be a k-irreducible component containing $\mathrm{im}\, b$. Clearly Z satisfies (1.3.3). Also, (1.3.3) implies (1.3.4) by taking $U = \mathbb{P}^1 \times Z$.

Let $\mathbb{P}^1 \times Z \dashrightarrow X$ be as in (1.3.3). We may assume that Z is affine. Let $Y \subset Z$ be a general $(\dim X - 1)$-dimensional linear space section of Z. It is clear that the restriction $\mathbb{P}^1 \times Y \dashrightarrow X$ is dominant, thus (1.3.3) implies (1.3.1).

Assume (1.3.4). Let $V \subset \mathrm{im}(U_K)$ be an open and dense subset. For every $v \in V(K)$ there is a $z \in Z(K)$ such that $v \in p(g^{-1}(z))$. $p(g^{-1}(z))$ is K-rational, hence (1.3.5) holds.

Assume (1.3.5). By (II.2.11) $\mathrm{RatCurves}^n(X)$ has only countably many components. Let $\mathrm{Univ} \to \mathrm{RatCurves}^n(X)$ be the universal family and $u : \mathrm{Univ} \to X$ the cycle map. The image of u is the union of countably many constructible subsets of X. Thus there is an irreducible component $W' \subset \mathrm{RatCurves}^n(X)$ such that $u : \mathrm{Univ}_{W'} \to X$ is dominant, giving (1.3.6).

Assume (1.3.6) and let $W \subset \mathrm{Chow}(X)$ be the closure of the image of $W' \to \mathrm{Chow}(X)$. This shows (1.3.7).

Finally assume (1.3.7). The geometric generic fiber of $\mathrm{Univ}_W \to W$ is a curve with $m \geq 1$ rational components, thus there is a generically finite morphism $Z \to W$ such that $\mathrm{Univ}_W \times_W Z$ is birational to a union of m copies of $\mathbb{P}^1 \times Z$. This shows (1.3.3). □

1.3.8 Remark. Let X be a variety over a countable algebraically closed field K. Assume that every K-point of X is contained in a rational curve in X. It is not known whether X is uniruled or not. The simplest example to investigate is degree four surfaces $X \subset \mathbb{P}^3$ over the closure of the prime field.

It is convenient to have a version of (1.3) which keeps track of the degrees of the occurring rational curves. This is automatic in the above proofs. We state only the three most important equivalent conditions:

1.4 Proposition – Definition. *Let X be a projective variety over an algebraically closed field k, M a nef $\mathbb{R}$-divisor on X and d a real number. We say that X is uniruled with curves of M-degree at most d if one of the following equivalent conditions is satisfied:*

(1.4.1) *There is a variety of dimension $n-1$ and a dominant morphism $f : \mathbb{P}^1 \times Y \to X$ such that $\deg f^*M|\mathbb{P}^1 \times \{y\} \leq d$ for every $y \in Y$.*

(1.4.2) *There is an uncountable algebraically closed field $K \supset k$ and a dense open subset $V \subset X_K$ such that every K-point $v \in V$ is contained in a K-rational curve $v \in C_v \subset X_K$ such that $M \cdot C_v \leq d$.*

(1.4.3) *There is a closed irreducible k-subvariety $W \subset \mathrm{Chow}(X)$ parametrizing 1-cycles with rational components of M-degree at most d such that the cycle map $\mathrm{Univ}_W \to X$ is dominant.* □

1.4.4 Corollary. *Let X be a proper uniruled variety over an algebraically closed field K. Then every K-point $x \in X$ is contained in a K-rational curve $x \in C_x \subset X$.*

If X is uniruled with curves of M-degree at most d, then we can find C_x such that $M \cdot C_x \leq d$.

Proof. Let $\mathrm{Univ}_W \to W$ be as in (1.3.7) (resp. (1.4.3)). Univ_W is proper and $\mathrm{Univ}_W \to X$ is dominant. Thus for every $x \in X(K)$ there is a $w \in W(K)$ such that x is contained in an irreducible component of $p(g^{-1}(z))$. Every irreducible component of $p(g^{-1}(z))$ is rational by (II.2.2). □

(1.4.4) can be generalized to schemes:

1.5 Proposition. *Assume that X has uniruled components over S. Let F be a field and $f : \mathrm{Spec}\, F \to S$ a morphism. Let $x \in X_F = \mathrm{Spec}\, F \times_S X$ be a point over F. Then there is a field extension $L \supset F$ and a rational curve $x \in C_x \subset X_L$.*

Proof. We may assume that X is irreducible. Let $p : \mathbb{P}^1 \times Y \to X$ show that X is uniruled. Let T be the spectrum of a DVR and $h : T \to X$ a morphism such that there are factorizations:

$$h_g : T_g \to \mathbb{P}^1 \times Y \xrightarrow{p} X \quad \text{and} \quad h_0 : T_0 \to \{x\} \hookrightarrow X_F \to X.$$

Thus we obtain a morphism

$$u_g : \mathbb{P}^1 \times T_g \to X \quad \text{such that} \quad \mathrm{im}\, h_g \in \mathrm{im}\, u_g.$$

Let $\Gamma \subset \mathbb{P}^1 \times T' \times_S X$ be the closure of the graph of u_g. The second projection $\Gamma \to X$ is proper and its image contains $h(T_g)$, therefore $h(T_0) \in \mathrm{im}\, \Gamma_0$. By (II.2.2) every component of Γ_0 is a rational curve. The image of one of these contains x. □

1.5.1 Corollary. *Let R be a DVR with quotient field K and residue field k, $T = \mathrm{Spec}\, R$. Let $f : X \to T$ be a morphism, X normal and irreducible. If X_K is uniruled, then X_k has uniruled components.* □

It is somewhat surprising that the analogous result for ruled varieties is almost true:

1.6 Theorem [Matsusaka68]. *Let R be a DVR with quotient field K and residue field k, $T = \mathrm{Spec}\, R$. Let $f : X \to T$ be a morphism, X normal and irreducible.*

(1.6.1) *If X_K is ruled over K, then X_k has ruled components over k.*

(1.6.2) *If X_K is geometrically ruled, then every reduced irreducible component of X_k is geometrically ruled.*

Proof. By assumption there is a variety Y_K and a birational map $g_K : X_K \dashrightarrow \mathbb{P}^1_K \times Y_K$. Let $Y_T \to T$ be a proper morphism extending $Y_K \to \operatorname{Spec} K$, Y_T normal and irreducible. There is a birational map

$$g_T : X_T \dashrightarrow \mathbb{P}^1_T \times Y_T.$$

Let $X' \subset X_k$ be an irreducible component and $x' \in X'$ the generic point. Let $p : \mathbb{P}^1_T \times Y_T \to Y_T$ be the second projection.

$p \circ g_T$ is defined at x'. By (VI.1.3) there is a proper birational morphism $h : Y'_T \to Y_T$ such that $y' := (h^{-1} \circ p \circ g_T)(x')$ is a codimension one point of Y'_T. By lifting g_T we obtain

$$X_T \overset{g'_T}{\dashrightarrow} \mathbb{P}^1_T \times Y'_T \overset{p'}{\longrightarrow} Y'_T.$$

Consider $z' := g'_T(x')$. If z' is a codimension one point of $\mathbb{P}^1_T \times Y'_T$, then $\overline{z'} = \mathbb{P}^1_k \times \overline{y'}$, thus $\overline{z'}$ is ruled hence so is $X' = \overline{x'}$.

Otherwise z' has codimension at least 2 and X' is an exceptional divisor of g'_T. By construction z' dominates y' and Y'_T is smooth at y' since y' is a codimension one point. Thus $\mathbb{P}^1_T \times Y'_T$ is smooth at z'. By (VI.1.2) this implies that X' is ruled over k.

The second part can be reduced to the first one. There is a finite extension $T' \to T$ with quotient field K' and residue field k' such that $X_{K'}$ is ruled over K'. Let X' be the normalization of $X \times_T T'$. By the first part, $X'_{k'}$ has ruled components.

The projection map $X'_{k'} \to X_k \times_{\operatorname{Spec} k} \operatorname{Spec} k'$ is birational over the components with multiplicity one, hence these components are geometrically ruled. □

1.6.3 Example. Let $Z \subset \mathbb{P}^4$ be a smooth cubic threefold. By [Clemens-Griffiths72] X is not rational and not even ruled by (3.12.4), but there is a (rational) involution τ on $\mathbb{P}^3$ such that $Z \sim \mathbb{P}^3/\tau$. Let X be the quotient of $\mathbb{P}^3 \times \mathbb{A}^1$ by the involution $(x,t) \mapsto (\tau(x), -t)$ and $f : X \to \mathbb{A}^1$ the morphism $f(x,t) = t^2$. The generic fiber of f is geometrically rational. The closed fiber has multiplicty 2 and its reduced part is birational to Z. Thus it is not geometrically ruled.

As an application and generalization we obtain the following result about specializations of degree d uniruled varieties:

1.7 Theorem. *Let R be a DVR with quotient field K and residue field k, $T = \operatorname{Spec} R$. Let $f : X \to T$ be a morphism, X normal and irreducible.*

(1.7.1) *Let* $g : \mathbb{P}^1_K \times Y_K \dashrightarrow X_K$ *be a degree* d *uniruling of* X_K *and* $X' \subset X_k$ *any irreducible component. Then there are degree* d_i *unirulings* $g_i : \mathbb{P}^1_k \times Y_i \dashrightarrow X'$ *and natural numbers* e_i *such that* $d = \sum d_i e_i$.

(1.7.2) *Let* $g : \mathbb{P}^1_{\bar{K}} \times Y_{\bar{K}} \dashrightarrow X_{\bar{K}}$ *be a degree* d *uniruling of* $X_{\bar{K}}$. *Let* $X' \subset X_{\bar{k}}$ *be any reduced irreducible component. Then there are degree* d_i *unirulings* $g_i : \mathbb{P}^1_{\bar{k}} \times Y_i \dashrightarrow X'$ *and natural numbers* e_i *such that* $d = \sum d_i e_i$.

Proof. Let us consider the first case. Extend Y_K to a normal scheme Y_T, proper over T and let $Z_T \to X_T \times (\mathbb{P}^1_T \times Y_T)$ be the normalization of the closure of the graph of g. Z_K is birational to $\mathbb{P}^1_K \times Y_K$, hence ruled. Therefore by (1.6.1) every irreducible component of Z_k is also ruled.

Let $Z_i \subset Z_k$ be those irreducible components which dominate X'. Let $d_i = \deg(Z_i/X')$ and e_i the ramification index of $Z_T \to X_T$ at the generic point of Z_i. Then $\sum d_i e_i = \deg(Z_T/X_T) = d$.

The second part reduces to the first one after a suitable base change as in the proof of (1.6.2). □

The following definition enables us to transform (1.7) to an upper semi continuity statement.

1.7.3 Definition. Let X_k be a variety over a field k, $\bar{k} \supset k$ an algebraic closure. The following invariants contain information about all unirulings of a variety.

$$u(X) := \gcd\{d | \exists \text{ degree } d \text{ uniruling of } X_k\},$$
$$\bar{u}(X) := \gcd\{d | \exists \text{ degree } d \text{ uniruling of } X_{\bar{k}}\}.$$

1.7.4 Corollary. *Let* R *be a DVR with quotient field* K *and residue field* k, $T = \operatorname{Spec} R$. *Let* $f : X \to T$ *be a morphism,* X *normal and irreducible. Let* $X' \subset X_k$ *any irreducible component. Then* $u(\operatorname{red} X') | u(X_K)$. *If* X' *is reduced, then also* $\bar{u}(X') | \bar{u}(X_K)$. □

The next result describes how uniruledness varies for fibers of morphisms.

1.8 Theorem. *Let* $f : X \to S$ *be an equidimensional proper morphism with irreducible fibers.*

(1.8.1) *There are countably many closed subvarieties* $U_i \subset S$ *such that* X_s *is uniruled iff* $s \in \cup U_i$.

(1.8.2) *Assume in addition that* S *is over a field of* char $= 0$. *There is a closed subvariety* $U \subset S$ *such that* X_s *is uniruled iff* $s \in U$.

(1.8.3) *Assume in addition that* f *is a well defined family of proper algebraic varieties such that every cycle theoretic fiber is geometrically irreducible and reduced (for instance,* f *is flat with geometrically irreducible and reduced fibers). There are countably many closed subvarieties* $R_i \subset S$ *such that* X_s *is geometrically ruled iff* $s \in \cup R_i$.

(1.8.4) *Assumptions as in* (1.8.3) *and let* m *be a natural number. There are countably many closed subvarieties* $M_i \subset S$ *such that* $\bar{u}(X_s)|m$ *iff* $s \in \cup M_i$.

Proof. In order to see (1.8.1) let $R/S \subset \mathrm{Chow}(X/S)$ be the subset parametrizing 1-cycles with rational components. R/S is proper by (II.2.2). Let $R/S = \cup T_i$ be the irreducible components and $g_i : C_i \to T_i$ the universal families. Let $Z_i \subset X$ be the image of $u_i : C_i \to X$ and

$$U_i = \{s \in S | Z_i \times \{s\} = X_s\}.$$

By upper semi continuity of fiber dimensions, U_i is closed. This shows (1.8.1).

(1.8.2) is proved by induction on $\dim S$. We may assume that S is irreducible and that $\cup U_i$ is dense in S. Let $X' \to X$ be a resolution of singularities, $f' : X' \to S$ the composition. There is an open $S^0 \subset S$ such that f' is smooth over S^0 and the fibers of f and f' are birational over S^0. By assumption there is an $s \in S^0$ such that $f^{-1}(s)$ is uniruled. By (1.10) every fiber of f' over S^0 is uniruled. Thus every fiber of $f : X \to S$ is uniruled.

In order to see (1.8.3) let H be the closure of $\mathrm{Hom}(\mathbb{P}^1_S, X)$ inside $\mathrm{Hilb}(\mathbb{P}^1_S \times X)$, and $V \subset \mathrm{Hilb}(H)$ the open subset parametrizing irreducible and reduced subvarieties which intersect $\mathrm{Hom}(\mathbb{P}^1_S, X)$ nontrivially. Let $U \to V$ and $\mathrm{Univ} \to \mathrm{Hilb}(\mathbb{P}^1_S \times X)$ be the universal families. The cycle map gives $u : U \to \mathrm{Hilb}(\mathbb{P}^1_S \times X)$ and set

$$P(U) := U \times_u \mathrm{Univ}.$$

Let $p : P(U) \to U \to V$ denote the projection. We have a commutative diagram

$$\begin{array}{ccc} P(U) & \xrightarrow{u} & X \times_S V \\ \downarrow & & \downarrow \\ V & = & V \end{array}$$

By (1.8.6) there is a locally closed subset $W \subset V$ such that u is birational exactly on the fibers over W. Let $W = \cup W_i$ be the irreducible components and $R_i \subset S$ the closure of $p(W_i)$. By (1.8.5) the geometrically ruled fibers of f are exatly those that lie over $p(W_i)$ for some i. By (1.7) every fiber over R_i is geometrically ruled.

The proof of (1.8.4) requires only slight changes. Assume that $\bar{u}(X_k)|m$. Then there are unirulings $g_i : Y_i \times \mathbb{P}^1 \dashrightarrow X_{\bar{k}}$ and $g_i' : Y_i' \times \mathbb{P}^1 \dashrightarrow X_{\bar{k}}$ such that $m = \sum \deg g_i - \sum \deg h_i$. For fixed values of $\deg g_i, \deg h_i$ we can encode the graphs of these morphisms as a point in a product of several copies of $\mathrm{Hilb}\big(P(U) \times (X \times_S V)\big)$.

The rest follows as before. □

1.8.5 Lemma. *Let* Z *be a variety. Then* Z *is ruled iff there is a subvariety* $Y' \subset \mathrm{Hom}(\mathbb{P}^1, Z)$ *such that the cycle map* $u : \mathbb{P}^1 \times Y' \to X$ *is birational.*

Proof. Let $g : \mathbb{P}^1 \times Y \dashrightarrow X$ be a birational map. g induces a map $c : Y \dashrightarrow \operatorname{Hom}(\mathbb{P}^1, Z)$, let $Y' \subset \operatorname{Hom}(\mathbb{P}^1, Z)$ denote its image. There is a factorization

$$g : \mathbb{P}^1 \times Y \overset{1\times c}{\dashrightarrow} \mathbb{P}^1 \times Y' \overset{g'}{\dashrightarrow} X.$$

Therefore g' is birational. □

1.8.6 Exercise. Let $f_i : X_i \to S$ be equidimensional proper morphisms with irreducible fibers and $g : X_1 \to X_2$ an S-morphism. Show that there is a locally closed subset $W \subset S$ such that $g_s : \operatorname{red} f_1^{-1}(s) \to \operatorname{red} f_2^{-1}(s)$ is birational iff $s \in W$.

1.8.7 Remarks. [Levine81b] gives an example as in (1.8.3) where it is necessary to have countably many subvarieties $R_i \subset S$.

(III.2.5.2) gives an example of a family of smooth projective surfaces over $\operatorname{Spec} \mathbb{Z}$ whose generic fiber is not uniruled but whose fiber over p is uniruled for $p \equiv -1 \bmod 4$.

The following result is the key to understanding smooth separably uniruled varieties:

1.9 Theorem. *Let X be a smooth proper variety over an algebraically closed field. Then X is separably uniruled iff there is a free morphism $f : \mathbb{P}^1 \to X$.*

Proof. If X is separably uniruled, then (II.3.10) guarantees the existence of a free morphism.

Conversely, let $f : \mathbb{P}^1 \to X$ be free. By replacing f with a generic deformation we may assume that f is separable (II.3.14.1). Consider

$$\bar{F} : \mathbb{P}^1 \times \operatorname{Hom}^{\text{free}}(\mathbb{P}^1, X) \to X.$$

By (II.3.5.4) $\bar{F}$ is smooth. Let $p \in \mathbb{P}^1$ be a point such that $df(p)$ is not zero. We can choose a subvariety $[f] \in Y \subset U$ of dimension $n-1$ such that Y is smooth at $[f]$ and the composite map

$$T_Y \otimes k([f]) \subset H^0(\mathbb{P}^1, f^*T_X) \to T_X \otimes k(x)$$

has rank $n-1$ and its image does not contain $df(p)$. $\mathbb{P}^1 \times Y$ is smooth at $(p, [f])$ and the restriction

$$F = \bar{F}|\mathbb{P}^1 \times Y : \mathbb{P}^1 \times Y \to X$$

has invertible differential at $(p, [f])$. Thus F is separable. □

1.10 Corollary [Fujiki81; Levine81a]. *Let $g : X \to S$ be a smooth and proper morphism, S connected. Assume that X_s is separably uniruled for some $s \in S$. Then there is an open neighborhood $s \in U \subset S$ such that X_u is separably uniruled for every $u \in U$.*

If $\operatorname{char} S = 0$, *then X_s is uniruled for every $s \in S$.*

Proof. Let us consider $\mathrm{Hom}(\mathbb{P}^1_S, X)$. By assumption there is an $[f] \in \mathrm{Hom}(\mathbb{P}^1_s, X_s)$ such that $f^*T_{X_s}$ is semi positive. By (I.2.17) this implies that $\pi : \mathrm{Hom}(\mathbb{P}^1_S, X) \to S$ is smooth at $[f]$. By (II.3.9.2.2) there is an open neighborhood $[f] \in V \subset \mathrm{Hom}(\mathbb{P}^1_S, X)$ such that $h^*T_{X/S}$ is semi positive for every $[h] \in V$. Since π is smooth at $[f]$, it is open, hence there is an open neighborhood $s \in U \subset \pi(V)$. By (1.9) X_u is separably uniruled for every $u \in U$.

In characteristic zero uniruled is the same as separably uniruled and being uniruled is a closed condition by (1.5.1). □

1.10.1 Comment. The char $S = 0$ assumption is probably necessary in the last claim. It is not hard to construct examples of separably uniruled varieties degenerating to one which is not separably uniruled, but I was unable to write down smooth examples.

1.11 Corollary. *Let X be a smooth, proper and separably uniruled variety. Then $H^0(X, K_X^m) = 0$ for every $m > 0$.*

Proof. Let $f : \mathbb{P}^1 \to X$ be a free morphism. By (II.3.14.1) $K_X \cdot_f \mathbb{P}^1 \le -2$, thus every section of K_X^m vanishes along $f(\mathbb{P}^1)$ for $m > 0$. The images of free morphisms are dense in X, hence every section of K_X^m is identically zero. □

The following partial converse of (1.11) is a very difficult open problem:

1.12 Conjecture. *Let X be a smooth, proper variety such that $H^0(X, K_X^m) = 0$ for every $m > 0$. Then X is uniruled.*

1.12.1 Comments. The result is true if $\dim X \le 3$ and char $X = 0$ by [Miyaoka88a]. The proof of Shepherd-Barron in [Kollár et al.92] is the simplest known.

It is not true that X is necessarily separably uniruled. An example is given in (1.13.5).

Asume that X is separably uniruled. By (1.9) for general $x \in X$ there is a free rational curve $x \in f_x(\mathbb{P}^1) \subset X$. Since f is free, $K_X \cdot f_x(\mathbb{P}^1) \le -2$. Conversely, assume that for every $x \in X$ there is a curve $x \in C_x \subset X$ such that $K_X \cdot C_x < 0$. It was hoped that this implies that X is separably uniruled, but this is unfortunately not quite true (1.13.5). The following result asserts that under the above conditions, X is at least uniruled.

1.13 Theorem [Miyaoka-Mori86]. *Let X be a projective variety over an uncountable algebraically closed field. Assume that for a general $x \in X$ there is a smooth proper curve C and a morphism $f : C \to X$ such that*

(1.13.1) $x \in f(C)$;

(1.13.2) *X is smooth along $f(C)$;*

(1.13.3) $\deg_C f^*K_X < 0$.

Then X is uniruled.

Proof. By (II.5.8) there is a rational curve $L_x \subset X$ containing x. Thus by (1.3.5) X is uniruled. □

1.13.4 Remark. It is not known whether the uncountability of the field is necessary in (1.13).

1.13.5 Exercise [Sato93]. Let char $= 2$. In $\mathbb{P}^2 \times \mathbb{P}^2$ with homogeneous coordinates $(x : y : z, u : v : w)$ consider the smooth hypersurface $X = (xu^2 + yv^2 + zw^2 = 0)$. Projection to the first factor makes it into a conic bundle and every fiber is a double line.

Using X construct a normal threefold Z which is a conic bundle over a nonuniruled smooth surface S such that the general fiber is a double line. Show that Z satisfies the assumptions of (1.13) and it is not separably uniruled.

Construct similar higher dimensional examples in higher characteristics.

Construct an example of a family of threefolds X_t such that X_t is separably uniruled for $t \neq 0$ but X_0 is not separably uniruled.

1.13.6 Exercise. Let X be a smooth and proper variety and $f : C \to X$ a nonconstant morphism from a smooth projective curve to X. Let $g : \mathbb{P}^1 \to X$ be a free morphism such that $\operatorname{im} g$ intersects $\operatorname{im} f$. Show that there is a morphism $f' : C \to X$ such that $K_X \cdot f'(C) < 0$. Hint: use (II.7.9).

1.14 Corollary. *Let X be a normal projective variety of dimension n. Assume that there is a nef $\mathbb{Q}$-Cartier Weil divisor L, an effective Weil divisor D and a natural number m such that $-mK_X \sim L + D$. Then either X is uniruled or $mK_X \sim 0$.*

Proof. The assumptions are unchanged under field extensions, thus we may assume that X is over an uncountable algebraically closed field. Let H be very ample on X. For a smooth point $x \in X - \operatorname{Supp} D$ let $x \in C \subset X$ be the intersection of $n-1$ general members of $|H|$. Since X is normal, C is smooth and $C \subset X - \operatorname{Sing} X$. If $\deg_C K_X < 0$, then X is uniruled by (1.13).

On $X - \operatorname{Sing} X$ all divisors are Cartier, thus

$$-\deg_C K_X = \frac{1}{m}\left(\deg_C L + \deg_C \mathcal{O}_X(D)\right).$$

Since H is ample, $H^{n-1} \cdot D \geq 0$ and the inequality is strict if D is nonempty. Thus $\deg_C K_X < 0$ unless $D = \emptyset$ and $\deg_C L = L \cdot H^{n-1} = 0$.

Assume that $L \cdot H^{n-1} = 0$. Let $B \subset X$ be an arbitrary irreducible curve. One can find $k > 0$ and $H_1, \ldots, H_{n-1} \in |kH|$ such that

$$H_1 \cap \ldots \cap H_{n-1} = B \cup B'$$

for some effective 1-cycle B'. Thus

$$0 = m^{(n-1)} L \cdot H^{n-1} = L \cdot (B + B') \geq L \cdot B \geq 0.$$

Thus $-mK_X \sim L$ is numerically trivial. □

In some cases one can get more information about the family of rational curves:

1.15 Corollary. *Let X be a normal projective variety of dimension n. Assume that $-K_X$ is $\mathbb{Q}$-Cartier, nef and not numerically trivial. Then X is uniruled with curves of $(-K_X)$-degree at most $2n$.*

If X is smooth, then X is uniruled with curves of $(-K_X)$-degree at most $n+1$.

Proof. We may assume that X is over an uncountable algebraically closed field. The proof of (1.14) shows that the conditions of (1.13) are satisfied. We apply (II.5.8) with $M = -K_X$. We obtain a rational curve $x \in L_x \subset X$ such that $-K_X \cdot L_x \leq 2n$.

If X is smooth, then by (II.5.7) the curve L_x can be replaced by another rational curve D_x such that $-K_X \cdot D_x \leq n+1$. Finally apply (1.4). □

1.15.1 Exercise. Let X be a normal projective variety and $E \subset X$ an irreducible Weil divisor. Assume that $-(E+K_X)$ is $\mathbb{Q}$-Cartier and $-(E+K_X)|E$ is nef and not numerically trivial. Show that E is uniruled.

Hint. First do the case when X is smooth in codimension two. In the general case use a minimal resolution of X in codimension two.

The following result is one of the most powerful uniruledness criteria. Unfortunately its proof requires techniques not discussed in this book. The proof given by Shepherd–Barron in [Kollár et al.92, 9] is the simplest known. This result should be viewed as a generalization of (1.13).

1.16 Theorem [Miyaoka87]. *Let X be a smooth and proper variety,* char $=$ 0. *Then X is uniruled iff there is a quotient sheaf $\Omega^1_X \twoheadrightarrow F$ and a family of curves $\{C_t\}$ covering an open subset of X such that $F|C_t$ is locally free and* $\deg(F|C_t) < 0$ *for every t.*

1.17 Exercises (Specialization of rational maps). Let R be a DVR with quotient field K and residue field k, $T = \operatorname{Spec} R$. Let $f_i : X^i \to T$ be proper morphisms, X^i normal and irreducible and X^1 regular. Assume that the X^i_k are irreducible. Let $\Gamma \subset X^1 \times_T X^2$ be a closed, irreducible subscheme such that Γ_K is the graph of a dominant rational map $h_K : X^1_K \dashrightarrow X^2_K$. Let $\Gamma'_k \subset \operatorname{red} \Gamma_k$ be the union of those irreducible components which dominate X^2_k and $h_k : \Gamma'_k \to X^2_k$ the projection.

(1.17.1) Asume that h_K is generically finite. Show that $\deg h_k \leq \deg h_K$. Furthermore, if h_K is birational (resp. purely inseparable), then h_k is birational (resp. purely inseparable).

(1.17.2) Show that either X_k^2 is uniruled or Γ_k' is the graph of a dominant rational map $X_k^1 \dashrightarrow X_k^2$.

(1.17.2) (Artin) Assume that h_K is birational. Show that either X_k^2 is ruled or Γ_k' is the graph of a birational map $X_k^1 \dashrightarrow X_k^2$.

(1.17.3) Assume that $f : X \to T$ is smooth and X_k is not ruled. Show that there is a specialization homomorphism

$$\{\text{birational selfmaps of } X_K\} \to \{\text{birational selfmaps of } X_k\}.$$

(1.17.4) Find examples where the above specialization homomorphism is not injective and Γ_k has several irreducible components. (Hint: The general Enriques surface has infinite automorphism group, but some special ones have only finite automorphism group [Barth-Peters83].)

(1.17.5) [Rosenlicht56] Let X_k be a proper variety. Assume that the additive group $G_a(k)$ or the multiplicative group $G_m(k)$ acts on X_k nontrivially. Show that X_k is ruled.

IV.2 Minimizing Families of Rational Curves

If X is uniruled, then it is covered by a family of rational curves. The aim of this section is to show that X is also covered by a family of low degree rational curves. The main technical result is (2.4) which shows that if a subvariety $Z \subset X$ is covered by a family of rational curves, then it can be covered by a "rather small" family of rational curves. This is especially simple for uniruled varieties where the normal bundle can be made as nice as possible (2.10). (2.12) gives an example of a Fano variety which is covered by a family of rational curves of anticanonical degree 2 whose general member is a rational curve with many nodes.

2.1 Definition. (2.1.1) Let X be proper over S and

$$\mathrm{Hom}^n_{bir}(\mathbb{P}^1_S, X/S) \xrightarrow{u} \mathrm{RatCurves}^n(X/S)$$

as in (II.2.15). We say that $V' \subset \mathrm{Hom}_{bir}(\mathbb{P}^1_S, X/S)$ is *closed under* $\mathrm{Aut}\,\mathbb{P}^1$ iff $V' = u^{-1}(u(V))$.

(2.1.2) Let $V \subset \mathrm{RatCurves}^n(X/S)$ be a closed irreducible subvariety and $V' = u^{-1}(V)$. We say that V is an *unsplit family of rational curves* (or that V' is an *unsplit family of morphisms*) if V is proper over S.

(2.1.3) Let $V \subset \mathrm{RatCurves}^n(X/S)$ be a closed irreducible subvariety and $V' = u^{-1}(V)$. Let

$$\Pi : V' \to X \times_S X \quad \text{be given by} \quad \Pi : [f] \mapsto (f(0), f(\infty)).$$

We say that V is a *generically unsplit family of rational curves* (or that V' is a *generically unsplit family of morphisms*) if the fiber of Π over the generic point of $\mathrm{im}\,\Pi$ has dimension at most one.

2.2 Explanation (Notation as above). Let $W \subset \mathrm{Chow}(X/S)$ be the image of V under the natural morphism RatCurves $\to$ Chow and $[C_g]$ the 1-cycle corresponding to the generic point of W. By (I.4.13) V is proper over S iff W is closed in $\mathrm{Chow}(X/S)$. Let $w \in \bar{W} - W$ be a point and $\sum a_i[C_i]$ the corresponding 1-cycle. By (II.2.2) every C_i is rational and $\sum a_i \geq 2$ since $w \notin W$. Informally speaking, C_g degenerates and in the limit it splits up into a reducible cycle $\sum a_i[C_i]$.

The motivation for the name generically unsplit should become clear from the proof of the next results.

2.3 Proposition (Notation as above). *If V (resp. V') is unsplit then it is also generically unsplit.*

Proof. This is done during the proof of the next result.

2.4 Theorem. *Let X be projective over S, $V \subset \mathrm{RatCurves}^n(X/S)$ a closed irreducible subvariety. Then there is a closed irreducible subvariety $W \subset \mathrm{RatCurves}^n(X/S)$ such that*

(2.4.1) *W is a generically unsplit family of rational curves;*

(2.4.2) $\mathrm{Fam}\, W \overset{e}{\prec} \mathrm{Fam}\, V$ (II.4.1.9);

(2.4.3) $\overline{\mathrm{Locus}(V)} = \overline{\mathrm{Locus}(W)}$ (III.1.5.1).

Proof. Let M be a line bundle on X. If $U \subset \mathrm{Chow}(X/S)$ is a connected subvariety, then $M \cdot Z$ is independent of $Z \in U$. This common value is denoted by $\deg_M \mathrm{Fam}\, U$.

Fix an ample line bundle H on X/S. We show that if $V = V_1$ is generically split, then we can find another irreducible family V_2 such that

$$\overline{\mathrm{Locus}(V_1)} = \overline{\mathrm{Locus}(V_2)},\ \mathrm{Fam}\, V_2 \overset{e}{\prec} \mathrm{Fam}\, V_1 \quad \text{and}$$
$$\deg_H \mathrm{Fam}\, V_2 < \deg_H \mathrm{Fam}\, V_1.$$

Thus after finitely many steps we get a generically unsplit family.

For general $(x, x') \in \mathrm{im}\, \Pi$ the fiber $Y := \Pi^{-1}(x, x')$ has dimension at least two. Let $g : \{0, \infty\} \to X$ be given by $g(0) = x, g(\infty) = x'$. With the above notation $Y \subset \mathrm{Hom}_{bir}(\mathbb{P}^1, X, g)$. Every $[f] \in Y$ is birational onto its image, thus by (II.1.9)

$$\dim_{[f]} \mathrm{Hom}_{bir}(\mathbb{P}^1, f(\mathbb{P}^1), g) = 1.$$

Therefore we can choose a smooth curve $B^0 \to Y$ such that

$$\dim \mathrm{im}[\mathbb{P}^1 \times B^0 \to X] = 2.$$

The conditions of (II.5.5.2) are satisfied, hence there is a reducible 1-cycle with rational components $\sum a_i(x)[C_i(x)] \overset{e}{\approx} f_*[\mathbb{P}^1]$ such that $x \in \cup C_i(x)$.

(2.3) is now easy. Assume that $V \subset \mathrm{RatCurves}^n(X/S)$ is generically split. By construction, $\sum a_i(x)[C_i(x)] \in \bar{V} - V$, thus V is split.

By suitable indexing we may assume that $x \in C_1(x)$. Also

$$H \cdot C_1(x) < \deg_H \operatorname{Fam} V.$$

Let $W_1 \subset \mathrm{RatCurves}^n(X/S)$ be the closure of

$$\bigcup_{x \in \mathrm{Locus}(V)} [C_1(x)].$$

By construction

$$\overline{\mathrm{Locus}(W_1)} = \overline{\mathrm{Locus}(V)}.$$

Thus there is an irreducible component $V_2 \subset W_1$ such that

$$\overline{\mathrm{Locus}(V_2)} = \overline{\mathrm{Locus}(V)} \quad \text{and} \quad \deg_H \operatorname{Fam} V_2 < \deg_H \operatorname{Fam} V. \qquad \square$$

2.5 Proposition. *Assume that $V \subset \mathrm{Hom}_{bir}(\mathbb{P}^1, X)$ is a closed, irreducible subvariety which is closed under* $\mathrm{Aut}\,\mathbb{P}^1$. *Let $x \in \mathrm{Locus}(V)$ be a point. Then*

$$\dim V \geq \dim \mathrm{Locus}(V) + \dim \mathrm{Locus}(V, 0 \mapsto x) + 1.$$

Equality holds for general $x \in \mathrm{Locus}(V)$ iff V is generically unsplit.

Proof. By upper semi continuity of the fiber dimension,

$$\dim\{[f] \in V | f(0) = x\} \geq \dim V - \dim \mathrm{Locus}(V). \tag{2.5.1}$$

Similarly, if $y \in \mathrm{Locus}(V, 0 \mapsto x)$, then

$$\begin{aligned} &\dim\{[f] \in V | f(0) = x,\ f(\infty) = y\} \\ &\qquad \geq \dim V - \dim \mathrm{Locus}(V) - \dim \mathrm{Locus}(V, 0 \mapsto x). \end{aligned} \tag{2.5.2}$$

The inequalities in (2.5.1–2) are equalities for general x and y. Let Π be as in (2.1). By construction

$$\Pi^{-1}(x,y) = \{[f] \in V | f(0) = x,\ f(\infty) = y\}$$

is not empty. Since V is closed under $\mathrm{Aut}\,\mathbb{P}^1$, $\dim \Pi^{-1}(x,y) \geq 1$ and equality holds for general x, y iff V is generically unsplit. Substituting into (2.5.2) we are done. $\square$

Using (II.1.2) to estimate $\dim V$ from below we obtain:

2.6 Corollary [Ionescu86; Wiśniewski91b]. *Let X be a smooth and proper variety and $V \subset \mathrm{Hom}_{bir}(\mathbb{P}^1, X)$ a generically unsplit irreducible component. Then for a general $x \in \mathrm{Locus}(V)$*

(2.6.1) $\dim X + \deg_{(-K)} \operatorname{Fam} V \le \dim \operatorname{Locus}(V) + \dim \operatorname{Locus}(V, 0 \mapsto x) + 1$;
(2.6.2) $\dim X + \deg_{(-K)} \operatorname{Fam} V \le 2 \dim \operatorname{Locus}(V) + 1 \le 2 \dim X + 1$; *and*
(2.6.3) $\deg_{(-K)} \operatorname{Fam} V \le \dim \operatorname{Locus}(V, 0 \mapsto x) + 1 \le \dim X + 1$. □

2.7 Proposition. *Let X be a smooth variety and $V \subset \operatorname{Hom}_{bir}(\mathbb{P}^1, X)$ an irreducible component such that $a_i(\operatorname{Fam} V) \ge -1$ for every i. Then*

$$\dim \operatorname{Locus}(V) \ge \#\{i | a_i(\operatorname{Fam} V) \ge 0\},$$

and equality holds in characteristic zero.

Proof. Let $F : \mathbb{P}^1 \times V \to \operatorname{Locus}(V)$ be the natural morphism. By (I.2.16) V is generically smooth. Thus

$$\dim \operatorname{Locus}(V) \ge \operatorname{rank} dF \text{ at a general point,}$$

and equality holds in characteristic zero. The rest follows from (II.3.10). □

2.7.1 Exercise. Use the example in (1.13.5) to show that the char $= 0$ assumption is necessary in (2.7).

2.8 Definition. A free morphism $f : \mathbb{P}^1 \to X$ is called *minimal* if

$$f^* T_X \cong \mathcal{O}(2) + \underbrace{\mathcal{O}(1) + \ldots + \mathcal{O}(1)}_{(d-2)\text{-times}} + \mathcal{O} + \ldots + \mathcal{O},$$

where $d = \deg_{(-K)}(f)$.

2.9 Corollary. *Let X be a smooth proper variety over a field of characteristic zero. Let $V \subset \operatorname{Hom}_{bir}(\mathbb{P}^1, X)$ be an irreducible component. Assume that f is free for general $[f] \in V$. Then V is generically unsplit iff f is minimal for general $[f] \in V$.*

Proof. Pick a general $x \in X$. By (2.5) and (2.7)

$$\begin{aligned} \dim X + \sum a_i(\operatorname{Fam} V) = \dim V &\ge \dim X + \dim \operatorname{Locus}(V, 0 \mapsto x) + 1 \\ &= \dim X + \#\{i | a_i(\operatorname{Fam} V) \ge 1\} + 1. \end{aligned} \tag{2.9.1}$$

Assume that V is generically unsplit. Then by (2.5) equality holds, hence

$$\sum a_i(\operatorname{Fam} V) = \#\{i | a_i(\operatorname{Fam} V) \ge 1\} + 1.$$

Therefore at most one of the $a_i(\operatorname{Fam} V)$ is at least two and f is minimal for general $[f] \in V$. Conversely, if f is minimal, then it forces equality in (2.9.1), thus by (2.5) f is generically unsplit. □

2.10 Theorem. *Let X be a smooth projective variety over a field of characteristic zero. Assume that there is a free morphism $g : \mathbb{P}^1 \to X$ (i.e. X is uniruled). Then there is a minimal free morphism $f : \mathbb{P}^1 \to X$.*

Proof. Let $V \subset \operatorname{Hom}_{bir}(\mathbb{P}^1, X)$ be an irreducible component containing $[g]$. By (II.3.5) $\operatorname{Locus}(V)$ is dense in X. Thus by (2.4) there is a $W \subset \operatorname{Hom}_{bir}(\mathbb{P}^1, X)$ such that $\operatorname{Locus}(W)$ is dense in X and W is generically unsplit. We may also assume that W is an irreducible component of $\operatorname{Hom}_{bir}(\mathbb{P}^1, X)$. By (2.9), f is minimal for general $[f] \in W$. □

2.10.1 Remark. The proof of (2.10) uses the projectivity of X since in (2.4) the ample divisor H is used to ensure that the procedure stops. I do not know if (2.4) remains true for smooth proper algebraic spaces or not.

I also do not know what happens in positive characteristic.

2.11 Proposition. *Let $f : \mathbb{P}^1 \to X$ be a minimal free morphism. Then f is an immersion. Furthermore, a general deformation of f is an embedding if $\deg_{(-K)}(f) \geq 4$.*

Proof. By (II.3.13) f is separable. $df : T_{\mathbb{P}^1} \to f^*T_X$ is an isomorphism of $T_{\mathbb{P}^1}$ and the unique $\mathcal{O}(2)$ summand of f^*T_X. Thus $df(p)$ is nonzero for every $p \in C$. The second claim follows from (II.3.14.3). □

2.12 Examples. (2.12.1) Let $X \subset \mathbb{P}^3$ be a smooth cubic surface. The intersection of X with a general tangent plane section is a nodal plane cubic. Let $f : \mathbb{P}^1 \to X$ be the normalization. $f^*T_X \cong \mathcal{O}(2) + \mathcal{O}(1), \deg_{(-K)} f = 3$ and $\dim_{[f]} \operatorname{Hom}(\mathbb{P}^1, X) = 5$.

The automorphisms of $\mathbb{P}^1$ account for three dimensions, and the movement of the tangent plane accounts for the remaining two. This shows that every deformation of f has a nodal image. Thus (2.11) is sharp. Observe, however, that X is covered by a family of conics, thus there is a degree two free morphism $\mathbb{P}^1 \to X$. See (II.2.13) for a similar example.

(2.12.2) Another interesting example is the following.(For simplicity take $\operatorname{char} k = 0$.) Let $Y \subset \mathbb{P}^n$ be a smooth hypersurface of degree $2n$ and $p : X \to \mathbb{P}^n$ the double cover of $\mathbb{P}^n$ ramified along Y. Then

$$K_X = p^*(K_{\mathbb{P}^n} + \frac{1}{2}[Y]) = p^*\mathcal{O}(-1).$$

Thus X is a Fano variety.

2.12.3 Claim. *There is a free morphism of degree two $f : \mathbb{P}^1 \to X$ such that $f(\mathbb{P}^1) \subset X$ is a rational curve with $(n-1)$ nodes, and the same holds for generic deformations of f.*

Proof. We start with a simple lemma on general position of tangent lines:

2.12.4 Lemma. *Let $p \in \mathbb{P}^n$ be a point and $Y \subset \mathbb{P}^n$ a general hypersurface of degree $k \geq 2n-2$. Then there are finitely many lines through p which are tangent to Y at $(n-1)$ points.*

Proof. Let $P \cong \mathbb{P}^{n-1}$ be the space of lines through p and $G = \mathbb{P}(H^0(\mathbb{P}^n, \mathcal{O}(k)))$ the space of degree k hypersurfaces. Consider the subset

$$Z \subset P \times G$$

consisting of those pairs (l, H) such that l is an $(n-1)$-fold tangent of H. Clearly Z is a closed subset. The fibers of the first projection $Z \to P$ are those hypersurfaces that are $(n-1)$-fold tangent to a given line. This is an irreducible variety of codimension $(n-1)$ in G. Thus

$$\operatorname{codim}(Z, P \times G) = n - 1.$$

Choose affine coordinates such that p is the origin and consider the hypersurface H given by the equation

$$\prod_{i=1}^{n-1} \left[(x_n - a_i)^2 + x_i\right] \cdot \prod_{i=2n-1}^{k} (x_n - b_i) = 0,$$

where a_i, b_i are different constants. The line $x_1 = \ldots = x_{n-1} = 0$ is tangent to H at $(n-1)$ points, and it is isolated in the fiber of $Z \to G$ over $[H]$. Thus $Z \to G$ is generically finite and surjective. □

The preimage of an $(n-1)$-fold tangent on X is a rational curve with nodes at the tangent points. Thus we obtain an $(n-1)$-dimensional family $\{C_\lambda\}$ of singular rational curves of degree 2 which cover X. Thus we have a dominant morphism

$$F : \mathbb{P}^1 \times V^{n-1} \to X \quad \text{such that} \quad \deg_{\mathbb{P}^1} F^*(-K_X) = 2.$$

Let $f_v = F|\mathbb{P}^1 \times \{v\}$. For general $v \in V$, $f_v^* T_X$ is semi positive by (II.3.11). Thus by (II.3.8.3)

$$f_v^* T_X \cong \mathcal{O}(2) + \mathcal{O} + \ldots + \mathcal{O}.$$

By (II.1.2) we obtain that

$$\dim_{[f_v]} \operatorname{Hom}(\mathbb{P}^1, X) = n + 2.$$

Counting automorphisms, we see that the family $\{C_\lambda\}$ accounts for all deformations of f_v. □

2.12.5. It is interesting to note that X has another covering family of degree two curves where the general member is smooth. A similar argument shows that there are finitely many conics Q_μ through p which are $2n$-fold tangents to Y. $p^{-1}(Q_\mu)$ has two irreducible components, both smooth rational curves. This gives the other family which is probably irreducible.

The same ideas lead to other interesting examples as well:

2.13 Exercise. Let $q : W \to \mathbb{P}^n$ be a double cover ramified along a smooth hypersurface H of degree $2m$.

(2.13.1) The preimages of $(m-1)$-fold tangents to H are rational curves with $(m-1)$ nodes. Show that for general W this gives a $(2n-1-m)$-dimensional family of degree two rational curves on W.

By (II.1.2) the expected dimension of the family of degree two rational curves is $(3n-1-2m)$. Thus if $m > n$, then we have more degree two rational curves than expected.

If $m < n$, then we seem to get fewer degree two rational curves than expected. Explain what happens.

(2.13.2) Assume that $m = n+1$, hence $K_W = 0$. The preimages of $(n-1)$-fold tangents to H are elliptic curves with $(n-1)$ nodes. These curves generically cover W.

Constant count suggests that for very general H and for every even d there is an $(n-1)$-dimensional (resp. $(n-2)$-dimensional) family of degree d rational curves in $\mathbb{P}^n$ which are tangent to H at $d(n+1)/2-2$ (resp. $d(n+1)/2-1$) points. Thus we get a family of degre $2d$ elliptic curves covering W and a family of degree $2d$ rational curves covering a divisor in W. Also, there should not be such families of odd degree on W.

IV.3 Rationally Connected Varieties

Probably the most important algebraic variety is $\mathbb{P}^n$. In the classification theory of varieties it is therefore important to isolate a class of varieties that "look like $\mathbb{P}^n$". It is also desirable that this class be stable under various operations. The definitions are given in (3.2) and their elementary properties are listed in (3.3–6). In analogy with (1.9) we obtain that a smooth variety X is separably rationally connected iff there is a morphism $f : \mathbb{P}^1 \to X$ such that f^*T_X is ample (3.7). This implies that a separably rationally connected variety does not have any holomorphic tensors (3.8). The converse is a very interesting open problem (3.8.1).

One of the main results is (3.10) which says that in characteristic zero a smooth variety is rationally connected iff it is rationally chain connected. The effective version of this result (3.10.1) is crucial for the applications in (V.2). The deformation theory of separably rationally connected varieties is discussed in (3.11). Finally we prove that if X is rationally chain connected, then the vector space of 1-cycles modulo algebraic equivalence is finite dimensional.

3.1 Definition. Let X_k be a variety of dimension n over a field k.

(3.1.1) We say that X_k is *rational* if there is a birational map $\mathbb{P}^n_k \dashrightarrow X$.

(3.1.2) We say that X_k is *unirational* if there is a dominant map $\mathbb{P}^n_k \dashrightarrow X$.

(3.1.3) Let $\bar{k} \supset k$ be an algebraic closure. We say that X_k is *geometrically rational* or *rational over* $\bar{k}$ (resp. *geometrically unirational* or *unirational over* $\bar{k}$) if there is a map $\mathbb{P}^n_{\bar{k}} \dashrightarrow X_{\bar{k}}$ which is birational (resp. dominant).

3.1.4 Comment. This terminology is not universally accepted. Frequently k-rational (or rational over k) is used in (3.1.1) instead of rational and "rational" may mean "geometrically rational". In some situations I use the "safe" version to avoid any possibility of confusion.

The class of all rational surfaces has very good properties. Being rational is deformation invariant among smooth surfaces, and the image of a rational surface by a map is again rational, at least in char $= 0$ (III.2.5).

In higher dimensions it is not known if being rational is deformation invariant or not. Any smooth cubic hypersurface $X_3 \subset \mathbb{P}^4$ has a 2-sheeted (ramified) cover which is rational, but X_3 itself is not rational [Clemens-Griffiths72].

The notion of rationally connected varieties was introduced in [KoMiMo92b] to remedy the situation. The idea is that instead of emphasizing global properties of $\mathbb{P}^n$, we try to concentrate on one special property: there are lots of rational curves in $\mathbb{P}^n$.

3.2 Definition. Let X be a variety (not necessarily proper).

(3.2.1) We say that X is *rationally chain connected* if there is a family of proper and connected algebraic curves $g : U \to Y$ whose geometric fibers have only rational components with cycle morphism $u : U \to X$ such that

$$u^{(2)} : U \times_Y U \to X \times X \quad \text{is dominant.}$$

(The image of $u^{(2)}$ consists of those pairs $x_1, x_2 \in X$ such that $x_1, x_2 \in u(U_y)$ for some $y \in Y$.)

(3.2.2) We say that X is *rationally connected* if there is a family of proper algebraic curves $g : U \to Y$ whose geometric fibers are irreducible rational curves with cycle morphism $u : U \to X$ such that $u^{(2)}$ is dominant.

(3.2.3) We say that X is *separably rationally connected* if there is a variety Y and a morphism $u : Y \times \mathbb{P}^1 \to X$ such that $u^{(2)}$ is dominant and smooth at the generic point.

3.2.4 Remark. The notion of rational chain connectedness makes sense for any (possibly reducible) scheme X.

3.2.5 Exercise. Show that the above notions do not depend on the field over which X is defined.

3.2.6 Examples. (3.2.6.1) The family of all lines in $\mathbb{P}^n$ shows that $\mathbb{P}^n$ satisfies all of the above definitions.

(3.2.6.2) If $p : \mathbb{P}^n \dashrightarrow X^n$ is dominant, then X^n is rationally connected and rationally chain connected. If p is separable, then X^n is also separably rationally connected.

(3.2.6.3) A cone over a variety Z is rationally chain connected but not rationally connected unless Z is. A cone minus its vertex is usually not rationally chain connected.

(3.2.6.4) Find an example of an irreducible, proper, rationally chain connected variety X whose normalization is not rationally chain connected.

The following properties are fairly straightforward from the definitions:

3.3 Proposition. (3.3.1) *If X is separably rationally connected, then X is rationally connected, and the converse holds if the characteristic is zero.*

(3.3.2) *If X is rationally connected, then X is rationally chain connected.*

(3.3.3) *Let X and X' be proper varieties, birational to each other. Then X is (separably) rationally connected iff X' is (separably) rationally connected.*

(3.3.4) *If X is rationally chain connected, then it is uniruled. If X is separably rationally connected, then it is separably uniruled.*

Proof. In order to see (3.3.1) let $u : U \to X$ show that X is rationally connected. Set $Y' = \operatorname{red}\operatorname{Hom}_{bir}(\mathbb{P}^1 \times Y/Y, U/Y)$. Then

$$\mathbb{P}^1 \times \mathbb{P}^1 \times Y' \to U \times_Y U \to X \times X$$

is dominant and in characteristic zero smooth at the generic point. The rest of the first two claims are clear from the definitions.

In order to see (3.3.3) assume that $g : U \to Y$ and $u : U \to X$ show that X is (separably) rationally connected. Let $u' : U \to X \dashrightarrow X'$ be the composition. There is an open subset $Y' \subset Y$ such that $u' : g^{-1}(Y') \to X'$ is a morphism. Thus X' is (separably) rationally connected.

Finally (3.3.4) is clear from the definitions. □

3.3.5 Exercise. Let X be a smooth proper surface over an algebraically closed field. Then X is separably rationally connected iff X is rational.

3.4 Lemma. *Let X be a proper scheme. X is rationally chain connected iff there is a closed subscheme $Y \subset \operatorname{Chow}(X)$ which parametrizes connected 1-cycles with rational geometric components such that $u^{(2)} : U \times_Y U \to X \times X$ is dominant, where $g : U \to Y$ is the universal family and $u : U \to X$ the cycle map.*

Proof. Let $g' : U' \to Y'$ show that X is rationally chain connected. Let $R \subset \operatorname{Chow}(X)$ be the closed subset parametrizing 1-cycles with rational components (II.2.2), $R = \cup R_i$ the irreducible components. Let $g_i : U_i \to R_i$ be the universal family. For some i there is a rational map $h : Y' \dashrightarrow R_i$ such that $h(y) = [u(U_y)]$ for general $y \in Y'$. Thus $u_i^{(2)} : U_i \times_{R_i} U_i \to X \times X$ is also dominant. □

3.5 Corollary. (3.5.1) *Let X be a proper and rationally chain connected variety over an algebraically closed field. Then for arbitrary closed points $x_1, x_2 \in X$ there is a connected curve $C \subset X$ which contains x_1 and x_2 such that every irreducible component of C is rational.*

(3.5.2) *Let $f : X \to S$ be a proper and equidimensional morphism. There are countably many closed subvarieties $S_i \subset S$ such that X_s is rationally chain connected iff $s \in \cup S_i$.*

Proof. Let $g_i : U_i \to R_i$ be as in (3.4). U_i is proper (II.2.2) hence $u_i^{(2)}$ is onto. This shows (3.5.1).

In order to see the second part let $R/S \subset \mathrm{Chow}(X/S)$ be the closed subset parametrizing geometrically connected 1-cycles with rational components (II.2.2). Let $R/S = \cup R_i$ be the irreducible components and $g_i : U_i \to R_i$ the universal families. Let $Z_i \subset X \times_S X$ be the image of $U_i \times_{R_i} U_i$. Z_i is closed and if $s \in S$, then

$$(U_i \times \{s\}) \times_{R_i} (U_i \times \{s\}) \to X_s \times X_s$$

is dominant iff $Z_i \times \{s\} = X_s \times X_s$. Let

$$S_i = \{s \in S | Z_i \times \{s\} = X_s \times X_s\}.$$

By upper semi continuity of fiber dimensions, S_i is closed. This shows the second part. □

In general we need countably many subvarieties in (3.5.2), cf. (1.8.7). As in (1.8.2), this can be improved in characteristic zero. The proof is given in (5.10).

3.5.3 Theorem. *Let $f : X \to S$ be a proper and equidimensional morphism with normal fibers. Assume that S is over a field of characteristic zero. There is a closed subvariety $R \subset S$ such that X_s is rationally chain connected iff $s \in R$.*

Over sufficiently large fields the above notions are equivalent to the corresponding set theoretic notions (cf. [KoMiMo92b]). As in (1.3.5), it is not known if uncountability of the field is essential or not.

3.6 Proposition. *Let X be a variety over an uncountable algebraically closed field. Then*

(3.6.1) *X is rationally chain connected iff for very general closed points $x_1, x_2 \in X$ there is a connected curve $C \subset X$ which contains x_1 and x_2 such that every irreducible component of C is rational.*

(3.6.2) *X is rationally connected iff for very general closed points $x_1, x_2 \in X$ there is an irreducible rational curve $C \subset X$ which contains x_1 and x_2.*

Proof. Let $X \subset \bar{X}$ be a compactification. Let $R \subset \mathrm{Chow}(\bar{X})$ be the locally closed subset parametrizing cycles C with rational components such that $C \subset$

X, $g : U \to R$ the universal family. (3.6.1) holds iff $u^{(2)} : U \times_R U \to X \times X$ is generically surjective on closed points. Since R has only countably many irreducible components this implies that there is an irreducible component $Y \subset R$ such that $U_Y \times_Y U_Y \to X \times X$ is dominant. This shows (3.6.1).

Arguing with $\mathrm{Hom}(\mathbb{P}^1, X)$ instead of R as above shows (3.6.2). □

The following result is the counterpart of (1.9).

3.7 Theorem. *Let X be a smooth variety over an algebraically closed field. Then X is separably rationally connected iff there is a morphism $f : \mathbb{P}^1 \to X$ such that f^*T_X is ample.*

Proof. Assume that $f : \mathbb{P}^1 \to X$ is a morphism such that f^*T_X is ample, that is $f^*T_X = \sum \mathcal{O}(a_i)$ and $a_i \geq 1$ for every i. Let $[f] \in Y \subset \mathrm{Hom}(\mathbb{P}^1, X)$ be an open and smooth neighborhood with cycle map $u : \mathbb{P}^1 \times \mathrm{Hom}(\mathbb{P}^1, X) \to X$. The evaluation map

$$\phi^{(2)}(p, q, f) : H^0(\mathbb{P}^1, f^*T_X) \to f^*T_X \otimes k(p) + f^*T_X \otimes k(q)$$

is surjective for $p \neq q \in \mathbb{P}^1$, thus by (II.3.5.2) $u^{(2)} : \mathbb{P}^1 \times \mathbb{P}^1 \times Y \to X \times X$ is smooth at $(p, q, [f])$.

Conversely, if $u^{(2)}$ is smooth at $(p, q, [f])$, then by (II.3.5.2) $\phi^{(2)}(p, q, f)$ is surjective. Thus $f^*T_X = \sum \mathcal{O}(a_i)$ and $a_i \geq 1$ for every i. Therefore f^*T_X is ample. □

3.7.1 Exercise. Let X be smooth and separably rationally connected, $Z \subset X$ a closed subscheme of codimension at least 2. Show that $X - Z$ is separably rationally connected.

The following result is the analog of (1.11):

3.8 Corollary. *Let X be a smooth projective variety. Assume that X is separably rationally connected. Then*

$$H^0(X, (\Omega^1_X)^{\otimes m}) = 0 \quad \textit{for every } m > 0.$$

Proof. Let $f : \mathbb{P}^1 \to X$ be a morphism such that f^*T_X is ample. Then $f^*\Omega^1_X$ is the sum of line bundles of negative degree. Therefore every section of $(\Omega^1_X)^{\otimes m}$ vanishes along $f(\mathbb{P}^1)$. Such curves cover a dense set of X, hence every section of $(\Omega^1_X)^{\otimes m}$ vanishes. □

3.8.1 Conjecture (Mumford). *The converse of (3.8) is also true, at least in* char $= 0$.

By [KoMiMo92b] the conjecture is true for $\dim X \leq 3$. Some aspects of this conjecture are studied in (5.7).

Separably rationally connected varieties have lots of rational curves.[1]

[1] See also [Kollár98, 4.1.2].

3.9 Theorem [KoMiMo92b, 2.1]. *Let X be a smooth proper separably rationally connected variety over an algebraically closed field, $x_1, \ldots, x_m \in X$ distinct closed points. Then there is a morphism $f = f_{x_1,\ldots,x_m} : \mathbb{P}^1 \to X$ such that*

(3.9.1) $x_1, \ldots, x_m \in f(\mathbb{P}^1)$;

(3.9.2) f^*T_X *is ample;*

(3.9.3) f *is an immersion if* $\dim X = 2$ *and an embedding if* $\dim X \geq 3$.

Proof. As a first step we prove that for every $x \in X$ there is a morphism $f_x : \mathbb{P}^1 \to X$ such that $x \in f_x(\mathbb{P}^1)$ and $f_x^*T_X$ is ample.

Let $Y \subset \mathrm{Hom}(\mathbb{P}^1, X)$ be the subset of all morphisms g such that g^*T_X is ample. Y is open by (II.3.9.2) and the cycle map $\mathbb{P}^1 \times Y \to X$ is smooth hence open (II.3.5.3). Let $X^0 \subset X$ be the image. We need to show that $X^0 = X$. Assume the contrary and pick $x \in X - X^0$. By (3.6.1) there is a connected rational chain joining x with a point of X^0. Thus there is a rational curve $h : D \cong \mathbb{P}^1 \to X$ such that $\mathrm{im}\, h \cap X^0 \neq \emptyset$ and $\mathrm{im}\, h \cap (X - X^0) \neq \emptyset$. Pick $p, x_1, \ldots, x_m \in D$ such that $h(p) \in X - X^0$ and $h(x_i) \in X^0$. By assumption there are morphisms $f_i : C_i \cong \mathbb{P}^1 \to X$ such that $f_i^*T_X$ is ample and $x_i = f_i(y_i)$ for some $y_i \in C_i$.

Out of these we can assemble a comb (II.7.7) $C = D + \sum C_i$ by identifying the pairs (x_i, y_i). The morphisms h and f_i glue together to a morphism $f : C \to X$. By (II.7.9) there is a subcomb $C' \subset C$ which can be smoothed fixing $f(p)$. By (II.7.10) if $m \gg 1$, then a general smoothing $f_t : \mathbb{P}^1 \to X$ is free over $f_t(p)$. This means that $f_t^*T_X(-p)$ is semi positive, hence $f_t^*T_X$ is ample. By construction $f_t(p) = h(p) \in X - X^0$. This is a contradiction, thus $X = X^0$.

In the second step pick an extra point $x_{m+1} \in X$. As in the construction of X^0 above, for every x_i there is an open set $U_i \subset X$ such that for every $u \in U_i$ there is a morphism $f_{i,u} : C_i \cong \mathbb{P}^1 \to X$ such that $f_{i,u}^*T_X$ is ample, $x_i = f_{i,u}(0)$ and $u = f_{i,u}(\infty)$. Let $h : D \cong \mathbb{P}^1 \to X$ be a morphism such that h^*T_X is ample and $\mathrm{im}\, h$ intersects $\cap U_i$. Choose points $y_i \in D$ such that $u_i = h(y_i) \in \cap U_i$. We can construct a comb $C = D + \sum C_i$ by identifying the point $y_i \in D$ with $\infty \in C_i$ for every i. The morphisms h and f_{i,u_i} glue together to a morphism $f : C \to X$. By (II.7.6) f can be smoothed fixing the points $x_i = f_{i,u_i}(0)$.

Let $f_t : \mathbb{P}^1 \to X$ be a general smoothing. Then $f_t^*T_X(-(p_1 + \ldots + p_{m+1}))$ is semi positive, thus $f_t^*T_X(-(p_1 + \ldots + p_m))$ is ample. By (II.3.14) a general deformation of f_t fixing $x_1 = f_t(p_1), \ldots, x_m = f_t(p_m)$ is an immersion if $\dim X = 2$ and an embedding if $\dim X \geq 3$. This gives the required morphism $f_{x_1,\ldots,x_m} : \mathbb{P}^1 \to X$. □

If X is not proper, then the first step in the above proof does not work since there may not be any curve $h : D \to X$. The rest of the proof works and we obtain the following:

3.9.4 Theorem. *Let X be a smooth separably rationally connected variety over an algebraically closed field. There is a unique largest open subset $\emptyset \neq X^0 \subset X$ such that if $x_1, \dots, x_m \in X^0$ are distinct closed points, then there is a morphism $f : \mathbb{P}^1 \to X^0$ such that*

(3.9.4.1) $x_1, \dots, x_m \in f(\mathbb{P}^1)$;

(3.9.4.2) f^*T_X *is ample;*

(3.9.4.3) f *is an immersion if* $\dim X = 2$ *and an embedding if* $\dim X \geq 3$.

Moreover, if $C \subset X$ is a rational curve such that $C \cap X^0 \neq \emptyset$, then $C \subset X^0$. □

3.9.5 Exercise. Find an example where $X^0 \neq X$. (I could not.)

Finally we prove that all the definitions given in (3.2) are equivalent for smooth proper varieties in char = 0.

3.10 Theorem [KoMiMo92b, 2.1]. *Let X be a smooth proper variety in characteristic zero. Then X is rationally chain connected iff X is rationally connected.*

For later applications we prove a version of (3.10) which measures the degrees of the curves involved (cf. [KoMiMo92c, Sect. 4]):

3.10.1 Complement. *Let X be a smooth, proper variety over an algebraically closed field of characteristic zero and H an ample divisor such that $H - K_X$ is nef. Assume that any two points $x_1, x_2 \in X$ can be connected by a chain of rational curves C_{x_1,x_2} such that $H \cdot C_{x_1,x_2} \leq d$. Then for any two points $x_1, x_2 \in X$ there is a morphism $f_{x_1,x_2} : \mathbb{P}^1 \to X$ such that $x_1, x_2 \in f_{x_1,x_2}(\mathbb{P}^1)$, $f^*_{x_1,x_2}T_X$ is ample and $H \cdot f_{x_1,x_2}(\mathbb{P}^1) \leq N(d, \dim X)$ where $N(d, \dim X)$ depends only on d and $\dim X$.*

Proof. We may assume that the field is uncountable. Let $C = C_1 + \ldots + C_n$ be a chain connecting x_1 and x_2, p_i the point where C_i and C_{i+1} join, $p_0 = x_1$ and $p_n = x_2$. $n \leq d$ since $H \cdot C_i \geq 1$ for every i. Let $f_i : \mathbb{P}^1 \to C_i \subset X$ be the normalization. If $\dim \operatorname{Hom}_{[f_i]}(\mathbb{P}^1, X) > 2\dim X + 1$, then we can deform C_i into a reducible curve keeping p_i and p_{i+1} fixed. Repeating this procedure (at most d times) we may assume that our chain satisfies

$$\dim \operatorname{Hom}_{[f_i]}(\mathbb{P}^1, X) \leq 2\dim X + 1 \ \forall i. \tag{3.10.2}$$

First consider the case when $p_0 = x_1$ is a very general point of X. This implies that $f_1 : \mathbb{P}^1 \to C_1 \subset X$ is free (II.3.11).

We prove inductively that there is a free rational curve $g_i : \mathbb{P}^1 \to X$ which connects p_{i-1} and p_i such that $\deg_H g_i \leq N_i(d, \dim X)$. We can start the induction with $g_1 = f_1$ and $N_1 = d$.

By inductive assumption $\operatorname{im} g_i$ passes through p_i and g_i is free. Thus deformations of g_i cover an open subset of X containing p_i. Pick very general $x_1, \dots, x_m \in C_{i+1}$ such that for every x_j there is a deformation g_i^j of g_i

which passes through x_j and a very general point of X. This is possible by (II.3.11.3). As in the proof of (3.9), using these morphisms we can assemble a comb with m teeth. By (II.7.9) it has a subcomb C' with at least one tooth which can be smoothed fixing p_i and p_{i+1} if

$$m > K_X \cdot C_{i+1} + 2\dim X - \dim X + 2\dim X + 1,$$

where we used the inequality (3.10.2). C' contains at least one of the curves $\operatorname{im} g_i^j$, hence it passes through a very general point of X. Thus by (II.3.11.3) so does a very general smoothing $f_t : \mathbb{P}^1 \to X$ of C'. $\operatorname{im} f_t$ is a free rational curve connecting p_{i-1} and p_i. Since $K_X \cdot C_i \le H \cdot C_i \le d$, we can take

$$N_{i+1}(d, \dim X) = d + (d + 3\dim X + 2)N_i(d, \dim X).$$

Once p_i can be connected to p_{i+1} by a free rational curve for every i, the resulting chain can be smoothed fixing $p_n = x_2$. Since x_1 is arbitrary, we obtain that x_2 can be connected to a very general point by a free rational curve $h : \mathbb{P}^1 \to X$ of H-degree at most $N_d(d, \dim X)$. By (II.3.10.11) this implies that h^*T_X is ample.

Let $x_1, x_2 \in X$ be arbitrary and $x_0 \in X$ very general. There are rational curves $f_i : \mathbb{P}^1 \to X$ connecting x_0 with x_i for $i = 1, 2$ such that $f_i^*T_X$ is ample. The length two chain consisting of f_1 and f_2 can be smoothed fixing x_1 and x_2 by (II.7.6). □

In the nonproper case we can not perform the first step degenerating the chain further. This part is used in the proof only through the explicit estimate (3.10.2). The proof still works without this estimate, but I do not see how to control the degree of the resulting curves. This way we obtain:

3.10.3 Theorem. *Let X be a smooth variety in characteristic zero. Then X is rationally chain connected iff X is rationally connected.* □

3.10.4 Remark. In positive characteristic it may happen that any two general points can be connected by a free rational curve but X is not separably rationally connected (V.5.19).

3.11 Theorem [KoMiMo92b, 2.4]. *Let $g : X \to S$ be a smooth morphism, S connected. Assume that X_s is separably rationally connected for some $s \in S$. Then there is an open neighborhood $s \in U \subset S$ such that X_u is separably rationally connected for every $u \in U$.*

If g is proper and char $= 0$, *then X_u is rationally connected for every $u \in S$.*

Proof. Let us consider $\operatorname{Hom}(\mathbb{P}^1_S, X)$. By assumption there is an $[f] \in \operatorname{Hom}(\mathbb{P}^1_s, X_s)$ such that $f^*T_{X_s}$ is ample. By (I.2.17) this implies that $\pi : \operatorname{Hom}(\mathbb{P}^1_S, X) \to S$ is smooth at $[f]$. By (II.3.9.2.2) there is an open neighborhood $[f] \in V \subset \operatorname{Hom}(\mathbb{P}^1_S, X)$ such that $h^*T_{X/S}$ is ample for every $[h] \in V$.

Since π is smooth at $[f]$, it is open, hence there is an open neighborhood $s \in U \subset \pi(V)$. By (3.7) X_u is separably rationally connected for every $u \in U$.

Next assume that g is proper and char $= 0$. It is sufficient to show that being rationally connected is also a closed condition. By (3.5.2) X_u is rationally chain connected for every $u \in \overline{U}$. Hence by (3.10) X_u is rationally connected for every $u \in \overline{U}$. □

3.12 Exercises. Let R be a DVR with quotient field K and residue field k, $T = \operatorname{Spec} R$. Let $f : X \to T$ be a proper, smooth morphism, cf. (1.17).

(3.12.1) Show that if X_K is purely inseparably uniruled, then X_k is also purely inseparably uniruled. (We say that a variety X is *purely inseparably uniruled* (resp. *purely inseparably unirational*) if there is a generically finite and purely inseparable map $Y \times \mathbb{P}^1 \dashrightarrow X$ (resp. $\mathbb{P}^n \dashrightarrow X$).)

(3.12.2) Let Z be a smooth and proper variety over an algebraically closed field which is ruled and rationally chain connected. Show that Z is rational if $\dim Z = 2$ or $\dim Z = 3$ and char $= 0$.

(3.12.3) [Timmerscheidt82] Show that if X_K is geometrically rational, then X_k is also geometrically rational if $\dim X_k = 2$ or $\dim X_k = 3$ and $\operatorname{char} X_k = 0$.

(3.12.4) [Blass-Levine82] Show that if X_K is a geometrically purely inseparably unirational surface, then X_k is also geometrically purely inseparably unirational.

(3.12.5) Give examples showing that the smoothness assumption is necessary in (3.12.2–4).

The low dimensional Chow groups of a rationally chain connected variety are well behaved:

3.13 Theorem. *Let X be a proper and rationally chain connected scheme over an algebraically closed field. Then*

(3.13.1) $A_0(X) \cong \mathbb{Z}$.

(3.13.2) *$A_1(X)_{\mathbb{Q}}$ is generated by rational curves and $B_1(X)_{\mathbb{Q}}$ is finite dimensional.*

Proof. Let $x_1, x_2 \in X$ be closed points. By assumption there is a connected curve C with rational components containing x_1, x_2. Any two points of C are rationally equivalent, thus any two points of X are rationally equivalent.

The second part is implied by the following more precise version:

3.13.3 Proposition. *Let $Y \subset \operatorname{Chow}(X)$ be a closed subvariety which parametrizes connected 1-cycles with geometrically rational components such that $u^{(2)} : U \times_Y U \to X \times X$ is surjective, where $g : U \to Y$ is the universal family and $u : U \to X$ the cycle map.*

Then $A_1(X)_{\mathbb{Q}}$ is generated by the irreducible components of fibers of g.

Proof. Pick a point $x_0 \in X$. By assumption $u^{-1}(x_0) \times_Y U \to \{x_0\} \times X$ is dominant. Let

$$Y_0 := g(u^{-1}(x_0)), U_0 := g^{-1}(Y_0) \quad \text{and} \quad Z_0 = u^{-1}(x_0).$$

We obtain a diagram

$$\begin{array}{ccccc} Z_0 & \xrightarrow{i} & U_0 & \xrightarrow{u} & X \\ & & {\scriptstyle g}\downarrow & & \\ & & Y_0 & & \end{array}$$

where u and $g \circ i$ are dominant and $(u \circ i)(Z_0) = x_0$. Thus (II.4.20) implies (3.13.3). □

IV.3.13.4 Exercise. Let X be a proper and rationally chain connected scheme over a field k. Show that $A_0(X) \cong \mathbb{Z} +$ (torsion group).

The following result is a considerable strenghtening of (3.13) for smooth varieties. The proof uses intersection theory on smooth varieties. We need only a few basic properties, see e.g. [Fulton84, Chap. 8.1].

3.14 Theorem [Bloch-Srinivas83]. *Let X be a smooth, proper and rationally chain connected variety over a field k. Then $B_1(X)_{\mathbb{Q}} \to N_1(X)_{\mathbb{Q}}$ is an isomorphism.*

Proof. By (II.4.3) it is sufficient to prove this for k algebraically closed.

Let $p_i : X \times X \to X$ be the two projections and $\Delta \subset X \times X$ the diagonal. Then $t : A_i(X) \to A_i(X)$ given by $z \mapsto (p_2)_*(\Delta \cdot p_1^*(z))$ is an isomorphism.

Pick a closed point $x \in X$ and let $F = k(X)$ be the field of rational functions on X. The generic fiber of p_1 can be identified with X_F. Δ and $X \times \{x\} \subset X \times X$ give F-points of X_F. By (3.13.4) $[\Delta] - [X \times \{x\}] \in A_0(X_F)$ is torsion. That is, there is a natural number n and a diagram

(3.14.1)
$$\begin{array}{ccc} W_F & \xrightarrow{g_F} & X_F \\ {\scriptstyle p_F}\downarrow & & \\ \mathbb{P}^1_F & & \end{array}$$

such that $(g_F)_*(p_F^*[0]) = n[\Delta]$ and $(g_F)_*(p_F^*[\infty]) = n[X \times x]$ (as cycles on X_F). Since Spec F is the generic point of X, we can take closures and extend (3.14.1) to a diagram

(3.14.2)
$$\begin{array}{ccc} W & \xrightarrow{g} & X \times X \\ {\scriptstyle p}\downarrow & & \\ \mathbb{P}^1 \times X & & \end{array}$$

such that

$$g_*(p^*([0 \times X] - [\infty \times X])) = n([\Delta] - [X \times \{x\}]) - Z, \tag{3.14.3}$$

where Z is an n-cycle on $X \times X$ that is disjoint from the generic fiber of p_1. Thus there is a divisor $D \subset X$ such that $p_1(\operatorname{Supp} Z) \subset D$. By (3.14.3) we see that

$$n[\Delta] \overset{r}{\approx} n[X \times \{x\}] + [Z]. \tag{3.14.4}$$

Define $T : A_i(X) \to A_i(X)$ by $z \mapsto (p_2)_*([n(X \times \{x\}) + Z] \cdot p_1^*(z))$. By (3.14.4) $nt - T : A_i(X) \to A_i(X)$ is the zero map.

If $z \in A_i(X)$ and $i > 0$, then $(p_2)_*([X \times \{x\}] \cdot p_1^*(z))$ is supported on $\{x\}$, thus this is the zero cycle. Therefore $T(z) = (p_2)_*(Z \cdot p_1^*(z))$.

Let $z \in B_1(X)$ be a 1-cycle which is numerically equivalent to zero. One can compute $Z \cdot p_1^*(z)$ in two steps: first intersect $D \times X = p_1^*(D)$ with $p_1^*(z)$ and, then further intersect by Z. $p_1^*(D) \cdot p_1^*(z) = p_1^*(D \cdot z)$. $D \cdot z$ is a zero-cycle which is numerically zero, hence also algebraically zero. Thus $Z \cdot p_1^*(z)$ is algebraically zero. Therefore $T(z) \overset{a}{\approx} 0$ and hence $nt(z) = T(z) \overset{a}{\approx} 0$. This shows that the kernel of $B_1(X)_{\mathbb{Z}} \to N_1(X)_{\mathbb{Z}}$ is n-torsion. □

3.14.5 Exercise. The aim of this exercise is to prove a relative version of (3.13). Let $f : X \to Y$ be a proper and surjective morphism of schemes.

(3.14.5.1) Let $C_1, C_2 \in Z_1(Y)_{\mathbb{Q}}$ such that $C_1 \overset{a}{\approx} C_2$. Show that there are $C_1', C_2' \in Z_1(X)_{\mathbb{Q}}$ such that $C_1' \overset{a}{\approx} C_2'$ and $f_*(C_i') = C_i$ for $i = 1, 2$.

(3.14.5.2) Assume that every fiber of f is rationally chain connected. Show that if $D_1, D_2 \in X$ are irreducible curves such that $f_*(D_1) = f_*(D_2)$ then

$$[D_1] - [D_2] \in \operatorname{im}[B_1(X/Y)_{\mathbb{Q}} \to B_1(X)_{\mathbb{Q}}].$$

(3.14.5.3) Assume that every fiber of f is rationally chain connected. Show that there is an exact sequence

$$B_1(X/Y)_{\mathbb{Q}} \to B_1(X)_{\mathbb{Q}} \to B_1(Y)_{\mathbb{Q}} \to 0.$$

(3.14.5.4) Use the method of (3.13) to prove that $B_1(X/Y)_{\mathbb{Q}}$ is finite dimensional.

(3.14.5.5) Remark. Even if X is smooth it may happen that $B_1(X/Y)_{\mathbb{Q}} \neq N_1(X/Y)_{\mathbb{Q}}$ as shown by the following example of [Ceresa - Collino83].

Let $Y \subset \mathbb{P}^4$ be a generic hypersurface of degree at least five with a single ordinary node locally given by the equation $xy - uv = 0$. Let $p : X \to Y$ be the blow-up of this node. The exceptional divisor of p is $E \cong \mathbb{P}^1 \times \mathbb{P}^1$. Let $L_1, L_2 \subset E$ be two intersecting lines. Show that L_1 and L_2 are numerically equivalent on X.

One can prove that the lines L_i are not algebraically equivalent.

The following exercises explore some properties of nonproper rationally connected varieties.

3.15 Exercise. Let X be a rationally chain connected scheme over a field k and $x \in X$ a general closed point.

(3.15.1) Show that there is a sequence of subschemes $\{x\} = X_0 \subset X_1 \subset \dots \subset X_m = X$ and irreducible subschemes $V_i \subset \operatorname{Hom}(\mathbb{P}^1, X_i)$ such that $u_i : \mathbb{P}^1 \times \operatorname{Hom}(\mathbb{P}^1, X_i) \to X_i$ is dominant and $f(0) \in X_{i-1}$ for every $[f] \in V_i$.

(3.15.2) If X satisfies the conditions of (3.15.1), then X is rationally chain connected.

(3.15.3) Show that if F is a torsion free sheaf on X, then $H^0(X, F)$ is finite dimensional.

3.16 Exercise. Let S be a smooth and separably rationally connected surface, $S' \supset S$ a smooth compactification and $C = \cup C_i := S' - S$.

(3.16.1) Show that the intersection matrix of the curves C_i is negative definite.

(3.16.2) Using the sequence

$$0 \to \omega_{S'} \to \omega_{S'}(mC) \to \omega_{mC} \to 0$$

show that $h^1(mC, \mathcal{O}_{mC}) = 0$ for every $m > 0$.

(3.16.3) By [Artin62] this implies that the connected components of C can be contracted to points. This way we obtain a compactification $\bar{S} \supset S$ such that $\bar{S}$ has rational singularities and $\bar{S} - S$ consists of finitely many points.

(3.16.4) The converse of (3.16.3) is very hard even in the simplest cases. For instance, let F be a proper surface with quotient singularities such that $-K_F$ is ample (i.e. F is a log Del Pezzo surface). Then $F - \operatorname{Sing} F$ is rationally connected by [Keel-McKernan99].

IV.4 Growing Chains of Rational Curves

Let X be a proper variety. If X is rationally chain connected, then any two points can be connected by a chain of rational curves. In general we can define an equivalence relation on closed points by declaring two points equivalent if they can be connected by a chain of rational curves. The aim of this section is to study this equivalence relation. It is not harder, and very useful, to study a more general class of algebraic relations. The main results of the section are (4.13) and (4.16). Some applications are discussed in the next section.

4.1 Definition. Let S be a set. A *relation* on S is a subset of $R \subset S \times S$. We say that $s_1, s_2 \in S$ are in relation if $(s_1, s_2) \in R$. If R_1, R_2 are relations on S, then we can define their *product* $R_1 * R_2$ by

$$(s_1, s_2) \in R_1 * R_2 \Leftrightarrow \exists s_3 \text{ such that } (s_1, s_3) \in R_1,\ (s_3, s_2) \in R_2.$$

If X is a scheme and S the set of points of X, then it is of interest to study those relations which are compatible with the scheme structure of X. Instead of the most straightforward definition, the following version seems more convenient:

4.2 Definition. Let X/S be a scheme.

(4.2.1) An *algebraic relation* on X is a scheme R together with a pair of morphisms $p : R \to X$ and $u : R \to X$. We frequently say that R is a relation, when the morphisms p, u are understood. We always assume that all occuring schemes are of finite type.

(4.2.2) A *proalgebraic relation* is a union of countably many algebraic relations.

(4.2.3) By Chevalley's theorem (see e.g. [Hartshorne77, Exercise II.3.19]), the images

$$\tilde{R} := \operatorname{im}[R \xrightarrow{(p,u)} X \times X] \subset X \times X \quad \text{and} \quad \tilde{R}(x) := u(p^{-1}(x)) \subset X$$

are constructible subsets (countable unions of constructible subsets in case of a proalgebraic relation). $\tilde{R}(x)$ is called the *relation class* of x. R is called an *algebraic realization* of the set theoretic relation $\tilde{R}$.

It is possible to work always with constructible subsets of $X \times X$ as relations, but it is easier to do constructions with the above definition.

(4.2.4) A (pro)algebraic relation R is called an *equivalence relation* if $\tilde{R}$ is an equivalence relation.

(4.2.5) We say that an algebraic relation (R, p, u) is *open* (resp. *proper*) if both of the morphisms p, u are open (resp. proper).

(4.2.6) We say that an algebraic relation (R, p, u) is *irreducible* if R is irreducible.

(4.2.7) Two (pro)algebraic relations R_1, R_2 are called *set theoretically equivalent* if $\tilde{R}_1 = \tilde{R}_2$. If R_1 and R_2 are set theoretically equivalent and R_1 is open (resp. proper or irreducible), then in general R_2 need not possess the same property.

(4.2.8) If $S' \to S$ is a morphism, then by base change we obtain the relation $R' = R \times_S S'$.

4.3 Definition–Lemma. Let $X \xleftarrow{p_i} R_i \xrightarrow{u_i} X$ be algebraic relations. Consider the following fiber product diagram:

$$\begin{array}{ccccc} R_3 & = & R_1 \times_X R_2 & \xrightarrow{p_2^* u_1} & R_2 \\ & & \downarrow{\scriptstyle u_1^* p_2} & & \downarrow{\scriptstyle p_2} \\ & & R_1 & \xrightarrow{u_1} & X. \end{array}$$

The scheme R_3 together with the morphisms

$$p_3 := p_1 \circ u_1^* p_2 : R_3 \to X \quad \text{and} \quad u_3 := u_2 \circ p_2^* u_1 : R_3 \to X$$

define an algebraic relation. It is called the *product* of R_1 and R_2 and is denoted by $R_1 * R_2$.

It is clear from the definition that

$$\widetilde{R_1 * R_2} = \tilde{R}_1 * \tilde{R}_2.$$

4.3.1 Lemma. *Let (R_i, p_i, u_i), $i = 1, 2$ be algebraic relations. If both of them are open (resp. proper), then $R_1 * R_2$ is also open (resp. proper).*

Proof. Follows from the fact that fiber products of open (resp. proper) morphisms are again open (resp. proper). □

In practice it is frequently convenient to describe a relation in a different way. The following motivating example explains how the definition arose:

4.4 Example. In applications we are interested in two types of relations obtained from families of subschemes of a scheme X. These are constructed as follows.

(4.4.1) Symmetric case. Let $U_v : v \in V$ be a family of subschemes of X. We define a set theoretic relation $\tilde{R}$ such that $(x_1, x_2) \in \tilde{R}$ iff there is a $v \in V$ such that $x_1, x_2 \in U_v$.

(4.4.2) Asymmetric case. Let $U_v^\ell : v \in V$ and $U_v^r : v \in V$ be two families of subschemes of X. We define a set theoretic relation $\tilde{R}$ such that $(x_1, x_2) \in \tilde{R}$ iff there is a $v \in V$ such that $x_1 \in U_v^\ell$ and $x_2 \in U_v^r$. This includes the symmetric case if $U_v^\ell = U_v^r$ for every $v \in V$.

(4.4.3) Algebraic case. We are interested in cases when $U_v : v \in V$ is an algebraic family. More generally, we consider cases when the U_v are not subschemes of X but admit a morphism to X. This situation can be codified by diagrams of schemes

$$V \xleftarrow{s} U \xrightarrow{w} X \quad \text{resp.} \quad V \xleftarrow{s^\ell} U^\ell \xrightarrow{w^\ell} X \text{ and } V \xleftarrow{s^r} U^r \xrightarrow{w^r} X.$$

(4.4.4) Reflexive relations. We are mainly interested in relations which are close to being reflexive, that is, $(x, x) \in \tilde{R}$ whenever x is in relation with something. For the relation constructed in (4.4.2) this means that $U_v^\ell \subset U_v^r$ for every $v \in V$. At the scheme level this can be guaranteed by a morphism $\sigma : U^\ell \to U^r$ such that $s^\ell = s^r \circ \sigma$ and $w^\ell = w^r \circ \sigma$.

(4.4.5) Normal form. It is very convenient to break the left/right symmetry in the above definition as follows. Instead of looking at pairs (U_v^ℓ, U_v^r) we look at pairs (u, U_v^r) such that $u \in U_v^\ell$. The set of all such pairs defines the same set theoretic relation. In the scheme setting we obtain the new relation

$$V_n := U^\ell, U_n^\ell := U^\ell, U_n^r := U^\ell \times_V U^r,$$

called the *normal form* of the relation given in (4.4.3). σ becomes a section $\sigma_n : V_n \to U_n^r$ such that $w_n^\ell = w_n^r \circ \sigma_n$. The schemes and morphisms $U_n^\ell, s_n^\ell, w_n^\ell$ are expresseble in terms of the other data and they are dropped from the notation in the final definition.

As in (4.3.1) we obtain the following:

4.5 Lemma (Notation as above). *Assume that we have* $V \xleftarrow{s^\ell} U^\ell \xrightarrow{w^\ell} X$, $V \xleftarrow{s^r} U^r \xrightarrow{w^r} X$ *and* $\sigma : U^\ell \to U^r$ *as in (4.4.3–4). If the morphisms* s^ℓ, s^r, w^ℓ, w^r *are all proper (resp. open) then the morphisms of the normal form* $s^r_n, w^r_n, w^r_n \circ \sigma_n$ *are also proper (resp. open).* □

4.6 Definition. Let U, V, X be schemes.

(4.6.1) The collection of schemes and morphisms

$$(V \xleftarrow{s} U \xrightarrow{w} X, V \xrightarrow{\sigma} U) \quad \text{such that} \quad s \circ \sigma = id_V$$

is called a *prerelation.* We frequently say that U is a prerelation if the other data are clear from the context.

(4.6.2) A prerelation is called *proper* (resp. *open*) if the morphisms $s, w, w \circ \sigma$ are all proper (resp. open).

(4.6.3) A prerelation is called *connected* if s has connected fibers.

(4.6.4) As before, $\tilde{U}(v) := w(s^{-1}(v))$ is called the *relation class* of $v \in V$.

(4.6.5) If $S' \to S$ is a morphism, then by base change we obtain the prerelation $U' = U \times_S S'$ etc.

4.7 Definition. Let $(V_i \xleftarrow{s_i} U_i \xrightarrow{w_i} X, V_i \xrightarrow{\sigma_i} U_i)$, $i = 1, \ldots, k$ be a collection of prerelations and $U_0 = X, s_0 = w_0 = \sigma_0 = id$ the identity prerelation. Fix an integer $m > 0$. Let $x_1, x_2 \in X$ be points. We say that x_1 and x_2 can be *connected by a* $(U_1, \ldots, U_k)$*-chain of length* m iff there are

(4.7.1) points $y_1 = x_1, y_2, \ldots, y_m, y_{m+1} = x_2 \in X$,

(4.7.2) a function $\tau : \{1, \ldots, m\} \to \{0, 1, \ldots, k\}$, and

(4.7.3) points $v_{\tau(j)} \in V_{\tau(j)}$ such that

$$y_j = w_{\tau(j)} \circ \sigma_{\tau(j)}(v_{\tau(j)}) \quad \text{and} \quad y_{j+1} \in \tilde{U}_{\tau(j)}(v_{\tau(j)}) \quad \forall j = 1, \ldots, m.$$

We say that x_1 and x_2 can be *connected by a* $(U_1, \ldots, U_k)$-chain iff x_1 and x_2 can be connected by a $(U_1, \ldots, U_k)$-chain of length m for some m.

4.8 Theorem. *Let* X/S *be a scheme and* $(V_i \xleftarrow{s_i} U_i \xrightarrow{w_i} X, V_i \xrightarrow{\sigma_i} U_i)$, $i = 1, \ldots$ *prerelations.*

(4.8.1) *For every* k, m *there is an algebraic relation* $\mathrm{Chain}_m(U_1, \ldots, U_k)$ *such that for every* $x_1, x_2 \in X$, x_1 *and* x_2 *can be connected by a* $(U_1, \ldots, U_k)$*-chain of length at most* m *iff*

$$(x_1, x_2) \in \widetilde{\mathrm{Chain}}_m(U_1, \ldots, U_k).$$

(4.8.2) *There is a proalgebraic relation* $\mathrm{Chain}(U_1, U_2, \ldots)$ *such that for every* $x_1, x_2 \in X$, x_1 *and* x_2 *can be connected by a* $(U_1, U_2, \ldots)$*-chain iff*

$$(x_1, x_2) \in \widetilde{\mathrm{Chain}}(U_1, U_2, \ldots).$$

4.8.3 Notation. For the rest of the section, $\mathrm{Chain}(U_1, U_2, \ldots)$ and $\mathrm{Chain}_m(U_1, \ldots, U_k)$ always stands for the particular relation constructed during the proof. This is important since the construction has further pleasant properties (4.8.4).

$\widetilde{\mathrm{Chain}}(U_1, U_2, \ldots)$ is also denoted by $\langle U_1, U_2, \ldots \rangle$. It is called the *set theoretic relation generated by the* U_i.

4.8.4 Complement (Notation as above). *If every prerelation* U_i *is proper (resp. open), then* $\mathrm{Chain}_m(U_1, \ldots, U_k)$ *is proper (resp. open).*

Proof. First we construct $R_i := \mathrm{Chain}_1(U_i)$. Set $R_i = U_i$ with morphisms $p_i = w_i \circ \sigma_i \circ s_i$ and $u_i = w_i$. R_0 denotes the identity relation.

Let $\tau : \{1, \ldots, m\} \to \{0, 1, \ldots, k\}$ be any function. By induction on m we obtain that there is an algebraic relation

$$R_{\tau(1)} * R_{\tau(2)} * \ldots * R_{\tau(m)}.$$

It is clear that the following relation has the required properties:

$$\mathrm{Chain}_m(U_1, \ldots, U_k) := \bigcup_{\tau:\{1,\ldots,m\}\to\{0,1,\ldots,k\}} R_{\tau(1)} * R_{\tau(2)} * \ldots * R_{\tau(m)}.$$

(4.8.2) holds since the class of proper (resp. open) morphisms is closed under fiber products and compositions.

$$\mathrm{Chain}(U_1, U_2, \ldots) := \bigcup_m \mathrm{Chain}_m(U_1, \ldots, U_m)$$

defines a proalgebraic relation on X as required. □

4.9 Definition–Lemma (Notation as in (4.8)). Let $X \xleftarrow{p} \mathrm{Chain}_m(U_1, \ldots, U_k) \xrightarrow{u} X$ be the projections.

(4.9.1) We say that *two general points of* X *can be connected by a* $(U_1, \ldots, U_k)$*-chain of length* m if

$$\mathrm{Chain}_m(U_1, \ldots, U_k) \xrightarrow{p \times u} X \times_S X$$

is dominant. As in (IV.2.6) we see that if S is the spectrum of a field, then two general points of X can be connected by a $(U_1, \ldots, U_k)$-chain of length m iff there is an open subset $W \subset X \times X$ such that if $(x_1, x_2) \in W$ is a closed point, then x_1 and x_2 can be connected by a $(U_1, \ldots, U_k)$-chain of length m.

(4.9.2) Let $\pi : X \to Z$ be a morphism. We say that the relation classes of U_i are *contained in fibers of* π if there is a factorization

$$U_i \times U_i \xrightarrow{w_i \times w_i} X \times_Z X \hookrightarrow X \times X.$$

If S is the spectrum of a field, then an equivalent formulation is:

$\pi(\tilde{U}_i(x))$ is a point for every closed point $x \in X$.

(4.9.3) Assume that the relation classes of every U_i are contained in fibers of π. We say that *two general points of general fibers of π can be connected by a $(U_1, \ldots, U_k)$-chain of length m* if

$$\mathrm{Chain}_m(U_1, \ldots, U_k) \xrightarrow{p\times u} X \times_Z X$$

is dominant. If S is the spectrum of a field, then this is equivalent to the following:

There is an open subset $W \subset X \times_Z X$ such that if $(x_1, x_2) \in W$ is a closed point, then x_1 and x_2 can be connected by a $(U_1, \ldots, U_k)$-chain of length m.

4.10 Example: Chains of Rational Curves. Let X be a proper scheme, $V' \subset \mathrm{Chow}_1(X)$ the subscheme parametrizing connected 1-cycles with rational components, $V_i' \subset V'$ the irreducible components, $V_i' \xleftarrow{s_i'} U_i' \xrightarrow{w_i'} X$ the universal families. The morphisms s_i' and w_i' are proper by (II.2.2).

Taking the normal form (4.4.5) we obtain proper prerelations $(V_i \xleftarrow{s_i} U_i \xrightarrow{w_i} X, V_i : \xrightarrow{\sigma_i} U_i)$.

By (4.8) there is a proalgebraic relation RC_X such that $(x_1, x_2) \in \widetilde{\mathrm{RC}}_X$ iff x_1 and x_2 can be connected by a chain of rational curves. By (4.8.4) RC_X is the disjoint union of proper algebraic relations.

(4.10.1) If X is rationally chain connected, then $\widetilde{\mathrm{RC}}_X$ has a single relation class.

(4.10.2) Let S be a K3 surface which contains infinitely many rational curves and no (-2)-curves. Then $\widetilde{\mathrm{RC}}_S$ has one relation class which is the union of countably many rational curves. All the other classes consist of a single point.

(4.10.3) Let X be a smooth variety which is birational to an Abelian variety A. The relation classes of $\widetilde{\mathrm{RC}}_X$ are in one-to-one correspondence with the points of A.

4.11 Example: Chains of Free Rational Curves. Let X be a smooth scheme, $V' = \mathrm{Hom}^{\mathrm{free}}(\mathbb{P}^1, X)$ the subscheme parametrizing free morphisms, $V_i' \subset V'$ the irreducible components, $V_i' \xleftarrow{s_i'} V_i' \times \mathbb{P}^1 \xrightarrow{w_i'} X$ the universal families. The morphisms s_i' and w_i' are smooth hence open by (II.3.5.4).

Taking the normal form (4.4.5) we obtain open prerelations $(V_i \xleftarrow{s_i} V_i \times \mathbb{P}^1 \xrightarrow{w_i} X, V_i : \xrightarrow{\sigma_i} V_i \times \mathbb{P}^1)$.

By (4.8) there is a proalgebraic relation FRC_X such that $(x_1, x_2) \in \widetilde{\mathrm{FRC}}_X$ iff x_1 and x_2 can be connected by a chain of free rational curves. By (4.8.4) FRC_X is the disjoint union of open algebraic relations.

In general it can easily happen that the relation classes of Chain $(U_1, \ldots, U_m)$ are countable unions of algebraic subvarieties:

4.12 Exercises. Let $t \in \mathrm{Aut}\,\mathbb{P}^1$ be an automorphism of infinite order.

(4.12.1) The relation

$$\mathbb{P}^1 \leftarrow \mathbb{P}^1 \times \{0,1\} \xrightarrow{w} \mathbb{P}^1 \quad w(x,0) = x,\ w(x,1) = t(x),$$

is proper. It generates an equivalence relation R. The relation classes of R are the orbits of t, thus most of them are dense in $\mathbb{P}^1$.

(4.12.2) Let C be any smooth proper curve with two marked points $a, b \in C$. Let τ be the isomorphism

$$\tau : \mathbb{P}^1 \times \{a\} \cong \mathbb{P}^1 \xrightarrow{t} \mathbb{P}^1 \cong \mathbb{P}^1 \times \{b\},$$

and X the surface obtained from $\mathbb{P}^1 \times C$ by identifying the above two sections via τ. It is easy to see that X is projective.

$\mathbb{P}^1 \leftarrow \mathbb{P}^1 \times C \to X$ defines a proper and connected prerelation on X which generates an equivalence relation R. The equivalence classes of R are in one-to-one correspondence with the orbits of t on $\mathbb{P}^1$. Thus most of the equivalence classes are dense in X.

It is quite surprising that the above examples account for most of the possible pathological behaviour of proalgebraic equivalence relations. Statements of this type have appeared in the literature many times. For the applications the most important results are (4.13) and (4.16–17). In various forms they are contained in [Campana81; Kollár87a; Campana91,92; KoMiMo92abc; EKL95].

4.13 Theorem. *Let X be a normal variety, $(V_i \xleftarrow{s_i} U_i \xrightarrow{w_i} X, V_i \xrightarrow{\sigma} U_i)$, $i = 1, \dots, m$ open prerelations on X such that every s_i is flat with irreducible fibers. Then there is an open subvariety $X^0 \subset X$ and a morphism $\pi : X^0 \to Z^0$ with connected fibers such that:*

(4.13.1) *$\langle U_1, \dots, U_m\rangle$ restricts to an equivalence relation on X^0.*

(4.13.2) *For every $z \in Z^0$ the closure of $\pi^{-1}(z)$ in X coincides with the closure of a suitable $\langle U_1, \dots, U_m\rangle$-equivalence class.*

(4.13.3) *Any two general points of $\pi^{-1}(z)$ can be connected by a $(U_1, \dots, U_m)$-chain of length at most* $\dim X - \dim Z^0$.

The scheme theoretic version is the following:

4.13′ Theorem. *Let X/S be a normal scheme, $(V_i \xleftarrow{s_i} U_i \xrightarrow{w_i} X, V_i \xrightarrow{\sigma} U_i)$, $i = 1, \dots, m$ open prerelations on X such that every s_i is flat with irreducible fibers. Then there is an open subscheme $X^0 \subset X$ and a morphism $\pi : X^0 \to Z^0$ with connected fibers such that*

$$(4.13'.1) \qquad \operatorname{im}\left[\operatorname{Chain}_r(U_1, \dots, U_m) \xrightarrow{(s \times w)} X \times_S X\right] \cap (X^0 \times_S X) \quad \textit{and} \quad X^0 \times_{Z^0} X^0$$

have the same closure for $r \geq$ $\dim X - \dim Z^0$.

Proof. We may assume that X and S are irreducible and reduced. Let F be the field of rational functions on S and $X_F = X \times_S \operatorname{Spec} F$. Assume that we find $X_F^0 \subset X_F$ and $\pi_F : X_F^0 \to Z_F^0$ satisfying (4.13). We can extend Z_F^0 to an S-scheme of finite type Z' and we obtain a rational map $\pi' : X \dashrightarrow Z'$. The two schemes in (4.13'.1) agree over the generic point of S, thus they agree for suitable choice of X^0. By further shrinking X^0 we may assume that π' restricts to a morphism π on X^0. □

In some cases (4.13) implies that two general points of X can be connected by a $(U_1, \dots, U_m)$-chain of length at most $\dim X$:

4.14 Corollary. *Let X be a smooth projective variety with $N_1(X) \cong \mathbb{R}$. Let $s : U \to V$ be proper, open with irreducible fibers of dimension ≥ 1 and $w : U \to X$ an open and equidimensional morphism. Assume that w is finite on the fibers of s. There is an open set $W \subset X \times X$ such that if $(x_1, x_2) \in W$, then x_1, x_2 can be connected by a U-chain of length at most* $\dim X$.

Proof. The conditions of (4.13) are satisfied, hence there is an open subset $X^0 \subset X$ and a morphism $\pi : X^0 \to Z^0$ such that two general points of the same fiber can be connected by a U-chain of length at most $\dim X$. We are done if Z^0 is a point.

Assume the contrary and let $H_Z^0 \subset Z^0$ be a divisor, $H^0 = \pi^{-1}(H_Z^0)$ and $H \subset X$ the closure of H^0. $w^{-1}(H - H^0)$ has codimension at least 2 in U since w is equidimensional. Thus $w^{-1}(H - H^0) \cap U(v)$ has codimension at least 2 in $U(v)$ for general $v \in V$. Therefore if $v \in V$ is general, then

$$\dim(\tilde{U}(v) \cap (H - H^0)) \leq \dim \tilde{U}(v) - 2 \quad \text{and} \quad \tilde{U}(v) \cap H^0 = \emptyset.$$

H is an effective Cartier divisor on X. Since $N_1(X) \cong \mathbb{R}$, any effective divisor is ample. In particular, $H \cap \tilde{U}(v) \neq \emptyset$ and

$$\dim(H \cap \tilde{U}(v)) = \dim \tilde{U}(v) - 1.$$

Thus $\tilde{U}(v)$ intersects H^0 but is not contained in H. This is a contradiction, thus Z^0 is a point. □

4.15 Proof of (4.13). We may asume that X is irreducible.

Let $X \xleftarrow{p_{ij}} R_{ij} \xrightarrow{u_{ij}} X$ be the irreducible components of $R_i := \operatorname{Chain}_1(U_i)$ and $\Delta \subset X \times X$ the diagonal.

Assume first that for every i, j, if Δ is contained in the closure of $\tilde{R}_{ij}$, then $\Delta \supset \tilde{R}_{ij}$. This means that there is an open subset $X^0 \subset X$ such that if $x \in X^0$ and $(x, x') \in \tilde{R}_i$ for some i, then $x = x'$. Thus $Z^0 = X^0$ and $\pi = id$ satisfy the requirements.

Otherwise there is $Q_1 := R_{ij}$ such that Δ is contained in the closure of $\tilde{Q}_1$, $\tilde{q}_1 := \tilde{p}_{ij} : \tilde{Q}_1 \to X$ is dominant and has relative dimension at least 1.

Let k be the largest natural number with the following property:

There is irreducible component

$$(4.15.1) \quad Q_k \subset \operatorname{Chain}_k(U_1, \ldots, U_m) \quad \text{with morphisms} \quad X \xleftarrow{q_k} Q_k \xrightarrow{v_k} X$$

such that Δ is contained in the closure of $\tilde{Q}_k$, $\tilde{q}_k : \tilde{Q}_k \to X$ is dominant and has relative dimension $d(k) \geq k$. If $d(k) > k$ then we can set $Q_{k+1} = Q_k$, thus in fact $d(k) = k$.

For simplicity, let $\bar{Q}_k$ (resp. $\bar{Q}_k(x)$) denote the closure of $\tilde{Q}_k$ (resp. of $\tilde{Q}_k(x)$). Let $\bar{q}_k : \bar{Q}_k \to X$ and $\bar{v}_k : \bar{Q}_k \to X$ be the projections.

4.15.2 Lemma (Notation as above). *Then there is a dense open subset $X^0 \subset X$ with the following properties:*

(4.15.2.1) *$\bar{Q}_k(x)$ is irreducible for $x \in X^0$.*

(4.15.2.2) *If $x \in X^0$ and $v \in V_i$ such that $w_i \circ \sigma_i(v) \in \bar{Q}_k(x)$ then $\tilde{U}_i(v) \subset \bar{Q}_k(x)$.*

(4.15.2.3) *If $x \in X^0$ and $x' \in \bar{Q}_k(x)$ then*

$$X^0 \cap \bar{Q}_k(x) \subset \langle U_1, \ldots, U_m\rangle(x') \subset \bar{Q}_k(x).$$

Proof. In the course of the proof we gradually shrink X^0.

By assumption $\bar{Q}_k$ is not contained in any larger dimensional irreducible component of the closure of $\widetilde{\operatorname{Chain}}_{k+1}(U_1, \ldots, U_m)$. Thus by shrinking X^0 we may assume that if $x \in X^0$ then none of the irreducible components of $\bar{Q}_k(x)$ is contained in a larger dimensional component of the closure of $\widetilde{\operatorname{Chain}}_{k+1}(U_1, \ldots, U_m)(x)$.

Let $\bar{Q}'_k(x) \subset \bar{Q}_k(x)$ be an irreducible component containing x. Since $\bar{Q}_k(x)$ is connected, if (4.15.2.1) or (4.15.2.2) fail then there is a $v \in V_i$ such that $w_i \circ \sigma_i(v) \in \bar{Q}'_k(x)$ but $\tilde{U}_i(v) \not\subset \bar{Q}'_k(x)$.

$w_i \circ \sigma_i : V_i \to X$ is open by assumption, hence there is an irreducible component $S \subset (w_i \circ \sigma_i)^{-1}(\bar{Q}_k(x))$ such that $w_i \circ \sigma_i : S \to \bar{Q}'_k(x)$ is dominant. Set $W := w_i(s_i^{-1}(S)) \subset X$. $\bar{W}$ is an irreducible subvariety of X which contains $\bar{Q}'_k(x)$. $\bar{W}$ also contains $\tilde{U}_i(v)$ which is not contained in $\bar{Q}'_k(x)$. Thus $\dim W > \dim \bar{Q}'_k(x) = k$. Every point $x' \in W$ can be connected to a point of $\tilde{Q}'_k(x)$ by a U_i-chain of length one, thus W is contained in the closure of $\widetilde{\operatorname{Chain}}_{k+1}(U_1, \ldots, U_m)(x)$, a contradiction.

This also shows that if $x' \in \bar{Q}_k(x)$ then $\langle U_1, \ldots, U_m\rangle(x') \subset \bar{Q}_k(x)$. It remains to show that $\langle U_1, \ldots, U_m\rangle(x')$ always contains $X^0 \cap \bar{Q}_k(x)$.

Choose an open and dense subscheme $Q^0_k \subset \tilde{Q}_k$. By further shrinking X^0 we may assume that the following conditions are satisfied:

(4.15.2.4.1) $\Delta \cap (X^0 \times X^0) \subset Q_k^0$; and

(4.15.2.4.2) the coordinate projections $\bar{q}_k, \bar{v}_k$ are flat over X^0. By (4.15.4) they are also open and by (4.15.2.4.1) they are surjective when restricted to Q_k^0.

Using openness of $\bar{v}_k$ we conclude that if $x' \in X^0$, then

$$B(x') := \{x \in \bar{Q}(x') | x' \in Q_k^0(x)\} \quad \text{is open in} \quad \bar{Q}_k(x').$$

Pick any $x \in X^0$ and $x' \in X^0 \cap \bar{Q}_k(x)$. $Q_k^0(x)$ and $B(x')$ are both open in $\bar{Q}_k(x)$; pick $x'' \in Q_k^0(x) \cap B(x')$. Then x'' can be reached from x by a $(U_1, \dots, U_m)$-chain of length k and x' can be reached from x'' by a $(U_1, \dots, U_m)$-chain of length k. Thus x' can be reached from x by a $(U_1, \dots, U_m)$-chain of length $2k$. This proves (4.15.2.3). □

4.15.3 End of Proof of (4.13). By (4.15.2.4.2) $\bar{Q}_k \to X$ is flat over X^0. This gives a morphism

$$\rho : X^0 \to \mathrm{Hilb}(X);$$

let $Y^0 \subset \operatorname{im} \rho$ be a dense open subscheme in the image. Let $T \subset Y^0 \times X$ be the universal family with cycle map $t : T \to X$ and projection $s : T \to Y^0$.

The fibers of s are the subschemes $\bar{Q}_k(x)$ and by (4.15.2) any two such are either disjoint or they coincide. Thus $t : T \to X$ is one-to-one on closed points. In particular it is quasi finite hence by (4.15.4) open since X is normal. Replace X^0 by $X^0 \cap \operatorname{im} t$. Set $T^0 = t^{-1}(X^0)$ and let $t^0 : T^0 \to X^0$ be the restriction of t.

In characteristic zero t^0 is an open immersion hence $t^0 : T^0 \to X^0$ is an isomorphism. Set

$$\pi := s \circ t^{-1} : X^0 \to Z^0 := Y^0.$$

In positive characteristic there is a morphism $\tau : X^0 \to T^0$ such that $\tau \circ t^0 : T^0 \to T^0$ is purely inseparable and the identity on the set of points. Set $\pi := s \circ \tau$.

By (4.15.2) the fibers of $\pi : X^0 \to Z^0$ are precisely the relation classes of $\langle U_1, \dots, U_m \rangle$ restricted to X^0. Thus the restriction is an equivalence relation.

If $z \in Z^0$ and $x \in \pi^{-1}(z)$ is a point, then $x \in X^0$ so $\tilde{Q}(x)$ is dense in $\pi^{-1}(z)$. Thus a general point of $\pi^{-1}(z)$ can be connected to x by a $(U_1, \dots, U_m)$-chain of length $k = \dim X^0 - \dim Z^0$. □

4.15.4 Exercise. Let $f : X \to Y$ be a dominant morphism of finite type between irreducible schemes. Let $y \in Y$ be a point such that Y is normal at y. Assume that $\dim_x f^{-1}(y) = \dim X - \dim Y$. Show that f is open at x. ([Mumford76, 3.10] calls this the *fundamental openness principle.*)

A similar result holds for proper prerelations but the chains get longer:

4.16 Theorem. *Let X be a normal, proper variety and $(V_i \xleftarrow{s_i} U_i \xrightarrow{w_i} X$, $V_i \xrightarrow{\sigma} U_i)$, $i = 1, \dots, m$ proper and connected prerelations on X. Then there* *is an open subvariety $X^0 \subset X$ and a proper morphism $\pi : X^0 \to Z^0$ with connected fibers such that*

(4.16.1) *$\langle U_1, \dots, U_m \rangle$ restricts to an equivalence relation on X^0.*

(4.16.2) *$\pi^{-1}(z)$ coincides with a $\langle U_1, \dots, U_m \rangle$ equivalence class for every $z \in Z^0$,*

(4.16.3) *for every $z \in Z^0$ any two points of $\pi^{-1}(z)$ can be connected by a $(U_1, \dots, U_m)$-chain of length at most $2^{\dim X - \dim Z} - 1$.*

4.16.4 Exercise. Let $c_i \in C_i$ be pointed proper curves and $X = \prod_{1 \le i \le n} C_i$. Let U_i, $i = 1, \dots, n$ be the prerelation

$$C_{i+1} \times \cdots \times C_n \leftarrow C_i \times \cdots \times C_n \cong \{c_1\} \times \cdots \times \{c_{i-1}\} \times C_i \times \cdots \times C_n \to C_1 \times \cdots \times C_n.$$

Show that any two points of X can be connected by a $(U_1, \dots, U_n)$-chain. For general points the minimal length of such a chain is $2^n - 1$.

The scheme theoretic version is again not hard to obtain. The proof is the same as for (4.13').

4.16' Theorem. *Let X/S be a normal scheme, proper over S and $(V_i \xleftarrow{s_i} U_i \xrightarrow{w_i} X, V_i \xrightarrow{\sigma} U_i)$, $i = 1, \dots, m$ proper and connected prerelations on X. Then there is an open subscheme $X^0 \subset X$ and a proper morphism $\pi : X^0 \to Z^0$ with connected fibers such that*

$$\text{(4.16'.1)} \quad \operatorname{im}\left[\operatorname{Chain}_r(U_1, \dots, U_m) \xrightarrow{(s \times w)} X \times_S X\right] \cap (X^0 \times_S X) = X^0 \times_{Z^0} X^0$$

for every $r \ge 2^{\dim X - \dim Z} - 1$. □

The following version with countably many prerelations is used frequently:

4.17 Theorem. *Let X be a proper and normal variety and $(V_i \xleftarrow{s_i} U_i \xrightarrow{w_i} X, V_i \xrightarrow{\sigma} U_i)$, $i = 1, \dots$ proper and connected prerelations on X. Then there is an open subset $X^0 \subset X$ and a proper morphism $\pi : X^0 \to Z^0$ with connected fibers such that*

(4.17.1) *$\pi^{-1}(z)$ coincides with a $\langle U_1, U_2, \dots \rangle$-relation class for very general $z \in Z^0$,*

(4.17.2) *every relation class of $\langle U_1, U_2, \dots \rangle$ intersects X^0 in a union of fibers of π,*

(4.17.3) *for every $z \in Z^0$ any two points of $\pi^{-1}(z)$ can be connected by an $(U_1, U_2, \dots)$-chain of length at most $2^{\dim X - \dim Z} - 1$.*

Proof. Applying (4.16) to the first t prerelations we obtain $\pi_t : X_t^0 \to Z_t^0$. Choose m such that the fiber dimension of π_m is maximal. I claim that $\pi := \pi_m$ satisfies the conclusions of (4.17). Clearly (4.16.3) implies (4.17.3).

Increasing t increases the equivalence classes, thus (4.16.2) implies (4.17.2). Every equivalence class is connected and by (4.16) general equivalence classes of $\langle U_1,\dots,U_t\rangle$ are irreducible. Thus if

$$\langle U_1,\dots,U_m\rangle(x) \neq \langle U_1,\dots,U_t\rangle(x)$$

for some $t > m$ and very general x, then

$$\dim\langle U_1,\dots,U_m\rangle(x) < \dim\langle U_1,\dots,U_t\rangle(x).$$

This contradicts the choice of m, showing (4.17.1). □

4.18 Proof of (4.16). We may asume that X is irreducible.

Let $\Delta \subset X \times X$ be the diagonal. If $w_i(U_i) \supset \Delta$ implies that $w_i(U_i) = \Delta$, then there is an open subset $X^0 \subset X$ such that if $x \in X^0$ and $(x,x') \in \tilde{U}_i$ for some i then $x = x'$. Thus $Z^0 = X^0$ and $\pi = id$ satisfy the requirements.

Otherwise there is an index, say $i = 1$ and an irreducible component $\sigma_1(V_1) \subset U_1^0 \subset U_1$ such that $w_1 : U_1^0 \to X$ is dominant and $w_1(U_1^0(v)) \subset X$ is positive dimensional for general $v \in V_1$. There are open subsets $V^0 \subset V_1, U^0 \subset U_1^0$ such that the induced prerelation

$$V^0 \xleftarrow{s_1|U^0} U^0 \xrightarrow{w_1|U^0} X$$

is open and $s_1|U^0$ is flat with irreducible fibers. Thus (4.13) applies and we obtain an open subvariety $X^* \subset X$ and a morphism $\tau : X^* \to Y^*$ whose properties are described in (4.13) (we use τ (resp. X^*, Y^*) instead of π (resp. X^0, Z^0) to avoid confusion with the notation of (4.16)).

Let $\Gamma \subset X \times Y^*$ be the closure of the graph of τ. By shrinking Y^* we may assume that the projection $\Gamma \to Y^*$ is flat. Thus we have a morphism $\rho^* : Y^* \to \mathrm{Hilb}(X)$. Let $Y \supset Y^*$ be a normal compactification of Y^* such that ρ^* extends to $\rho : Y \to \mathrm{Hilb}(X)$. Let X' be the unique irreducible component of the normalization of $Y \times_{\mathrm{Hilb}(X)} \mathrm{Univ}(X)$ which dominates X. Let $u : X' \to X$ and $\tau' : X' \to Y$ denote the projections. Both are proper and u is birational.

The prerelations $V_i \leftarrow U_i \to X$ induce proper and connected prerelations

$$V_i \leftarrow U_i' := U_i \times_X X' \to X'.$$

These have the following obvious property:

4.18.1 Claim (Notation as above).

(4.18.1.1) *Two points $x_1', x_2' \in X'$ can be connected by a $(U_1',\dots,U_m')$-chain of length r iff $u(x_1')$ and $u(x_2')$ can be connected by a $(U_1,\dots,U_m)$-chain of length r.*

(4.18.1.2) *If $\pi' : (X')^0 \to (Z')^0$ satisfies the requirements of (4.16), then $(X')^0 = u^{-1}(X^0)$ for some open set $X^0 \subset X$ and π' factors through $\pi : X^0 \to Z^0 = (Z')^0$.* □

The main advantage of replacing X with X' is the following:

4.18.2 Claim (Notation as above). *Let $x_1', x_2' \in X'$ be closed points such that $\tau'(x_1') = \tau'(x_2')$. Then x_1' and x_2' can be connected by a $(U_1', \dots, U_m')$-chain of length* $\dim X - \dim Y$.

Proof. In scheme theoretic language we assert that

$$\operatorname{im}\left[\operatorname{Chain}_r(U_1') \to X' \times X'\right] \supset X' \times_Y X' \quad \text{for } r \geq \dim X - \dim Y.$$

By construction $X' \to Y$ is equidimensional with connected fibers. Thus by (4.18.3) $X' \times_Y X'$ is irreducible. Therefore it is sufficient to check the containment over an open set. The latter is implied by (4.13.2). □

The prerelations U_i' induce prerelations

$$V_i \leftarrow U_i'' := U_i' \to Y.$$

$\dim Y < \dim X$, thus by induction on the dimension we may assume that (4.16) holds for Y. Thus there is an open subset $Y^0 \subset Y$ and a proper morphism $\pi_Y : Y^0 \to Z^0$. Let $(X')^0 := \pi^{-1}(Y^0)$, $\pi' := \pi_Y \circ \tau'$.

$\pi' : (X')^0 \to Z^0$ is proper by construction. Let $x_1', x_2' \in (X')^0$ be closed points which are connected by a $(U_1', \dots, U_m')$-chain. Then $\tau'(x_1')$ and $\tau'(x_2')$ are connected by a $(U_1'', \dots, U_m'')$-chain. Thus $\pi_Y(\tau'(x_1')) = \pi_Y(\tau'(x_2'))$. This means that x_1' and x_2' are in the same fiber of π'.

Conversely, assume that x_1' and x_2' are in the same fiber of π'. By induction $\tau'(x_1')$ and $\tau'(x_2')$ are connected by a $(U_1'', \dots, U_m'')$-chain of length d where $d = 2^{\dim Y - \dim Z} - 1$. The connecting chain is given by the data

$$a_j, b_j \in U_{i_j}''(v_j) = U_{i_j}'(v_j),\ j = 1, \dots, d \quad \text{such that}$$
$$\tau'(w_{i_1}'(a_1)) = \tau'(x_1'),\ \tau'(w_{i_j}'(b_j)) = \tau'(w_{i_{j+1}}'(a_{j+1})),\ \tau'(w_{i_d}'(b_d)) = \tau'(x_2').$$

We would like to connect x_1' and x_2' using the links $U_{i_j}'(v_j)$. In general $w_{i_1}'(a_1) \neq x_1'$, but they are contained in the same fiber of τ'. Thus by (4.18.2) they can be connected by a U_1-chain of length $\dim X - \dim Y$. Similarly, $w_{i_j}'(b_j)$ and $w_{i_{j+1}}'(a_{j+1})$ are contained in the same fiber of τ' and they can be connected by a U_1-chain of length $\dim X - \dim Y$. Thus x_1' and x_2' can be connected by a $(U_1', \dots, U_m')$-chain of length at most $(d+1)(\dim X - \dim Y) + d$. Elementary inequalities show that

$$(d+1)(\dim X - \dim Y) + d \leq 2^{\dim X - \dim Z} - 1. \qquad \square$$

4.18.3 Exercise. Let Y be normal, X irreducible and $f : X \to Y$ an equidimensional morphism with geometrically connected fibers. Show that $X \times_Y X$ is irreducible.

IV.5 Maximal Rationally Connected Fibrations

In this section we apply the results of the previous section to study the equivalence relation given by the family of all rational curves on a variety X. The main result (5.2) asserts that there is a map $\pi : X \dashrightarrow Z(X)$, called the maximal rationally chain connected fibration, such that the fibers are rationally chain connected and there are "very few" rational curves on X not contained in the fibers of π. For smooth varieties the general fiber of π is rationally connected and π is called the maximal rationally connected fibration or MRC-fibration. For smooth varieties the assignment $X \mapsto Z(X)$ is functorial for dominant maps (5.5).

The notion of MRC-fibrations clarifies the conjecture (3.8.1) and this leads to a proof of the conjecture in dimension 3 (5.7.1). The dimension of the target of the MRC-fibration can be characterized by free rational curves (5.8); this gives a common generalization of (1.9) and (3.7). MRC-fibrations behave very well in families (5.9).

5.1 Definition. Let X be a normal and proper variety, $X^0 \subset X$ an open subset.

(5.1.1) A proper morphism $\pi : X^0 \to Z^0$ is called a *rationally chain connected fibration* if the fibers of π are rationally chain connected and $\pi_*\mathcal{O}_{X^0} = \mathcal{O}_{Z^0}$. (The last condition is important only in positive characteristic where it implies that π can not be factored through a purely inseparable morphism $Z^1 \to Z^0$.)

Equivalently, there is an irreducible component $R_i \in \mathrm{RC}_X$ (4.11) such that

$$\mathrm{im}[R_i \to X \times X] \supset X^0 \times_{Z^0} X^0.$$

(5.1.2) A proper morphism $\pi : X^0 \to Z^0$ is called a *maximal rationally chain connected fibration* or *MRCC-fibration* if π is a rationally chain connected fibration and if $\pi' : X' \to Z'$ is any rationally chain connected fibration then there is a map $\tau : Z' \dashrightarrow Z^0$ such that $\pi = \tau \circ \pi'$.

(5.1.3) In the above definitions X^0 is not unique. Therefore we frequently say that a map $\pi : X \dashrightarrow Z$ is a (maximal) rationally chain connected fibration if there are open subsets $X^0 \subset X$ and $Z^0 \subset Z$ such that $\pi|X^0 : X^0 \to Z^0$ is a morphism satisfying the above properties.

5.2 Theorem [Campana92, KoMiMo92b]. *Let X be a normal and proper variety. Then the maximal rationally chain connected fibration $\pi : X \dashrightarrow Z$ exists.*

Proof. In (4.10) we defined a proalgebraic equivalence relation RC_X such that $(x_1, x_2) \in \widetilde{\mathrm{RC}}_X$ iff x_1 and x_2 can be connected by a chain of rational curves.

By (4.17) there is an open subset $X^0 \subset X$ and a proper morphism $\pi : X^0 \to Z^0$ such that if $x_1, x_2 \in X^0$ are very general, then $\pi(x_1) = \pi(x_2)$

iff x_1 and x_2 can be connected by a chain of rational curves. (In positive characteristic replace π by its Stein factorization. This is a purely inseparable morphism, thus does not change any set theoretic property.) π is a rationally chain connected fibration by construction.

Assume that $\pi' : X' \to Z'$ is another rationally chain connected fibration. Let $\Gamma := \operatorname{im}[X' \cap X^0 \to Z' \times Z^0]$. I claim that the projection $\Gamma \to Z'$ is birational, thus it is the graph of the map τ. The birationality of a morphism can be checked after a base change, thus we may assume that X is defined over an uncountable field.

If $x_1, x_2 \in X' \cap X^0$ such that $\pi'(x_1) = \pi'(x_2)$ then x_1 and x_2 can be connected by a chain of rational curves. If x_1 is very general, then this implies that $\pi(x_1) = \pi(x_2)$. Therefore a very general fiber of π' is contained in a fiber of π. Therefore $\Gamma \to Z'$ is generically one-to-one, hence birational in characteristic zero. In positive characteristic π' can not be factored through a purely inseparable morphism by assumption, hence again $\Gamma \to Z'$ is birational.

This gives the rational map $Z' \dashrightarrow Z^0$. □

By (4.17) we get the following extra information:

5.2.1 Complement (Notation as above). *If X is defined over an uncountable field, then π^0 is characterized by the following property:*

If $z \in Z^0$ is very general and $C \subset X$ is a rational curve intersecting $\pi^{-1}(z)$, then $C \subset \pi^{-1}(z)$. □

5.3 Definition. Let X be a smooth proper variety over a field of characteristic zero and $\pi : X^0 \to Z^0$ the maximal rationally chain connected fibration. By shrinking Z^0 we may assume that π is smooth. In characteristic zero a smooth and rationally chain connected variety is rationally connected by (3.10.3), hence every fiber of π is rationally connected. $\pi : X^0 \to Z^0$ is called the *maximal rationally connected fibration* of X or the *MRC-fibration* of X for short.

As in (5.1.3) we frequently use the notation $\pi : X \dashrightarrow Z$.

Reformulating (5.2) we obtain:

5.4 Theorem [Campana92, KoMiMo92b]. *Let X be a smooth proper variety over a field of characteristic zero. Then the MRC-fibration $\pi : X \dashrightarrow Z$ exists and it is unique up to birational equivalence.* □

The MRC-fibration is functorial:

5.5 Theorem. *Let X_1, X_2 be smooth proper varieties over a field of characteristic zero and $f_X : X_1 \dashrightarrow X_2$ a dominant map. Let $\pi_i : X_i \dashrightarrow Z_i$ be the MRC-fibrations. Then there is a rational map $f_Z : Z_1 \dashrightarrow Z_2$ such that $\pi_2 \circ f_X = f_Z \circ \pi_1$.*

Proof. The proof is very similar to the proof of (5.2). We may assume that everything is defined over an uncountable field. Let $x_1, x_2 \in \pi_1^{-1}(z)$ be very general points for very general $z \in Z_1$. Then there is an irreducible rational curve $C \subset X_1$ containing x_1, x_2. f_X is defined at x_1, x_2, thus $f_X(C) \subset X_2$ is a rational curve containing $f_X(x_1), f_X(x_2)$. Thus $f_X(\pi_1^{-1}(z))$ is contained in a fiber of π_2. This gives the rational map f_Z on the set theoretic level.

An argument as in the proof of (5.2) shows that f_Z is a rational map. □

5.5.1 Exercise. Find an example which shows that the analog of (5.5) fails for normal varieties and for the maximal rationally chain connected fibration.

The following is one of the main open problems concerning MRC-fibrations. We consider it in more detail in the next section.

5.6 Conjecture. *Let X, Y be smooth proper varieties over a field of characteristic zero.*

(5.6.1) *Let $\pi : X \dashrightarrow Z$ be the MRC-fibration. Then Z is not uniruled.*

(5.6.2) *Let $\rho : Y \to \mathbb{P}^1$ be a dominant morphism whose smooth fibers are rationally connected. Then Y is rationally connected.*

5.6.3 Proposition. *The conjectures* (5.6.1) *and* (5.6.2) *are equivalent.*

Proof. Assume that (5.6.1) is true and let $\pi : Y \dashrightarrow Z$ denote the MRC-fibration. By (5.5) there is a factorization $Y \dashrightarrow \mathbb{P}^1 \dashrightarrow Z$. By (5.6.1) Z is not uniruled, thus Z is a point and Y is rationally connected. Thus (5.6.1) implies (5.6.2).

Conversely, assume that (5.6.2) holds and let $\pi : X \dashrightarrow Z$ be the MRC-fibration. By base change assume that everything is defined over an uncountable field. Assume that Z is uniruled and let $f : \mathbb{P}^1 \to Z$ be a morphism whose image passes through a very general point of Z. Let Y be a desingularization of $X \times_Z \mathbb{P}^1$ and $\rho : Y \to \mathbb{P}^1$ the projection. The general fiber of ρ is birational to a general fiber of π, hence rationally connected. Thus Y is rationally connected. This gives rational curves on X which intersect a very general fiber of π but are not contained in it, a contradiction. Thus (5.6.2) implies (5.6.1). □

(5.6) is closely related to some earlier conjectures:

5.7 Proposition. *Assume that* (5.6) *and* (1.12) *are true in dimensions at most k. Then* (3.8.1) *is also true in dimensions at most k.*

Proof. Assume that $H^0(X, (\Omega_X^1)^{\otimes m}) = 0$ for every $m > 0$. Let $\pi : X \dashrightarrow Z$ be a map. By choosing birational models suitably we may assume that X and Z are smooth and π is a morphism. We have natural inclusions

$$\pi^*\omega_Z^r \subset \left(\wedge^{\dim Z}\Omega_X^1\right)^{\otimes r} \subset \left(\Omega_X^1\right)^{\otimes r \dim Z}.$$

Thus

$$h^0(Z,\omega_Z^r) \le h^0\big(X,(\Omega_X^1)^{\otimes r \dim Z}\big) = 0,$$

and if (1.12) holds for Z, then Z is uniruled. If $\pi : X \dashrightarrow Z$ is the MRC-fibration, then (5.6) implies that Z is a point, thus X is rationally connected. □

In dimension 3 one can be more precise:

5.7.1 Corollary [KoMiMo82b]. *Let X be a smooth threefold over a field of characteristic zero. The following are equivalent:*

(5.7.1.1) *X is rationally connected;*

(5.7.1.2) *$H^0\big(X,(\Omega_X^1)^{\otimes m}\big) = 0$ for every $m > 0$;*

(5.7.1.3) *$H^0(X,\omega_X^m) = 0$ for every $m > 0$, $h^0(X,\Omega_X^1) = 0$ and $h^0(X,S^2\Omega_X^2) = 0$.*

Proof. (5.7.1.1) implies (5.7.1.2) by (3.8) and (5.7.1.2) implies (5.7.1.3). Assume (5.7.1.3).

X is uniruled by (1.12.1). Let $\pi : X \to Z$ be the MRC-fibration (where we assume that π is a morphism and X, Z are smooth and proper by choosing suitable birational models). We are done if $\dim Z = 0$.

$h^0(Z,\Omega_Z^1) \le h^0(X,\Omega_X^1) = 0$, thus Z has irregularity zero. If $\dim Z = 1$, then $Z \cong \mathbb{P}^1$. We see in Sect. 6 that X is rationally connected.

We are left with the case when $\dim Z = 2$. $h^0(Z,\omega_Z^2) \le h^0(X,S^2\Omega_X^2) = 0$. By (III.2.4) Z is a rational surface hence uniruled. By (6.6.2) this again leads to a contradiction. □

The following result is a common generalization of (1.9) and (3.7):

5.8 Theorem. *Let X be a smooth proper variety over a field of characteristic zero and $\pi : X \dashrightarrow Z$ the MRC-fibration. Then $\dim X - \dim Z$ is the largest number r such that there is a free morphism $f : \mathbb{P}^1 \to X$ such that*

(5.8.1) $$f^*T_X = \sum \mathcal{O}(a_i) \quad \text{and} \quad \#\{i | a_i > 0\} \ge r.$$

Proof. Let f be a morphism as above. By (II.3.9.2) the condition (5.8.1) also holds for small deformations of f. Thus we may assume in addition that $\operatorname{im} f$ passes through a very general point of X. Thus for a suitable $z \in Z$ there is a factorization

$$f : \mathbb{P}^1 \to \pi^{-1}(z) \to X.$$

This gives an exact squence

(5.8.2) $$0 \to f^*T_{\pi^{-1}(z)} \to f^*T_X \to f^*N_{\pi^{-1}(z)|X} \cong \mathcal{O}_{\mathbb{P}^1}^{\dim Z} \to 0.$$

f^*T_X has at least r positive summands, thus $r \le \dim X - \dim Z$.

Conversely, since $\pi^{-1}(z)$ is rationally connected, we can choose $f : \mathbb{P}^1 \to \pi^{-1}(z)$ such that $f^*T_{\pi^{-1}(z)}$ is ample by (3.9). Looking at (5.8.2) again we see that f^*T_X has at least $\dim X - \dim Z$ positive summands. □

MRC-fibrations behave very well in families:

5.9 Theorem [KoMiMo92b]. *Let $f : X \to S$ be a smooth proper morphism of varieties over a field of characteristic zero, S normal. Then there is an open subset $X^* \subset X$, an S-scheme Z^* and a proper morphism $\pi^* : X^* \to Z^*$ such that for every $s \in S$ the induced morphism $\pi^*_s : X^*_s \to Z^*_s$ is the MRC-fibration of X_s.*

In particular, the dimension of the image of the MRC-fibration is constant in smooth families.

Proof. By (4.16′) there is an open set $X^0 \subset X$, an S-scheme Z^0 and a proper morphism $\pi^0 : X^0 \to Z^0$ such that for very general $s \in S$ the induced morphism $\pi^0_s : X^0_s \to Z^0_s$ is the MRC-fibration of X_s. There are two problems that need to be considered:

π^0 may not give the MRC-fibration on some fibers and X^0 may be disjoint from some fibers.

Let $\pi_s : X_s \dashrightarrow Z_s$ be the MRC-fibration of the fiber X_s. First we prove that

(5.9.1) $$\dim X_s - \dim Z_s \le \dim X^0 - \dim Z^0.$$

Indeed, let $r = \dim X_s - \dim Z_s$. By (5.8) there is a free morphism $f_s : \mathbb{P}^1 \to X_s$ such that $f^*T_{X_s}$ has r positive summands. By (I.2.17) we can deform f_s to a morphism $f_g : \mathbb{P}^1 \to X_g$ (where $g \in S$ is a general point) and by (II.3.9.2) $f^*T_{X_g}$ has at least r positive summands. Applying (5.8) again we obtain (5.9.1).

By shrinking X^0 we may assume that π^0 is flat. Let $\rho : Z^0 \to \mathrm{Hilb}(X/S)$ be the corresponding morphism. ρ is an immersion since Z^0 parametrizes fibers of a morphism. Let $Z \supset \rho(Z^0)$ be the closure of the image, $p : U \to Z$ the universal family and $u : U \to X$ the cycle map. u is birational by construction. Let $E \subset U$ be the set of points where u is not a local isomorphism. $p(E) \subset Z$ is closed thus $Z^* := Z - p(E)$ is open. Set $U^* = U - p^{-1}(p(E))$ and $X^* := u(U^*)$. By construction $u^* : U^* \to X^*$ is an isomorphism; set $\pi^* := p \circ (u^*)^{-1}$.

The fibers of $p : U \to Z$ are rationally connected over an open set, hence every fiber of p is rationally chain connected by (3.5.2). Since U is proper over S and $u : U \to X$ is dominant, we conclude that u is proper hence surjective.

Let us look at $Z_s \xleftarrow{p_s} U_s \xrightarrow{u_s} X_s$. u_s is dominant and every fiber of p_s is rationally chain connected. Thus if $z \in Z_s$ is very general, then $u_s(U_s(z)) \subset X_s$ is contained in a fiber of π_s. By (5.9.1) the fibers of π_s have dimension at most r, thus $u_s(U_s(z))$ is a fiber of π_s. Therefore $u_s : U_s \to X_s$ is an injection over very general points of X_s. This shows that X^* intersects every fiber X_s in a dense open set and π_s^* is birational to the MRC-fibration on every fiber. □

5.10 Proof of (3.5.3). Let $f : X \to S$ be as in (3.5.3). We may assume that the base field is uncountable. By (3.5.2) there are countably many closed subvarieties $S_i \subset S$ such that X_s is rationally chain connected iff $s \in \cup S_i$. By induction on $\dim S$ we may assume that S is irreducible and that $\cup S_i$ is dense in S. Let $X_1 \to X$ be a resolution of singularities, $f_1 : X_1 \to S$ the composition. By shrinking S we may assume that f_1 is smooth over S and the fibers $f^{-1}(s)$ and $f_1^{-1}(s)$ are birational for every $s \in S$.

Let $\pi_1 : X_1 \dashrightarrow Z_1$ be the MRC-fibration of $X_1 \to S$. As in (4.18.2), by choosing suitable birational models $X_2 \to X_1$ and Z_2 and shrinking S we may assume that $\pi_2 : X_2 \to Z_2$ is a morphism and any two points in the same fiber of π_2 can be connected by a chain of rational curves. Let $p_2 : X_2 \to X$ and $f_Z : Z_2 \to S$ be the natural morphisms.

$X \leftarrow X_2 \to Z_2$ defines a proper prerelation on Z_2, call it U. By construction, if $s \in S$ and $x, y \in f_2^{-1}(s)$ are very general then

$$(5.10.1) \qquad (\pi_2(x), \pi_2(y)) \in \widetilde{\mathrm{Chain}}(U) \Leftrightarrow (p_2(x), p_2(y)) \in \widetilde{\mathrm{Chain}}(\mathrm{RC}_{X_s}).$$

(The implication $\Rightarrow$ always holds.) Therefore if $s \in \cup S_i$, then any two points of $f_Z^{-1}(s)$ can be connected by a U-chain.

By (4.16′) there is an open subset $Z_2^0 \subset Z_2$ and a proper morphism $\pi_Z : Z_2^0 \to T$ such that two points of Z_2^0 can be connected by a U-chain iff they are in the same fiber of p. $T \to S$ has relative dimension zero over a dense set, thus it has relative dimension zero over on open subset $S^0 \subset S$. By (5.10.1) the fibers of f are rationally chain connected for every $s \in S^0$. □

Instead of developing a theory of MRC-fibrations for nonproper varieties, here is an example where everything behaves very nicely. These results are taken from [BBK96].

5.11 Exercise. Let $X = G/H$ be a homogeneous space under the action of a connected algebraic group G.

(5.11.1) Show that there is a unique G-equivariant smooth morphism $p : X \to X'$ with the following properties:

(5.11.1.1) If $f : \mathbb{P}^1 \to X$ is a morphism, then $f(\mathbb{P}^1)$ is contained in a fiber of p.

(5.11.1.2) If $x_1 \in X$ is arbitrary and $x_2 \in p^{-1}(p(x_1))$ general, then there is a morphism $f_{12} : \mathbb{P}^1 \to X$ such that $x_1, x_2 \in f_{12}(\mathbb{P}^1)$.

Hint. Use (V.1.4.3.1) to reduce to the case when H is reduced. If H is reduced, then by the proof of (V.1.4) T_X is generated by global sections, hence every morphism $f : \mathbb{P}^1 \to X$ is free.

(5.11.2) Show that there is a unique sequence of G-equivariant smooth morphisms

$$P : X = X_0 \xrightarrow{p_0} X_1 \to \cdots \to X_{n-1} \xrightarrow{p_{n-1}} X_n = Z$$

with the following properties:

(5.11.2.1) If $f : \mathbb{P}^1 \to X_i$ is a morphism, then $f(\mathbb{P}^1)$ is contained in a fiber of p_i.

(5.11.2.2) If $x_1 \in X_i$ is arbitrary and $x_2 \in p_i^{-1}(p_i(x_1))$ general, then there is a morphism $f_{12} : \mathbb{P}^1 \to X_i$ such that $x_1, x_2 \in f_{12}(\mathbb{P}^1)$.

(5.11.2.3) Every morphism $f : \mathbb{P}^1 \to Z$ is constant.

(5.11.3) Let $P : X \to Z$ be as above and F a coherent G-sheaf on X. Using (3.15) show that $P_* F$ is a coherent G-sheaf on Z.

IV.6 Rationally Connected Varieties over Nonclosed Fields

The aim of this section is to study the following problem which arose in Sect. 5:

6.1.1 Problem. Let $f : X \to C$ be a morphism from a smooth projective variety X to a smooth curve C. Assume that the general fiber of f is separably rationally connected. Is it true that f has a section?

Let $K(C)$ be the function field of C and $X_{K(C)}$ the generic fiber of f. There is a one-to-one correspondence between sections of f and $K(C)$-valued points of $X_{K(C)}$. Thus (6.1.1) is a special case of the following more general question:

6.1.2 Problem. Let F be a field and X_F a variety over F. Find conditions on F and on X_F which imply that X_F has a point in F.

Let us observe first that the answer to (6.1.2) does not depend on the birational model chosen, at least for smooth and proper varieties:

6.2 Proposition [Nishimura55]. *Let X_F be a variety over a field F. Assume that there is an F-point $P \in X_F$ such that X_F is smooth at P. Let Y_F be a proper variety and $g : X_F \dashrightarrow Y_F$ a rational map defined over F. Then Y_F has an F-point.*

Proof. We can replace X_F by an affine neighborhood of P, thus we may assume that X_F is quasi projective. Let $U \subset X_F$ be the open set where g is a morphism. By taking sufficiently general hypersurface sections through P we obtain an irreducible curve $P \in C_F \subset X_F$ such that C_F is smooth at

P and $C_F \cap U \neq \emptyset$. Let $n : \bar{C}_F \to C_F$ be the normalization. n is a local isomorphism over P, let $\bar{P} := n^{-1}(P) \in \bar{C}_F$ be the corresponding F-point. The map $h := f \circ n : \bar{C}_F \dashrightarrow Y_F$ is defined over a dense open set, hence extends to a morphism $\bar{h} : \bar{C}_F \to Y_F$. Thus $\bar{h}(\bar{P}) \in Y_F$ is an F-point. □

6.2.1 Exercise. Find an example of a normal variety X_F with a point in F such that none of the desingularizations of X_F have F-points.

The most classical versions of (6.1.1–2) come from number theory, where the cases $F = \mathbb{Q}$ and $F = \mathbb{Z}/p\mathbb{Z}$ were investigated intensively. The guiding principle that emerges from these studies can be summarized as follows:

6.3 Principle. *Let X_F be a smooth projective variety over F, $\bar{F}$ an algebraic closure of F.*

(6.3.1) *If $X_{\bar{F}}$ is rationally connected, then X_F should have lots of F-points, at least if F is "nice" (e.g. F is a finite field, a function field of a curve or a sufficiently large number field).*

(6.3.2) *If $X_{\bar{F}}$ is not rationally connected, then X_F should have "few" F-points.*

6.3.3 Comments. The situation is well understood for curves. Already for surfaces most of the conjectures are unsolved. The precise formulation of what we mean by "few" points is complicated by Abelian varieties, which can possess infinitely many points over number fields, and these points can even be Zariski dense. See [Lang86; Mazur86] for a general discussion of these questions.

6.3.4 Example. Let F be a field with a finite Galois extension $F' \supset F$ and C a smooth curve over F. Let $p_i \in \mathbb{P}^1$ and $c_j \in C$ be F'-points, conjugate over F, neither of which is in F. On $C \times \mathbb{P}^1$ the points (c_j, p_i) are conjugate over F but none of them is in F. Blow them up to obtain $\pi : S' \to C \times \mathbb{P}^1$. The birational transforms of the sections $C \times \{p_i\}$ become contractible on S'. Let $S' \to S$ be the surface obtained by contracting them. S is defined over F, it is rationally chain connected and S has an F-point iff C has.

Thus (6.3) has no analog for singular rationally chain connected varieties.

The first general result which has considerable interest for us is the following:

6.3.5 Exercise [Chevalley35]. Let F be a finite field and $f(x_0, \ldots, x_n)$ a homogeneous polynomial of degree $< n+1$. Show that f has a nontrivial zero in F. (Note that the condition $\deg f < n+1$ is precisely the one needed to assure that $(f = 0)$ is rationally chain connected (V.2).)

Hints. Let $|F| = q$. Show that

$$h := 1 - f^{q-1} - \prod_i (1 - x_i^{q-1})$$

is identically zero on F^{n+1}. Show that h has too many zeros compared with its degree.

See [Borevich-Shafarevich64, Ch. 1] for more details and related results.

6.3.6 Remark. Unfortuately, this is not a very convincing evidence for (6.3). The results of Weil assert that if f is irreducible over $\bar{F}$, then $f = 0$ has a solution if $|F|$ is sufficiently large (depending on the degree). Thus, for instance, a cubic form in three variables which defines a smooth cubic always has a nontrivial zero in a finite field.

The function field version of (6.3.5) was first investigated by [Tsen36] and later studied systematically by [Lang52] who introduced the following notions:

6.4.1 Definition. F is a *C_r-field* if every homogeneous polynomial $f(x_1,\dots,x_m)$ of degree d in $m > d^r$ variables has a nontrivial zero in F.

By (6.3.5) a finite field is a C_1 field.

It is already noted in [Lang52] that the above definition has several variants and the equivalence of these versions is not known. Instead of trying to figure out the optimal conditions, we consider an extra assumption which is satisfied in the cases that interest us.

6.4.2 Definition. A homogeneous polynomial $f \in F[x_1,\dots,x_m]$ of degree m is called a *normic form* if $f = 0$ has only the trivial solution in F.

6.4.3 Examples. (6.4.3.1) Let $F = K(X)$ be the field of rational functions of a normal algebraic variety. Let $t \in K(X)$ be a rational function whose zero set has an irreducible component of multiplicity one. Show that $f_m = \sum_{i=1}^{m} t^{i-1}x_i^m$ is normic. Thus any field finitely generated but not algebraic over another field has normic forms of arbitrary degree.

(6.4.3.2) Let $E \supset F$ be an extension of degree m and $\gamma_1,\dots,\gamma_m$ an F-basis of E. Set $f_m(x_1,\dots,x_m) := N_{E/F}(\sum \gamma_i x_i)$ where $N_{E/F}$ is the norm. Show that f_m is normic. Thus any field finitely generated over its prime field has normic forms of arbitrary degree.

The following result is usually called Tsen's theorem. It implies that the field of rational functions on an algebraic curve over an algebraically closed base field is a C_1-field:

6.5 Theorem. *Let F be a field of transcendence degree one over an algebraically closed field K. Let $f_1,\dots,f_k \in F[x_0,\dots,x_n]$ be homogeneous polynomials. Assume that $\sum \deg f_i \le n$. Then the equations $f_1 = \dots = f_k = 0$ have a nonzero solution in F.*

Proof. It is sufficient to prove this for finitely generated extensions. Pick $t \in F - K$ and let $R \subset F$ be the integral closure of $K[t]$ in F. R is free as a $K[t]$-module, let $c_1, \dots, c_r$ be a basis. Multiplication of the c_i is given by formulas

$$c_i c_j = \sum_k \gamma_{i,j,k} c_k \quad \text{where} \quad \gamma_{i,j,k} \in K[t].$$

Set $h := \max \deg \gamma_{i,j,k}$.

Multiplying the f_i by suitable constants, we may assume that they have coefficients in R. We can write every coefficient $f_{j,J}$ of f_j as

$$f_{j,J} = \sum_k f_{j,J,k} c_k \quad \text{where} \quad f_{j,J,k} \in K[t].$$

Set $m := \max \deg f_{j,J,k}$.

We are looking for a solution of the form $x_i = \sum_k u_{i,k} c_k$ were the $u_{i,k}$ are polynomials of degree at most d in $K[t]$. The coefficients are viewed as unknowns, they correspond to points in a projective space of dimension $r(d+1)(n+1)-1$.

Write

$$f_j\Big(\sum_k u_{0,k} c_k, \dots, \sum_k u_{n,k} c_k\Big) = \sum_k F_{j,k}[t] c_k$$

where $F_{j,k}[t] \in K[t]$. $f_j(x_0, \dots, x_n) = 0$ imposes $\sum_k (1 + \deg F_{j,k})$ equations on the coefficients of the $u_{i,k}$. Expand f_j to obtain terms of the form

$$\Big(f_{j,J} \prod_{1 \le s \le \deg f_j} c_{k_s}\Big)\Big(\prod_{1 \le s \le \deg f_j} u_{i_s,k_s}\Big).$$

The degree of the last term is at most $d \deg f_j$ and the degree of the previous product is at most $h \deg f_j + m$. Thus $\deg F_{j,k} \le (d+h) \deg f_j + m$. Therefore the total number of equations is at most

$$r\Big(\sum_j (d+h) \deg f_j + m + 1\Big) \le nr((d+h)+m+1).$$

For $d \gg 1$ we se that $r(d+1)(n+1)-1 > nr((d+h)+m+1)$. Thus we obtain a nonzero solution. □

6.5.1 Exercises. (6.5.1.1) From the proof it would seem that we always have lots of solutions. To disprove this, for every n find a polynomial $f \in K(t)[x_0, \dots, x_n]$ of degree n such that $(f = 0) \in \mathbb{P}^n_{K(t)}$ has exactly one solution in $K(t)$.

(6.5.1.2) Let $F = K(t)$. Find examples of homogeneous polynomials $f \in F[x_0, \dots, x_n]$ of degree $n+1$ such that $(f = 0)$ defines a smooth hypersurface which has no points in F or even in $K((t))$. This fits very well with the predictions of (6.3).

(6.5.1.3) [Lang52] Let F be a field of transcendence degree r over an algebraically closed field K. Show that F is a C_r-field.

6.5.2 Exercise. Let F be a field of transcendence degree one over an algebraically closed field K and $X = \mathbb{P}^{a_1}_F \times \cdots \times \mathbb{P}^{a_m}_F$. Let $f_i \in H^0(X, \mathcal{O}_X(d_{i,1}, \ldots, d_{i,m}))$. Assume that $\sum_i d_{i,j} \leq a_j$ for every j. Show that the subscheme $(f_1 = \ldots = f_k = 0) \subset X$ has a point in F.

The simplest application of (6.5) is to curves:

6.6 Proposition. *Let F be a C_1-field and X_F a curve. Assume that $X_{\bar F} \cong \mathbb{P}^1_{\bar F}$. Then $X_F \cong \mathbb{P}^1_F$.*

Proof. X_F is smooth hence $\omega^{-1}_{X_F}$ is a line bundle. A field extension does not change h^0, thus $h^0(X_F, \omega^{-1}_{X_F}) = 3$ and $\omega^{-1}_{X_F}$ is very ample. This gives a morphism $X_F \to \mathbb{P}^2_F$ whose image is a conic. The conic is given by a quadratic form of 3 variables, thus it has a point P in F. Projecting from P gives an isomorphism $X_F \cong \mathbb{P}^1_F$. □

6.6.1 Exercise. Let X_F be a k-dimensional variety such that $X_{\bar F}$ is isomorphic to a hypersurface of degree $k+1$ or $k+3$. Show that X_F is isomorphic to a hypersurface of degree $k+1$ or $k+3$.

6.6.2 Corollary. *Let S be a projective surface over an algebraically closed field K. Let $f : S \to C$ be a morphism whose general fiber is a smooth rational curve. Then f has a section and S is birational to $C \times \mathbb{P}^1$.*

Proof. Set $F = K(C)$ and let S_F be the generic fiber of f. By assumption S_F is smooth and $S_{\bar F} \cong \mathbb{P}^1_{\bar F}$. By (6.5) F is a C_1-field, hence by (6.6) $S_F \cong \mathbb{P}^1_F$. □

(6.5) can be generalized to weighted projective spaces. I thank Colliot–Thélène and Mori for pointing this out to me. I do not know if the assumption about the existence of normic forms is necessary or not.

6.7 Theorem. *Let F be a C_1-field and assume that F has normic forms of arbitrary degree. Consider the weighted polynomial ring $F[x_0, \ldots, x_n]$ where* weight$(x_i) = a_i$. *Let $f_1, \ldots, f_k \in F[x_0, \ldots, x_n]$ be weighted homogeneous polynomials. Assume that*

$$\sum \textit{weighted-deg}\, f_j < \sum \mathrm{weight}(x_i).$$

Then $f_1 = \ldots = f_k = 0$ has a nontrivial solution in F.

Proof. For every i let $g_i(y_{i,1}, \ldots, y_{i,a_i})$ be a normic form of degree a_i. Set

$$F_j := f_j(g_0, \ldots, g_n) \in F[y_{i,k_i} | 1 \leq i \leq n, 1 \leq k_i \leq a_i].$$

By construction $\deg F_i =$ weighted-deg f_i and the number of variables y_{i,k_i} is $\sum a_i$, thus $F_1 = \ldots = F_k = 0$ has a solution $\{b_{i,k_i}\}$ such that $b_{s,t} \neq 0$ for some s, t. Since g_s is normic, $x_s = g_s(b_{s,1}, \ldots, b_{s,a_s}) \neq 0$. Thus

$$x_j = g_j(b_{j,1}, \ldots, b_{j,a_j}) \quad j = 0, \ldots, n$$

is a nontrivial solution of $f_1 = \ldots = f_k = 0$. □

The following is one of the main results of this section. A smooth proper surface is separably rationally connected iff it is rational (3.7.3). Thus we have a complete solution of (6.1.1) for $\dim X \leq 3$.

6.8 Theorem [Manin72, Colliot–Thélène86]. *Let S_F be a smooth, proper and geometrically irreducible surface over a C_1-field F such that $S_{\bar{F}}$ is rational. Then S_F has a point in F.*

Proof. By (6.2) it is sufficient to prove this for a suitable birational model of S_F. By (III.2.2) S_F is birational to a surface X_F which satisfies one of the following assumptions:

(6.8.1) X_F is a conic bundle over a smooth curve C. Since X_F is rationally connected, C is a smooth rational curve.

(6.8.2.k) X_F is a Del Pezzo surface such that $k = (K^2_{X_F})$.

Assume first that we have (6.8.1). There is an F-point $P \in C$ by (6.6). The fiber X_P is thus a conic over F which again has an F-point by the definition of C_1. This gives an F-point in X_F.

We are left with the cases (6.8.2.k). Unfortunately the only known proof is a case-by-case analysis which makes essential use of the classification results from (III.3).

If $k \leq 4$, then X_F is isomorphic to a complete intersection in a weighted projective space as described in (III.3.5). Thus X_F has a point in F by (6.7), at least if F has normic forms of arbitrary degree.

If $k = 5$, then by (III.3.9.6) X_F has a point in F and is even rational over F.

If $k \geq 6$, then X_F is rational over F by (III.3.12) since the Brauer group of a C_1-field is trivial [Serre79, X.7].

We are left with the cases $k \leq 4$ when F does not have normic forms of arbitrary degree. For our purposes this case is not important, the solution is outlined in (6.8.3) □

6.8.3 Exercises. Let F be a field, S_F a Del Pezzo surface over F and $k = (K^2_{S_F})$.

(6.8.3.1) If $k = 1$, then S_F has a point in F.

(6.8.3.2) If $k = 2$, then S_F is defined by an equation $x_3^2 + x_3 a(x_0, x_1, x_2) + b(x_0, x_1, x_2) = 0$ in $\mathbb{P}^2(1,1,1,2)$. If F has a normic form of degree 2, then the proof of (6.7) applies to show that S_F has an F-point if F is a C_1-field. Otherwise show that if $x_0, x_1, x_2 \in F$ are arbitrary and $(x_0, x_1, x_2, x_3) \in S_{\bar{F}}$, then $x_3 \in F$.

(6.8.3.3) If $k = 3$, then and F is a C_1-field, then S_F has an F-point by definition.

(6.8.3.4) If $k = 4$, then S_F is the intersection of two quadrics Q_1, Q_2 defiend over F. If F has a normic form $H(u,v)$ of degree 2 then $H(Q_1, Q_2)$ is a quartic in 5 variables. If F is a C_1-field, this gives an F-point in S_F. Otherwise F does not have quadratic extensions. Show that $Q_1 = 0$ contains a line over F which intersects S_F in two F-points. ((6.8.3.4) also follows from general results of [Lang52].)

Another large class of rationally connected varieties where the answer to (6.1.1) is known is homogeneous spaces. The techniques of the proof rely on Galois cohomology and the theory of algebraic groups. We just state the result and refer to [Serre64, III-14-16] and [Borel69, 18.2] for proofs.

6.9 Theorem (Chevalley–Springer). *Let F be a perfect C_1-field and X_F a homogeneous space over a connected affine algebraic group G over F. Then X_F is unirational over F.* □

6.9.1 Corollary. *Let F be a perfect C_1-field with algebraic closure $\bar{F}$ and X_F a smooth proper variety over F. Assume that the connected component of* $\mathrm{Aut}(X_{\bar{F}})$ *is a linear group which has a dense orbit on $X_{\bar{F}}$. Then X_F is unirational over F.* □

6.9.2 Exercises. For nonperfect fields some strange things can happen.

(6.9.2.1) Let K be a field of characteristic $p > 0$ and set $F = K(t)$. Then $X_F := (x^p + ty^p + t^2 z^p = 0) \subset \mathbb{P}^2$ is a regular F-scheme which is homogeneous but has no points in F. X_F is not geometrically reduced.

(6.9.2.2) (cf. [Queen71, LangW79]) Let K be an algebraically closed field of characteristic 3 and $F = K(t)$. Pick $f(t) \in K[t]$ which is not a cube. Show that

$$C_F := (zy^2 + y^3 + x^3 + f(t)z^3 = 0)$$

is a plane cubic which is regular over F. Let C_F^0 denote the smooth affine cubic $(z + 1 + x^3 + f(t)z^3 = 0)$.

Show that C does not have any F-points. Also show that C_F^0 is a homogeneous space under $\mathrm{Aut}(C_F)$.

The method of (6.5) gives one F-point on certain varieties, but does not give more in general. In the case of function fields, a method of obtaining more points was discovered in [KoMiMo92b]. The result shows that once the existence of *one* F-point is established, we get lots of F-points under some mild conditions.

6.10 Theorem [KoMiMo92b, 2.13]. *Let X be a projective varety over an algebraically closed field and $f : X \to C$ a morphism onto a smooth curve. Assume that f has a section $\sigma : C \to X$. Let $c_1, \dots, c_k \in C$ be closed points such that $X_i := f^{-1}(c_i)$ are smooth and separably rationally connected. Pick arbitrary points $p_i \in X_i$.*

Then f has a section $s = s_{p_1,\dots,p_k} : C \to X$ such that $s(c_i) = p_i$.

6.10.1 Remark. We can relax the condition of the fibers X_i being smooth as follows. It is sufficient to assume that there is a morphism $h_i : \mathbb{P}^1 \to X_i$ passing through $\sigma(c_i)$ and p_i such that X_i is smooth along $\operatorname{im} h_i$ and $h_i^* T_{X_i}$ is ample.

Proof. We have not assumed that X is smooth, thus f need not be smooth along $\sigma(C)$. By [Artin69a, 4.6] after a suitable sequence of blowing ups we obtain a birational morphism $p : X' \to X$ such that σ lifts to a section $\sigma' : C \to X'$ and $X' \to C$ is smooth along $\sigma'(C)$. We change notation and assume that $\sigma(C) \subset X^0$ where $X^0 \subset X$ is the open set where X and f are smooth.

The basic idea of the proof is the following. Let $h_i : \mathbb{P}^1 \to X_i$ be morphisms connecting $\sigma(c_i)$ and p_i. We can assemble a comb with handle $\sigma : C \to X$ and teeth $h_i : \mathbb{P}^1 \to X$. If the comb is smoothable fixing the points p_i, then the resulting smooth curve gives the required section. Unfortunately the general smoothing results do not give this right away, so we do the smoothing in several steps. At each step we stay inside X^0.

Assume that we already found a section s_k such that $s_k(c_i) = p_i$ for $i \le k$ and f is smooth along $s_k(C)$. We would like to use this to find a section s_{k+1} such that $s_{k+1}(c_i) = p_i$ for $i \le k+1$ and f is smooth along $s_{k+1}(C)$. Let $X' \to X$ be the blow up of the points $p_i : 1 \le i \le k$, $E_i \subset X'$ the corresponding exceptional divisors and $\sigma' : C \to X'$ the lifting of σ.

If we assemble any comb using $\sigma'(C)$ as handle and teeth contained in fibers of $X' \to C$ and take a small deformation, then the resulting curve still passes through the divisors E_i. Thus its image in X is a section which passes through the points $p_1, \dots, p_k$. Therefore it is sufficient to find a section s of $X' \to C$ which is obtained from σ' by small deformation of combs and which passes through a given point $p = p_{k+1} \in X$.

Changing notation again, we solve this special case for $f : X \to C$ with section σ. As a first step, we obtain a section whose normal bundle is positive:

6.10.2 Claim. *Given any section $\sigma : C \to X$ there is a section $\rho : C \to X$ obtained by smoothing a suitable comb with handle σ such that $H^1(C, \rho^* N_{\rho(C)|X}) = 0$.*

Proof. Note that if $\tau : C \to X^0$ is any section, then $\tau^* N_{\tau(C)|X} \cong \tau^* T_{X^0/C}$.

Choose m such that $H^1(C, M \otimes \sigma^* N_{\sigma(C)|X}) = 0$ for every line bundle of degree at least $m-1$.

Pick sufficiently many points $c_j \in C$ such that the X_j are smooth and separably rationally connected and morphisms $f_j : \mathbb{P}^1 \to X$ passing through $\sigma(c_j)$ such that the $f_j^* T_{X_j}$ are ample. We can assemble a comb with handle σ and teeth f_j to obtain $h_0 : C + \sum \mathbb{P}^1_j \to X$. By (II.7.9) this has a subcomb $C + \sum' \mathbb{P}^1_j$ which is smoothable to $h_t : C \to X$, and the number of teeth in the subcomb is at least m. The smoothing is given by a diagram

$$\begin{array}{ccc} Y & \xrightarrow{H} & X \\ {\scriptstyle q}\downarrow & & \\ 0 \in T & & \end{array} \qquad \text{such that} \quad h_t = H|q^{-1}(t)$$

$h_0^*T_{X^o/C}|C \cong \sigma^* N_{\sigma(C)|X}$ and $h_0^*T_{X^o/C}|\mathbb{P}_j^1 \cong f_j^*T_{X_j}$. Thus $h_0^*T_{X^o/C}$ is ample on the teeth of $C + \sum' \mathbb{P}_j^1$ and therefore $H^1(C, h_t^* N_{h_t(C)|X}) = 0$ for $t \neq 0$ by (II.7.10.1). Set $\rho = h_t$. □

We change notation for the last time. Let $c = f(p)$ and $X_c = f^{-1}(c)$. Set $F \cong \mathbb{P}^1$ with marked points $0, \infty$. Let $f : F \to X_c$ be a free morphism such that $f(0) = \rho(c)$, $f(\infty) = p$ and define g by $g(\infty) = p$.

$C + F$ denotes the curve obtained by identifying the points $0 \in F$ and $c \in C$ and $h : C + F \to X$ the morphism given by (ρ, f). Our aim is to show that h is smoothable fixing p. The smoothing, then gives the required section of $f : X \to C$.

As in (II.1.11) we obtain a flat morphism $q : U \to (0 \in M)$ such that $q^{-1}(0) \cong C + F$, no other fiber is isomorphic to $C + F$ and

(6.10.3)

$$\dim M = -3\chi(\mathcal{O}_{C+F}) + \dim \operatorname{Aut}(C + F) = -3\chi(\mathcal{O}_C) + \dim \operatorname{Aut}(C, c) + 2.$$

We may also assume that there is a section $s : M \to U$ such that $s(0) = \infty \in F$. Let $G : s(M) \to X$ be the constant morphism with image p.

6.10.4 Claim. *Let $[h] \in Z \subset \operatorname{Hom}(U/M, M \times X/M, G)$ be a small open neighborhood. Assume that there is a $0 \neq m \in M$ and a morphism $h_m : q^{-1}(m) \to X$ such that $[h_m] \in Z$. Then $q^{-1}(m) \cong C$ and $h_m : C \to X$ is the sought after section.*

Proof. The image of h_m has an irreducible component which is a section of f, hence isomorphic to C. If $q^{-1}(m)$ is irreducible, then we are done. Otherwise $q^{-1}(m)$ is the union of C and of a smooth rational curve $f_m : \mathbb{P}^1 \to X$ joining at a point $x \in C$. By assumption $p \in \operatorname{im} f_m(\mathbb{P}^1)$, thus $\operatorname{im} f_m(\mathbb{P}^1) \subset X_c$. Therefore $c = x$ and $q^{-1}(m) \cong C + F$, a contradiction. □

This can be reformulated as:

6.10.5 Claim. *If $\dim_{[h]} \operatorname{Hom}(U/M, M \times X/M, G) > \dim_{[h]} \operatorname{Hom}(C+F, X, g)$, then h is smoothable fixing p.* □

The necessary estimates for the above dimensions are given by the following:

6.10.6 Claim.

(6.10.6.1)

$$\dim_{[h]} \operatorname{Hom}(C + F, X, g) \leq \chi(C + F, h^*T_{X^o/C}(-\infty)) + \dim \operatorname{Aut}(C, c);$$

(6.10.6.2)

$$\dim_{[h]} \operatorname{Hom}(U/M, M \times X/M, G) \geq \chi(C + F, h^*T_X(-\infty)) + \dim M.$$

Proof. The second estimate is a special case of (II.1.7). The key point is the first one, since it does not follow from general results about the dimension of spaces of morphisms.

ρ is an embedding, thus by (I.6.2) there is a natural morphism $\mathrm{Hom}(C, X) \to \mathrm{Hilb}(X)$, at least in a neighborhood of $[\rho]$. $\rho(C)$ is a section, thus every small deformation is again a section hence isomorphic to C. Thus $\mathrm{Hom}(C, X) \to \mathrm{Hilb}(X)$ is an $\mathrm{Aut}(C)$-bundle locally near $[\rho]$. $H^1(C, N_{\rho(C)|X}) = 0$ by (6.10.2), hence by (I.2.15.3)

$$\dim_{[\rho(C)]} \mathrm{Hilb}(X) = \chi(C, \rho^* N_{\rho(C)|X}) = \chi(C, \rho^* T_{X^0/C}).$$

The fiber of the restriction morphism $\mathrm{Hom}(C+F, X, g) \to \mathrm{Hom}(C, X) \to \mathrm{Hilb}(X)$ above $[\rho(C)]$ is the product of $\mathrm{Hom}(F, X, f|\{0,\infty\}) = \mathrm{Hom}(F, X_c, f|\{0,\infty\})$ and of $\mathrm{Aut}(C, c)$. By (II.1.7.1) we obtain that

$$\begin{aligned}
\dim_{[f]} \mathrm{Hom}(F, X, f|\{0,\infty\}) &= \chi(F, f^* T_{X_c}(-0-\infty)) \\
&= \chi(F, f^* T_{X^0/C}(-0-\infty)), \quad \text{thus} \\
\dim_{[h]} \mathrm{Hom}(C+F, X, g) &\le \chi(C, \rho^* T_{X^0/C}) + \chi(F, f^* T_{X^0/C}(-0-\infty)) \\
&\quad + \dim \mathrm{Aut}(C, c) \\
&= \chi(C+F, h^* T_{X^0/C}(-\infty)) + \dim \mathrm{Aut}(C, c),
\end{aligned}$$

where in the last line we used the exact sequence

$$0 \to f^* T_{X^0/C}(-0-\infty) \to h^* T_{X^0/C}(-\infty) \to \rho^* T_{X^0/C} \to 0.$$

Putting all of this together, we obtain that

$$\begin{aligned}
&\dim_{[h]} \mathrm{Hom}(U/M, M \times X/M, G) \ge \chi(C+F, h^* T_X(-\infty)) + \dim M \\
&\quad = \chi(C+F, h^* T_{X^0/C}(-\infty)) + \chi(C+F, (f \circ h)^* T_C(-\infty)) - 3\chi(\mathcal{O}_C) \\
&\qquad + \dim \mathrm{Aut}(C, c) + 2 \\
&\quad \ge \dim_{[h]} \mathrm{Hom}(C+F, X, g) - \dim \mathrm{Aut}(C, c) + 3\chi(\mathcal{O}_C) - 1 - 3\chi(\mathcal{O}_C) \\
&\qquad + \dim \mathrm{Aut}(C, c) + 2 \\
&\quad = \dim_{[h]} \mathrm{Hom}(C+F, X, g) + 1.
\end{aligned}$$

This shows that the assumption of (6.10.5) is satisfied and the proof of (6.10) is complete. □

Chapter V. Fano Varieties

A variety X is called Fano if the anticanonical bundle of X is ample. Thus Fano surfaces are the same as Del Pezzo surfaces. The importance of Fano varieties in the theory of higher dimensional varieties is similar to the significance of Del Pezzo surfaces in the two dimensional theory. The interest in Fano varieties increased recently since Mori's program predicts that every uniruled variety is birational to a fiberspace whose general fiber is a Fano variety (with terminal singularities).

From this point of view it is more important to study the general properties of Fano varieties with terminal singularities than to understand the properties of smooth Fano varieties. At the moment, however, we know much more about smooth Fano varieties, and their theory should serve as a guide to the more subtle questions of singular Fano varieties.

Fano varieties also appear naturally as important examples of varieties. In characteristic zero every projective variety which is homogeneous under a linear algebraic group is Fano (1.4), and their study is indispensable for the theory of algebraic groups.

Also, Fano varieties have a very rich internal geometry, which makes their study very rewarding. This is one of the reasons for the success of the theory of Fano threefolds. This is a beautiful subject, about which I say essentially nothing.

Section 1 is devoted to presenting the basic examples of Fano varieties and to the study of low degree rational curves on them. The largest class of examples are weighted complete intersections (1.2–3); these are probably the most accessible by elementary methods. Homogeneous spaces also provide many examples but their detailed study requires the machinery of algebraic groups (1.4). The most studied examples are the moduli spaces of stable vector bundles with fixed determinant on curves. Their theory deserves a monograph in itself; we mention them for sake of completeness only.

The cone of curves of a Fano variety X is generated by rational curves $C_i \subset X$ such that $-K_X \cdot C_i \leq \dim X + 1$ (1.6). This and many other examples lead to the following

Principle. The geometry of a Fano variety is governed by rational curves of low degree.

The rest of Sect. 1 is devoted to various assertions that support this principle.

The aim of Sect. 2 is to prove two general results about Fano varieties. Any smooth Fano variety is rationally chain connected (2.1, 2.13). For any dimension, there are only finitely many deformation types of Fano varieties, at least in characteristic zero (2.3, 2.14). This result implies that, in principle, it is possible to obtain a complete list of Fano varieties of any given dimension. In dimension three this has been accomplished, but the complexity of the arguments and the length of the list suggests that already in dimension 4 this is not a feasible project.

Section 3 discusses Mori's characterization of $\mathbb{P}^n$ as the only algebraic variety whose tangent bundle is ample (3.2). [Mori79] is the article where the bend-and-break technique was first introduced. Mori's arguments are very elegant and many of his ideas found later applications. I give a somewhat shortened version of the proof.

This result can also be considered as another example of the validity of the above principle. The canonical line bundle of $\mathbb{P}^n$ is $\mathcal{O}(-n-1)$, thus lines in $\mathbb{P}^n$ have anticanonical degree $n+1$ and there are no curves with smaller anticanonical degree. Ampleness of the tangent bundle of a variety X easily implies that there are no rational curves of anticanonical degree less than $\dim X + 1$ (3.6.1).

Section 4 is more like an overgrown exercise about lines of low degree hypersurfaces. The main technical result says that the family of all lines is connected when dimension count suggests that it might be so. This can be used to prove that the group of 1-cycles modulo algebraic equivalence is one dimensional (4.1). In some cases it also implies that the group of 1-cycles modulo rational equivalence is one dimensional (4.2). Even in this very concrete situation there are many interesting open problems.

There are many similarities between rational and rationally connected varieties. In fact, it is not easy to show that not every rationally connected variety is rational. This question is studied in Sect. 5. The method first produces rationally connected varieties in positive characteristic which are not separably uniruled. These can then be lifted to characteristic zero to produce examples of Fano hypersufaces which are not ruled. These techniques produce other interesting examples in positive characteristic as well.

V.1 Low Degree Rational Curves on Fano Varieties

The aim of this section is to present the basic examples of Fano varieties and to prove some simple results about low degree rational curves on them.

In this section everything is defined over an algebraically closed field k.

1.1 Definition. Let X be a smooth projective variety. X is called *Fano* if $-K_X$ is ample. Let X be a normal projective variety. X is called $\mathbb{Q}$-*Fano* (or just Fano) if $-K_X$ is $\mathbb{Q}$-Cartier and ample.

1.2 Example. Let $X \subset \mathbb{P}^n$ be a smooth complete intersection of k hypersurfaces of degrees $d_1, \ldots, d_k$. Then $-K_X = \mathcal{O}(n+1-\sum d_i)$. Thus X is Fano iff $\sum d_i < n+1$.

More generally, complete intersections in weighted projective spaces also provide lots of examples of Fano varieties. See [Dolgachev82] for a general introduction or [Fletcher89] for a series of examples of $\mathbb{Q}$-Fano threefolds. The basic properties of weighted projective spaces are recalled in the next series of exercises.

1.3 Exercises. Let k be a field and $S = k[x_0, \ldots, x_n]$ the polynomial ring in $n+1$ variables. Let $a_i \in \mathbb{N}$. Define a grading of S by $\deg x_i = a_i$. Proj S is called the *weighted projective space* of dimension n with weights a_i. It is denoted by $\mathbb{P}(a_0, \ldots, a_n)$. We may and do assume that $a_0, \ldots, a_n$ are relatively prime.

The following shortened version of this notation is frequently convenient:

$$\mathbb{P}(a_0^{r_0}, \ldots, a_n^{r_n}) \quad \text{denotes} \quad \mathbb{P}(\underbrace{a_0, \ldots, a_0}_{r_0\text{-times}}, \ldots, \underbrace{a_n, \ldots, a_n}_{r_n\text{-times}}).$$

Because of this convention, one should never use $\mathbb{P}(1,2,3^2)$ to denote $\mathbb{P}(1,2,9)$!

(1.3.1) Show that $\mathbb{P}(a_0, a_1, \ldots, a_n) \cong \mathbb{P}(a_0, da_1, \ldots, da_n)$ for every $d \in \mathbb{N}$. Thus in working with weighted projective spaces of dimension n we may always assume that any n weights are relatively prime. We say that $\mathbb{P}(a_0, \ldots, a_n)$ is *well formed* if this condition is satisfied. For the rest of the section we assume that all weighted projective spaces are well formed.

(1.3.2) Show that

$$\mathbb{P}(a_0, a_1, \ldots, a_n) \cong \mathbb{P}^n / \mu_{a_0} \times \ldots \times \mu_{a_n}$$

where μ_{a_j} denotes the group of a_j^{th} roots of unity and it acts on $\mathbb{P}^n$ via multiplication on the jth coordinate. (If you feel uneasy about group schemes, assume that the characteristic does not divide any of the a_i.)

(1.3.3) Describe a covering of $\mathbb{P}(a_0, \ldots, a_n)$ with affine charts. Show that $\mathbb{P}(a_0, \ldots, a_n)$ has only cyclic quotient sigularities.

(1.3.4) Let $\mathcal{O}(m)$ denote the coherent sheaf associated to the graded module $S(m)$. Show that $\mathcal{O}(m)$ is locally free iff $a_i | m$ for every i. Show that

$$(\mathcal{O}(1)^n) = \frac{1}{a_0 \cdots a_n}.$$

(1.3.5) Show that the dualizing sheaf $K_{\mathbf{P}}$ of $\mathbb{P}(a_0, \ldots, a_n)$ is isomorphic to $\mathcal{O}(-\sum a_i)$.

(1.3.6) Show that $\mathbb{P}(a_0, \ldots, a_n)$ has isolated singularities iff the weights are pairwise relatively prime.

(1.3.7) Assume that $\mathbb{P}(a_0, \ldots, a_n)$ has isolated singularities and $a_i | m$ for every i. Show that the smooth members of $|\mathcal{O}(m)|$ form a dense open set. If $H \in |\mathcal{O}(m)|$ is smooth, then $K_H = \mathcal{O}(m - \sum a_i)|H$.

(1.3.8) Let $d_0, \ldots, d_r$ be pairwise relatively prime natural numbers. Let H be a smooth member of $\mathcal{O}(d_0 \cdots d_r)$ on $\mathbb{P}(d_0, \ldots, d_r)$. $\mathcal{O}_H(1)$ is a locally free sheaf, which is ample and has selfintersection 1.

(1.3.9) Let $X = X_{d_1, \ldots, d_k} \subset \mathbb{P}(a_0, \ldots, a_n)$ be a smooth (or normal) complete intersection of k hypersurfaces of degrees $d_1, \ldots, d_k$. Then $-K_X = \mathcal{O}(\sum a_j - \sum d_i)$. Thus X is Fano iff $\sum d_i < \sum a_j$.

(1.3.10) Let $f : \mathbb{P}^1 \to \mathbb{P}(a_0, \ldots, a_n)$ be a morphism such that $\mathbb{P}(a_0, \ldots, a_n)$ is smooth along $\operatorname{im} f$. Show that f is given by a a collection of sections $f_i \in H^0(\mathbb{P}^1, \mathcal{O}(da_i))$, $i = 0, \ldots, n$ where $d = \deg f^*\mathcal{O}(1)$. d is called the *degree* of f.

(1.3.11) Find further examples of Fano manifolds which are complete intersections in products of weighted projective spaces.

Another large class of examples of Fano varieties is provided by homogeneous spaces:

1.4 Theorem. *Let G be a reduced and connected linear algebraic group and X a proper homogeneous space under G. Pick a point $x \in X$ and let $H < G$ be the stabilizer of x. Assume that H is reduced (which is always the case in characteristic zero). Then T_X is generated by global sections and K_X^{-1} is very ample.*

Proof. As we see in (1.4.2.4), all such homogeneous spaces can be classified in terms of Dynkin diagrams and the theorem can be read off from various assertions about algebraic groups.

Mori pointed out to me the following argument which shows that T_X is generated by global sections, K_X^{-1} is ample and generated by global sections. There are then various methods to show that K_X^{-1} is very ample.

We may assume that no subgroup of G acts trivially on X. (This needs the result that the quotient of an affine group by a normal subgroup is again affine, cf. [Borel69, 6.8].) Thus $T_1 G \to T_1 \operatorname{Aut}(X) = H^0(X, T_X)$ is injective

Sending $x \in X$ to $xg \in X$ defines a G-equivariant surjective morphism $m_x : G \to X$. By definition $H = m_x^{-1}(x)$ and every fiber is isomorphic to H. In characteristic zero the general fiber is smooth, so H is smooth. In positive characteristic m_x is smooth by assumption. In particular $dm_x : T_1 G \to T_x X$ is surjective.

By (II.2.16.5) there is a natural map $T_1 G \to H^0(X, T_X)$ and the resulting sections generate T_X. G is affine and X is projective, thus $\dim H > 0$ unless $\dim X = 0$. The image of $T_1 H \to H^0(X, T_X)$ gives sections of T_X which vanish at x but not everywhere.

Therefore we obtain that K_X^{-1} has a section which vanishes at x but is not everywhere zero. By homogeneity, K_X^{-1} is generated by global sections and K_X is not the trivial line bundle. The sections of K_X^{-1} define a G-equivariant morphism $p : X \to Y \subset \mathbb{P}$ to some projective space. Let $F \subset X$ be the fiber containing x with reduced scheme structure.

Any fiber of a G-equivariant morphism is again a homogeneous space under a subgroup $G' < G$ and the stabilizers of $x \in F$ in G' and in G are the same. Applying the above considerations to F we conclude that K_F^{-1} is not the trivial line bundle, unless $\dim F = 0$. On the other hand, $K_F = K_X|F$ is trivial. Therefore $p : X \to Y$ is finite and H has finite index in G'.

This shows that K_X^{-1} is ample and generated by global sections.

By a general result of [Ramanathan85], any ample line bundle on X is very ample. There are more elementary ways to complete the proof of (1.4), at least in characteristic zero. These are outlined in a series of exercises. □

1.4.1 Exercise. Let X be a smooth Fano variety over a field of characteristic zero.

(1.4.1.1) Use Kodaira vanishing to show that $\chi(\mathcal{O}_X) = 1$.

(1.4.1.2) Show that if X is homogeneous and L is a line bundle on X which is ample and generated by global sections, then L is very ample.

(1.4.1.3) Let X be a smooth Fano variety over a field of characteristic zero. Show that $\mathrm{Pic}(X)$ is torsion free.

1.4.2 Exercises. Here is a more group theoretic approach to the end of (1.4), still in characteristic zero.

(1.4.2.1) (Borel fixed point theorem, cf. [Borel69, 10.4]) Let H be a connected and solvable algebraic group acting on a proper variety X. Show that H has a fixed point.

(1.4.2.2) Let G be a reduced and connected linear algebraic group and X a proper homogeneous space under G. Let $H \triangleleft G$ be a connected, solvable normal subgroup. Show that H acts trivially on X. In particular, the connected component of $\mathrm{Aut}(X)$ is semi-simple.

(1.4.2.3) Notation as in (1.4). Show that H contains a Borel subgroup of G. By definition, $G' > H$ is parabolic, hence connected by [Borel69, 11.15]. This again shows that if X is homogeneous and L is a line bundle on X which is ample and globally generated, then L is very ample.

(1.4.2.4) Semisimple algebraic groups and their reduced parabolic subgroups can be completely classified in terms of Dynkin diagrams. By the above results, this gives a complete classification of the homogeneous spaces considered in (1.4).

1.4.3 Exercises. (1.4.3.1) Let G be a reduced and connected linear algebraic group and X a proper homogeneous space under G. Let $H < G$ be the stabilizer of a point $x \in X$ and set $H' = \mathrm{red}\, H$. G/H' is a homogeneous space such that $K_{G/H'}^{-1}$ is very ample. Furthermore there is a purely inseparable G-equivariant morphism $G/H' \to X$.

(1.4.3.2) In $\mathbb{P}^n \times \mathbb{P}^n$ let $(x_i) \times (y_j)$ be homogeneous coordinates. Let q be a power of the characteristic. Show that the hypersurface $Y_q := \sum x_i^q y_i$ is smooth. Y_q is Fano iff $q \leq n$.

Show that Y_q is a homogeneous space under $SL(n+1)$ where

$$g \in SL(n+1) \quad \text{acts by} \quad (x_i) \mapsto (x_i)g \quad \text{and} \quad (y_i)^t \to g^{-q}(y_i)^t.$$

(1.4.3.3) Show that not every deformation of Y_q is homogeneous for $q \geq 2$. Thus in positive characteristic there are nonrigid homogeneous varieties. In characteristic zero every proper homogeneous variety (under an affine group) is rigid by [Bott57].

(1.4.3.4) Proper homogeneous spaces with nonreduced stabilizers provide very nice examples of algebraic varieties over fields of positive characteristic. Many interesting questions are discussed in [Lauritzen96; Wentzel93].

The last set of examples of Fano varieties is more complicated. We do not use them in the future.

1.5 Example [Ramanan73]. Let C be a smooth curve of genus $g \geq 2$, r a positive integer and L a line bundle on C such that r and $\deg L$ are relatively prime. Let $M(r, L, C)$ denote the moduli space of stable vector bundles of rank r and determinant L on C. $M(r, L, C)$ is a smooth Fano variety of dimension $(r^2 - 1)(g - 1)$.

These Fano varieties have been studied extensively, see e.g. [Drezet-Narasimhan89] and the references there. Their study usually involves a very nice mixture of higher dimensional methods and curve theory.

The following summarizes the basic general results about rational curves on Fano varieties. The proofs are in [Mori79] but the explicit statement of the first part seemed to have been noticed only later [Kollár81]. The first part is a special case of (II.5.14) and the second part is a special case of (III.1.2).

1.6 Theorem. *Let X be a smooth Fano variety over an algebraically closed field k.*

(1.6.1) *Through every point $x \in X$ there is a rational curve $x \in C_x \subset X$ such that $-K_X \cdot C_x \leq \dim X + 1$.*

(1.6.2) *On X there are only finitely many families of rational curves C_μ such that $-K_X \cdot C_\mu \leq \dim X + 1$. Let $C_i : 1 \leq i \leq N$ be a set of representatives. Then*

$$\overline{NE}(X) = NE(X) = \sum \mathbb{R}^+[C_i]. \qquad \square$$

By (2.13) and (IV.3.14) the second part can be strengthened:

1.6.3 Corollary. *Let X be a smooth Fano variety over an algebraically closed field k. Then*

$$BE_1(X)_{\mathbb{Q}} = \sum \mathbb{Q}^+[C_i]. \qquad \square$$

1.6.4 Exercise [Szurek-Wiśniewski90; KoMiMo92c]. Let $f : X \to Y$ be a smooth morphism between smooth projective varieties. Assume that X is Fano. Using the following steps show that Y is also Fano.

(1.6.4.1) Let $g : Z \to \mathbb{P}^1$ be a smooth, projective morphism with a section σ. Use the bend-and-break technique to prove that there is a section $s : \mathbb{P}^1 \to Z$ such that $K_{Z/\mathbb{P}^1} \cdot_s \mathbb{P}^1 \geq 0$.

(1.6.4.2) Let $h : \mathbb{P}^1 \to X$ be a morphism. Show that $K_Y \cdot_{f \circ h} \mathbb{P}^1 < 0$.

(1.6.4.3) Use (VI.2.19) to conclude that Y is Fano.

(1.6.4.4) Remark. [Wiśniewski91b] contains an example of a Fano variety which is a quadric bundle over a variety which is not Fano.

In (1.6.1) the bound $\dim X + 1$ is optimal. The simplest Fano variety is $\mathbb{P}^n$. Since $-K_{\mathbb{P}^n} = \mathcal{O}_{\mathbb{P}^n}(n+1)$, $-K_{\mathbb{P}^n} \cdot C \geq n+1$ for every curve and equality holds only for lines.

The following conjecture asserts that $\mathbb{P}^n$ is the only such example. Even in low dimensions no conceptual proof is known. It can be read off from the classification of Fano varieties for $n \leq 2$ or $n = 3$, $\operatorname{char} k = 0$.

1.7 Conjecture.[1] *Let X be a smooth projective variety of dimension n over an algebraically closed field. Then $X \cong \mathbb{P}^n$ iff*

(1.7.1) *$-K_X$ is ample, and*

(1.7.2) *$-K_X \cdot C \geq n+1$ for every rational curve $C \subset X$.*

In general it seems that on "most" Fano varieties there are rational curves of low degree. The following result answers a conjecture raised by [Mukai88].

1.8 Theorem [Wiśniewski90]. *Let X be a smooth Fano variety. Assume that*

$$-K_X \cdot C \geq \tfrac{1}{2}(\dim X + 3)$$

for every rational curve $C \subset X$. Then $\dim B_1(X) = 1$.

Proof. By (1.6.1) there is an irreducible component $V \subset \operatorname{RatCurves}_d(X)$ with cycle morphism $V \stackrel{f}{\longleftarrow} U \stackrel{p}{\longrightarrow} X$ such that p is dominant and $d \leq \dim X + 1$. By (II.2.14) V is proper, hence p is surjective.

Let $D \subset X$ be a rational curve such that $-K_X \cdot D \leq \dim X + 1$. Pick a point $x \in D$ and let $W \subset \operatorname{RatCurves}(x, X)$ be an irreducible component containing $[D]$. By (II.2.14) W is proper. By (IV.2.6.3)

$$\dim \operatorname{Locus}(W, 0 \mapsto x) \geq \deg_{(-K)} D - 1 \geq \tfrac{1}{2}(\dim X + 1).$$

Since p is surjective, $x \in \operatorname{Locus}(V)$, thus again by (III.2.6.3)

$$\dim \operatorname{Locus}(V, 0 \mapsto x) \geq \deg_{(-K)} C - 1 \geq \tfrac{1}{2}(\dim X + 1).$$

Therefore

$$\dim \operatorname{Locus}(W, 0 \mapsto x) \cap \operatorname{Locus}(V, 0 \mapsto x) \geq 1.$$

[1] A solution is announced in [Cho-Miyaoka98].

Let B be an irreducible curve in the above intersection. By (II.4.21) there are rational numbers $c \neq 0 \neq d$ such that

$$B \overset{a}{\approx} cC \quad \text{and} \quad B \overset{a}{\approx} dD.$$

Thus D is algebraically equivalent to a multiple of C. By (1.6.3) this implies that $\dim B_1(X) = 1$. □

If $-K_X = mH$ for some ample divisor H, then $-K_X \cdot C \geq m$ for any curve C. This is one of the motivation for the following definition:

1.9 Definition. Let X be a smooth Fano variety. The *index* of X is the largest natural number m such that $-K_X \equiv mH$ for some Cartier divisor H. It is denoted by $\operatorname{index}(X)$. By (1.4.1.3) if char $= 0$, then $-K_X = mH$.

Assume that $N^1(X)_{\mathbb{Z}} \cong \mathbb{Z}$ and let H be the positive generator. $-K_X = \operatorname{index}(X)H$, thus the index measures how far $-K_X$ is from being the "simplest" divisor on X. The notion of index is less useful if $\operatorname{rank} N^1(X)_{\mathbb{Z}} \geq 2$. For instance, $\operatorname{index}(\mathbb{P}^n \times \mathbb{P}^{n+1}) = 1$.

1.9.1 Example. Let $X = X_{d_1,\ldots,d_k} \subset \mathbb{P}(a_0, \ldots, a_n)$ be a smooth weighted complete intersection of hypersurfaces of degrees $d_1, \ldots, d_k$. By (1.3.9) $-K_X = \mathcal{O}_X(\sum a_j - \sum d_i)$. The usual methods of Lefschetz theory can be modified to prove that $\mathcal{O}_X(1)$ generates $\operatorname{Pic}(X)$ if $\dim X \geq 3$, see e.g. [Dolgachev82, 3.2.4]. Thus if $\sum d_i < \sum a_j$ and $\dim X \geq 3$, then $\operatorname{index}(X) = \sum a_j - \sum d_i$.

These examples may describe all Fano varieties with large index:

1.10 Problem. Fix an integer c. Does there exist a bound $N(c)$ such that if X is a smooth Fano variety of dimension at last $N(c)$ and $\operatorname{index}(X) = \dim X + 1 - c$, then X is a weighted complete intersection?

Very little is known about this problem. $c \leq 1$ is the only case when a satisfactory solution is known:

1.11 Theorem [Kobayashi-Ochiai70]. *Let X be a smooth Fano variety of dimension n over a field of characteristic zero. Then*

(1.11.1) $\operatorname{index}(X) \leq n + 1$;
(1.11.2) $\operatorname{index}(X) = n + 1$ *iff* $X \cong \mathbb{P}^n$;
(1.11.3) $\operatorname{index}(X) = n$ *iff* $X \subset \mathbb{P}^{n+1}$ *is a smooth quadric.*

Proof. X contains rational curves C of anticanonical degree at most $n + 1$, thus $\operatorname{index}(X) \leq -K_X \cdot C \leq \dim X + 1$.

The rest is clear for $n = 1$, thus we may assume that $n \geq 2$. Let $m = \operatorname{index}(X)$ and $-K_X = mH$. First we claim that

$$\chi(X, \mathcal{O}_X(jH)) = \begin{cases} 1 & \text{if } j = 0 \\ 0 & \text{if } -m < j < 0 \\ (-1)^n & \text{if } j = -m. \end{cases} \tag{1.11.4}$$

By the Kodaira vanishing theorem

(1.11.5)
$$\begin{aligned} h^i(X,\mathcal{O}_X(jH)) &= h^i(X,K_X+(m+j)H) \\ &= 0 \quad \text{if} \quad -m<j \quad \text{and} \quad i>0. \end{aligned}$$

Since $h^0(X,\mathcal{O}_X(jH))=0$ for $j<0$ and $h^0(X,\mathcal{O}_X)=1$, we can compute χ for $-m<j\le 0$. By Serre duality $\chi(X,K_X)=(-1)^n\chi(\mathcal{O}_X)=(-1)^n$.

$\chi(X,\mathcal{O}_X(jH))$ is a polynomial in j of degree at most n (VI.2.3). Therefore it has at most n roots, which shows again that $m\le n+1$. Also, if $m\ge n$, then the formulas (1.11.4) compute $\chi(X,\mathcal{O}_X(jH))$ at $n+1$ places, thus (1.11.4) determines $\chi(X,\mathcal{O}_X(jH))$ for every j. We obtain that

$$\chi(X,\mathcal{O}_X(jH)) = \begin{cases} \binom{j+n}{n} & \text{if } m=n+1; \\ \binom{j+n+1}{n+1} - \binom{j+n-1}{n+1} & \text{if } m=n. \end{cases}$$

By (1.11.5) $h^0(X,\mathcal{O}_X(H))=\chi(X,\mathcal{O}_X(H))$, thus

(1.11.6)
$$h^0(X,\mathcal{O}_X(H)) = \begin{cases} n+1 \\ n+2 \end{cases} \quad \text{and} \quad (H^n) = \begin{cases} 1 & \text{if } m=n+1; \\ 2 & \text{if } m=n. \end{cases}$$

1.11.7 Claim (Notation as above). *Then $\mathcal{O}_X(H)$ is generated by global sections.*

Proof. Let $V\subset \operatorname{Hom}(\mathbb{P}^1,X)$ be the open subset parametrizing morphisms whose anticanonical degree is at most $n+1$. If $[f]\in V$, then $0>\deg f^*K_X\ge -n-1$. On the other hand $\deg f^*K_X=-m\deg f^*H$. Thus $\deg f^*K_X=-m$, $\deg f^*H=1$ and V is proper by (II.2.14).

Let $x\in X$ be a point. By (IV.2.6.3)

(1.11.8)
$$\dim \operatorname{Locus}(V,0\mapsto x)\ge m-1.$$

Assume that $x\in X$ is a base point of $H^0(X,\mathcal{O}_X(H))$. Fix a local trivialization of $\mathcal{O}_X(H)$ and consider the differential

$$d: H^0(X,\mathcal{O}_X(H))\to (T_{x,X})^*.$$

Comparing dimensions we see that there is a section $s_1\in H^0(X,\mathcal{O}_X(H))$ which is singular at x and if $m=n$, then there are two such sections s_1,s_2.

Assume first that $m=n+1$. Choose $[f]\in(V,0\mapsto x)$ such that f^*s_1 is not identically zero. This is possible by (1.11.8). Then

$$2\le \deg(f^*s_1)=\deg f^*H=1,$$

which is a contradiction.

Next let $m=n$. By (1.8) $N^1(X)_{\mathbb{Z}}\cong\mathbb{Z}$ and H is the positive generator. Therefore every member of $|H|$ is irreducible, hence $\operatorname{codim}((s_1=0)\cap(s_2=0),X)=2$. Choose $[f]\in(V,0\mapsto x)$ such that f^*s_j is not identically zero for some $j\in\{1,2\}$. This is possible by (1.11.8). Then

$$2\le \deg(f^*s_1)=\deg f^*H=1,$$

which is again a contradiction. □

Finally consider the morphism $p: X \to \mathbb{P}^n$ (resp. $p: X \to \mathbb{P}^{n+1}$) given by the base point free linear system $|H|$. Let $Y = \operatorname{im} p$. Then

$$(H^n) = \deg p \cdot \deg Y.$$

If $m = n+1$, then $\deg Y = 1$ hence $Y = \mathbb{P}^n$ and p is birational. If $m = n$, then Y can not be a linear space thus $\deg Y \geq 2$ hence $Y \subset \mathbb{P}^{n+1}$ is a quadric hypersurface and p is birational. $-K_X = p^*\mathcal{O}_Y(m)$ is ample, thus p is also finite. Therefore p is an isomorphism. □

1.11.9 Remark. If one assumes only that $-K_X = mH$ is nef and big, then essentially the same proof gives that $X \cong \mathbb{P}^n$ if $m = n+1$. If $m = n$, then either X is a smooth quadric in $\mathbb{P}^{n+1}$ or $p: X \to Y$ is a birational morphism onto a singular quadric of rank 2.

1.11.10 Exercise. Let X be a smooth Fano variety of dimension n (any characteristic) such that $-K_X \equiv (n+1)H$. Show that $(H^n) = 1$. Hint: use (VI.2.15.7).

1.11.11 Exercise [Shepherd-Barron97]. Let X be a smooth Fano variety over a field of characteristic $p > 0$.

(1.11.11.1) Use the method of (III.3.2.1) to show that $h^1(X, \mathcal{O}_X) \leq h^2(X, \mathcal{O}_X)$.

(1.11.11.2) Let $-K_X \equiv mH$. Show that $H^1(X, \mathcal{O}(-jH)) = 0$ if $(p-1)j + m > 1 + \dim X$.

(1.11.11.3) If $\dim X = 3$ and $\operatorname{index}(X) = 4$ or $\operatorname{index}(X) = 3$ and $p > 2$, then all the vanishings required for (1.11.4) hold. Thus $X \cong \mathbb{P}^3$ in the first case and X is isomorphic to a smooth quadric in the second case.

1.11.12 Exercise. (1.11.12.1) Let X be a smooth projective variety and L a line bundle on X. Assume that

$$L^{\dim X+1} \cong K_X^{-1} \quad \text{and} \quad h^0(X, L) \geq \dim X + 1, \quad \text{or}$$
$$L^{\dim X} \cong K_X^{-1} \quad \text{and} \quad h^0(X, L) \geq \dim X + 2.$$

Use induction on $\dim X$ to show that X is isomorphic to $\mathbb{P}^n$ (resp. to a quadric in $\mathbb{P}^{n+1}$).

(1.11.12.2) Let $f: X \to S$ be a smooth projective morphism, S connected. Assume that one geometric fiber of X is isomorphic to $\mathbb{P}^n$ (resp. to a quadric in $\mathbb{P}^{n+1}$). Show that every geometric fiber of X is isomorphic to $\mathbb{P}^n$ (resp. to a quadric in $\mathbb{P}^{n+1}$).

It is possible to push the above direction somewhat further. The following result, established by Fujita in a series of articles, gives a complete answer to (1.10) for $c = 2$. The proof is summarized in his book [Fujita90].[2] The proof is by induction on the dimension, using the classification of Del Pezzo surfaces. Therefore it does not serve as a model for a general approach to (1.10).

[2] The case $c = 3$ is treated in [Mukai89, Mella99].

1.12 Theorem [Fujita90, 8.11]. *Let X be a smooth Fano variety of dimension $n \geq 3$ over a field of characteristic zero such that* $\operatorname{index}(X) = n - 1$. *Assume that $N^1(X) \cong \mathbb{R}$ (by (1.8) this is automatic if $n \geq 5$). Let $-K_X = (n-1)H$. Then one of the following holds:*

(1.12.1) $(H^n) = 1$ *and* $X \cong X_6 \subset \mathbb{P}(1^{n-1}, 2, 3)$.
(1.12.2) $(H^n) = 2$ *and* $X \cong X_4 \subset \mathbb{P}(1^n, 2)$.
(1.12.3) $(H^n) = 3$ *and* $X \cong X_3 \subset \mathbb{P}(1^{n+1})$.
(1.12.4) $(H^n) = 4$ *and* $X \cong X_{2,2} \subset \mathbb{P}(1^{n+2})$.
(1.12.5) $(H^n) = 5$ *and X is a linear space section of the Grassmannian* $\operatorname{Grass}(2,5) \subset \mathbb{P}^9$. *(Thus $n \leq 6$).*

Remark. For $n = 2$ one recovers the models of Del Pezzo surfaces obtained in (III.3.5) in the first four cases.

1.12.6 Exercise. Let X be a smooth Fano variety of dimension n over a field of characteristic zero and H a line bundle on X such that $-K_X = \operatorname{index}(X)H$. Use (1.11.4) to show that

$$\chi(X, \mathcal{O}(xH)) = \begin{cases} (H^n)\binom{x+n-1}{n} + \binom{x+n-2}{n-2} & \text{if } \operatorname{index}(X) = n-1; \\ (H^n)\left(\frac{2x+n-2}{2n}\right)\binom{x+n-2}{n-1} + \binom{x+n-2}{n-2} + \binom{x+n-3}{n-2} & \text{if } \operatorname{index}(X) = n-2. \end{cases}$$

Therefore

$$h^0(X, \mathcal{O}(H)) = \begin{cases} (H^n) + n - 1 & \text{if } \operatorname{index}(X) = n-1; \\ \frac{1}{2}(H^n) + n & \text{if } \operatorname{index}(X) = n-2. \end{cases}$$

The precise version of the problem about low degree rational curves is the following:

1.13 Problem. Let X be a smooth Fano variety such that $N^1(X) \cong \mathbb{R}$. Does there exist a rational curve $C \subset X$ such that $-K_X \cdot C = \operatorname{index}(X)$?

In analogy with the case of hypersurfaces such a curve C is called a *line* on X. In general X cannot be embedded into projective space in such a way that C becomes a projective line.

I do not know any good reason why lines should exist, but I also do not know any counterexamples. The existence of lines can be established for most of the examples introduced in this section. The case of weighted complete intersections is considered in (4.10–11).

1.14 Exercise. Let X be a smooth Fano variety such that $2\operatorname{index}(X) > \dim X + 1$. Show that X contains lines.

1.15 Theorem. *Let G be a semisimple group, P a (reduced) parabolic subgroup and $X = G/P$ the corresponding homogeneous space. Assume that $N_1(X) \cong \mathbb{R}$. Then X contains a line.*

Proof. This requires some familiarity with the theory of algebraic groups. I thank I.-H. Tsai and D. Miličić for explaining the proof to me.

Let G be a simply connected semisimple algebraic group over an algebraically closed field and $B < G$ a Borel subgroup containing a maximal torus T. Let Δ be the set of corresponding simple roots, $S_\alpha \subset G: \alpha \in \Delta$ the corresponding subgroups, isomorphic to $SL(2)$. $B_\alpha := S_\alpha \cap B$ is a Borel subgroup of S_α, and $T_\alpha := S_\alpha \cap T$ a maximal torus of S_α.

Up to conjugation we may assume that $B < P$. Every parabolic subgroup containing B can be written as $P = P_I = \langle B, S_\alpha : \alpha \notin I \rangle$ where $I \subset \Delta$ is a subset. For every α there is an irreducible representation of G with highest weight α and the stabilizer of the highest weight line is P_α. Thus P_α has a one-dimensional representation

$$\rho_\alpha : P_\alpha \to GL(1) \quad \text{such that} \quad \rho_\alpha : T_\alpha \to GL(1) \quad \text{is an isomorphism.}$$

Restricting ρ_α to P yields a 1-dimensional representation $\rho_\alpha : P \to GL(1)$. Let L_α be the corresponding line bundle on $X = G/P$. $B_\alpha = S_\alpha \cap P$, thus the induced morphism

$$f : \mathbb{P}^1 \cong S_\alpha / B_\alpha \to G/P = X$$

gives a rational curve in X. f^*L_α is a line bundle on $\mathbb{P}^1$ which is obtained from $\rho_\alpha | B_\alpha$. This is the standard 1-dimensional representation of B_α which corresponds to $\mathcal{O}(1)$ on $\mathbb{P}^1$. Thus $\deg f^*L_\alpha = 1$.

If $\dim N_1(X) = 1$, then $-K_X$ is a multiple of L_α. Thus $\operatorname{im} f$ is a line on X. □

1.16 Exercise. Let X be a proper scheme over an algebraically closed field. Assume that a connected, solvable and linear algebraic group G acts on X.

(1.16.1) Show that G also acts on $\mathrm{Chow}(X)$. Use (1.4.2.1) and this action to show that every d-dimensional cycle on X is effectively rationally equivalent to a cycle $\sum_i a_i[Y_i]$ where every Y_i is G-invariant.

(1.16.2) Assume that G has only finitely many orbits on X. Show that every d-dimensional cycle on X is effectively rationally equivalent to a cycle $\sum_i a_i[\overline{Y_i}]$ where the sum is over d-dimensional G-orbits.

(1.16.3) Assume that G has only finitely many orbits on X. Conclude that $A_k(X)$ is a finitely generated Abelian group and $AE_k(X)$ is a finitely generated semigroup. Thus $N_k(X)$ is a finite dimensional vector space and $NE_k(X)$ is a polyhedral cone.

(1.16.4) Let X be a scheme and G a group acting on X. Define the notion of G-rational equivalence by requiring in (II.2.4.1) all schemes to be with a G-action and all morphisms to be G-equivariant. The resulting groups are denoted by $A_k^G(X)$.

(1.16.5) The following more precise result is proved in [FMSS94]. Let X be a scheme (not necessarily proper) and G a connected and solvable linear algebraic group acting on X. Then the natural map $A_k^G(X) \to A_k(X)$ is an isomorphism.

(1.16.6) Show that $\mathrm{Chow}_{k,d}(\mathbb{P}^n)$ is connected for any d, k, n.

(1.16.7) Let X be a proper homogeneous space under a linear algebraic group such that the stabilizer of a point is reduced. Show that every effective 1-cycle $D \subset X$ is rationally equivalent to an irreducible rational curve $C_D \subset X$. Hint: First show that D is rationally equivalent to a connected 1-cycle D' all of whose irreducible components are rational curves. Show that there is a tree of rational curves C and a morphism $f : C \to X$ such that $f_*[C] = D'$. Then use (II.7.6).

V.2 Boundedness of Fano Varieties

In the previous section we investigated low degree rational curves on Fano varieties and we proved that every Fano variety is uniruled. The aim of this section is to show that every Fano variety is rationally connected, at least in characteristic zero. Moreover, it is possible to give an explicit upper bound on the degree of a rational curve connecting two points.

The proofs are easier, and the resulting bound is much sharper, for Fano varieties with Picard number 1. Two different proofs of this case are given. Both are short and interesting (2.6–12). The general case is harder and the bounds are so large that I did not even write them down explicitly, though this is easy to do so (2.14). These bounds imply that there are only finitely many deformation types of smooth Fano varieties of any given dimension (2.3, 2.14).

We start with Fano varieties X such that $N_1(X) = \mathbb{R}$. By (1.4.1.3) this implies that $N_1(X)_{\mathbb{Z}} = \mathbb{Z}$.

2.1 Theorem [KoMiMo92a]. *Let X be a smooth Fano variety of dimension n over an algebraically closed field of characteristic zero. Assume that* $\mathrm{Pic}(X) \cong \mathbb{Z}$. *There is an open set $U \subset X \times X$ such that if $(x_1, x_2) \in U$, then there is an irreducible rational curve $C_{12} \subset X$ such that*

$$x_1, x_2 \in C_{12} \quad \textit{and} \quad -K_X \cdot C_{12} \le n(n+1).$$

2.2 Theorem [Nadel90; Campana91; Nadel91; KoMiMo92a]. *Let X be a smooth Fano variety of dimension n over an algebraically closed field of characteristic zero. Assume that* $\mathrm{Pic}(X) \cong \mathbb{Z}$. *Then*

$$0 < (-1)^n (K_X^n) \le (n(n+1))^n.$$

2.3 Corollary. *Let k be an algebraically closed field of characteristic zero. For every $n > 0$ there are ony finitely many deformation types of n-dimensional Fano varieties over k with* Pic $\cong \mathbb{Z}$.

2.4 History of the Results. Fano varieties of dimension at most 3 over $\mathbb{C}$ have been completely classified. For dim = 1 we have only $\mathbb{P}^1$. For dim = 2 these are the Del Pezzo surfaces (III.3). There are 10 deformation types.

For dim = 3 the classification is considerably harder. See [Iskovskikh83; Mori83] and the references there.[3] There are seventeen families with Pic $\cong \mathbb{Z}$, four score and seven with rank Pic > 1.

(2.3) for toric Fano varieties is due to [Batyrev82].

[Nadel90] proved (2.2–3) for dim = 4. His ideas are presented in (2.11–12). He used a weaker version of the results from (IV.4), this accounted for the dimension restriction. Based on his ideas, the general result was completed in three independent articles.

[Campana91] realized that his earlier works concerning equivalence relations can be used to complete Nadel's approach to (2.2–3) in all dimensions. [Nadel91] used a somewhat different version of (IV.4) to prove (2.2–3). [KoMiMo92a] contains a proof of (2.1–3); this approach is described in (2.5–9).

For rank Pic > 1 the above methods do not seem to work. Rational chain connectedness is proved in [Campana92; KoMiMo92c]. Boundedness is established in [KoMiMo92c]

Explicit estimates for the number of deformation types is given in [Kollár93]. The bounds are certainly very crude.

It is not clear how sharp the bounds given in (2.1–2) and (2.14) are. It has been conjectured that for a smooth Fano variety of dimension n the inequality $(-1)^n(K_X^n) \le (n+1)^n$ holds with equality only for $X = \mathbb{P}^n$. A counterexample with rank Pic > 1 is given in [Batyrev82]. It is still possible that this inequality holds if rank Pic = 1 but I do not see any reason to believe it.

2.5 Proof of (2.3). The proof requires some results that we are unable to prove here.

[Matsusaka70] proves that there are only finitely many deformation types of Fano varieties with fixed Hilbert polynomial $\chi(X, \mathcal{O}(tK_X))$. The two highest coefficients of $\chi(X, \mathcal{O}(tK_X))$ depend only on (K_X^n), hence bounded for fixed n by (2.2). The rest of the coefficients are also bounded by (VI.2.15.8.9).

More recent results of [Demailly93; Kollár93; Angehrn-Siu95; Tsuji96; Kollár97] imply that $-(n(n+1)(n+3)/2)K_X$ is very ample. Thus all Fano varieties of dimension n and with Picard number one can be realized as subvarieties of $\mathbb{P}^{2n+1}$ of degree at most

$$2^{-n}n^{2n}(n+1)^{2n}(n+3)^n.$$

[3] See [Shepherd-Barron97, Megyesi98] for positive characteristic.

Thus (2.3) follows from the general boundedness results about Chow varieties (I.3.21). (I.3.28) can be used to obtain an explicit bound for the number of deformation types. □

2.5.1 Remark. The above proof shows that Fano varieties with $(-1)^n(K_X^n)$ bounded from above form a bounded family.

The key result in the proof of (2.1) is the next application of (IV.4.14):

2.6 Proposition. *Let X be a smooth Fano variety of dimension n over an algebraically closed field of characteristic zero such that $N_1(X) = \mathbb{R}$. There is an open set $U \subset X \times X$ such that if $(x_1, x_2) \in U$, then x_1 and x_2 can be joined by a connected chain of free rational curves of anticanonical degree at most $n(n+1)$.*

Proof. By (IV.2.10) there is an irreducible component $V \subset \operatorname{Hom}(\mathbb{P}^1, X)$ parametrizing free morphisms of anticanonical degree at most $n+1$. Let $V \xleftarrow{s} U = V \times \mathbb{P}^1 \xrightarrow{w} X$ be the universal family. s is proper and open. w is smooth by (II.3.5.4) hence open and equidimensional. Thus by (IV.4.14) two general points can be connected by a U-chain of length at most n. The anticanonical degree of the chain is at most $n(n+1)$. □

2.7 Proof of (2.1). Let $W \subset \operatorname{Hom}(\mathbb{P}^1, X)$ be the closed subscheme parametrizing morphisms f such that $-K_X \cdot_f \mathbb{P}^1 \leq n(n+1)$. We obtain a morphism

$$u^{(2)} : W \to X \times X \quad \text{by} \quad [f] \mapsto (f(0), f(\infty)).$$

(2.1) is equivalent to the assertion that $u^{(2)}$ is dominant. Assume the contrary and let $Z \subsetneq X \times X$ be a closed subvariety containing the image of $u^{(2)}$.

Let $h : C \to X$ be a connected chain of rational curves joining two points $x_1, x_2 \in X$ such that $-K_X \cdot_h C \leq n(n+1)$. If C is smoothable (II.7.2), then $(x_1, x_2) \in Z$.

By (II.7.6) a chain of free morphisms can always be smoothed. Thus (2.6) implies (2.1). □

2.8 Remark. There are two parts of the proof which do not work in positive characteristic. In order to smooth the chains we need to use free rational curves. Some Fano varieties may not be separably uniruled. Even if a Fano variety X is separably uniruled, it is not clear that it is separably uniruled with curves of low degree.

The implication (2.1) $\Rightarrow$ (2.2) goes back to Fano:

2.9 Proposition. *Let X be a proper variety of dimension n, $x \in X$ a smooth point and L an nef and big line bundle on X. Choose $d > 0$ such that a general point $x' \in X$ can be connected to x by an irreducible curve $C_{x'}$ such that $L \cdot C_{x'} \leq d$. Then $(L^n) \leq d^n$.*

Proof. Fix $\epsilon > 0$. By (VI.2.15.7) there is a $k > 0$ and a divisor $D_k \in |kL|$ such that $\operatorname{mult}_x D_k \geq k\sqrt[n]{(L^n)} - k\epsilon$. Pick a general point $x' \notin \operatorname{Supp} D_k$. Then $C_{x'}$ is not contained in D_k hence

$$kd \geq D_k \cdot C_{x'} \geq \operatorname{mult}_x D_k \geq k\sqrt[n]{(L^n)} - k\epsilon.$$

Divide by k and let ϵ go to zero. □

2.10 First Proof of (2.2) [KoMiMo92a]. Pick a general $x \in X$. By (2.1) a general point $x' \in X$ can be connected to x by a rational curve $C_{x'}$ such that $-K_X \cdot C_{x'} \leq n(n+1)$. By assumption $-K_X$ is ample. Thus (2.9) implies (2.2). □

Another interesting argument showing that (2.6) implies (2.2) is given in [Nadel90]. Let $D \subset X$ be an effective divisor and $C \subset X$ an irreducible curve. If $C \not\subset D$, then $\operatorname{mult}_x D \leq C \cdot D$ for every $x \in C$. In other words, $\operatorname{mult}_x D$ is an upper semi continuous function of $x \in C$ and its jumps are at most $C \cdot D$, provided $C \not\subset D$. Nadel's observation is that this assertion remains true if $C \subset D$ and C is free.

2.11 Proposition [Nadel90,91]. *Let X be a smooth variety, $D \subset X$ an effective divisor, $f : \mathbb{P}^1 \to X$ a free morphism, and $a, b \in \mathbb{P}^1$ closed points. Then*

$$|\operatorname{mult}_{f(a)} D - \operatorname{mult}_{f(b)} D| \leq D \cdot_f \mathbb{P}^1.$$

Proof. By (II.3.5.4) there is a smooth variety $V \subset \operatorname{Hom}(\mathbb{P}^1, X)$ such that f can be extended to a smooth morphism $F : \mathbb{P}^1 \times V \to X$ and $F|\mathbb{P}^1_v = f$ for some $v \in V$. Let $D_V = F^*D$. Since F is smooth, $\operatorname{mult}_u D_V = \operatorname{mult}_{F(u)} D$ for every $u \in \mathbb{P}^1 \times V$.

Let $v \in C \subset V$ be a general curve section which is smooth at v. Let $D_C = D_V|\mathbb{P}^1 \times C$. Since C is general,

$$\operatorname{mult}_{(a,v)} D_V = \operatorname{mult}_{(a,v)} D_C \quad \text{and} \quad \operatorname{mult}_{(b,v)} D_V = \operatorname{mult}_{(b,v)} D_C.$$

Write $D_C = \alpha\mathbb{P}^1_v + D^h_C$ where D^h_C does not contain $\mathbb{P}^1_v$ as an irreducible component. Then

$$\operatorname{mult}_{(a,v)} D_C = \alpha + \operatorname{mult}_{(a,v)} D^h_C, \quad \text{and}$$
$$\operatorname{mult}_{(b,v)} D_C = \alpha + \operatorname{mult}_{(b,v)} D^h_C.$$

Therefore it is sufficient to prove that

$$(2.11.1) \qquad |\operatorname{mult}_{f(a)} D^h_C - \operatorname{mult}_{f(b)} D^h_C| \leq D \cdot_f \mathbb{P}^1.$$

$\mathbb{P}^1_v$ is not contained in D^h_C, thus

$$0 \leq \operatorname{mult}_{f(a)} D^h_C \leq \mathbb{P}^1_v \cdot D^h_C = \mathbb{P}^1_v \cdot D_C = \mathbb{P}^1_v \cdot D_V = D \cdot_f \mathbb{P}^1,$$

and similarly for $\operatorname{mult}_{f(b)} D^h_C$. This shows (2.11.1) which in turn implies (2.11). □

2.11.2 Remarks. (2.11.2.1) The above proof shows that if $d = \min_{b\in\mathbb{P}^1} \operatorname{mult}_{f(b)} D$, then

$$\sum_{a\in\mathbb{P}^1} |\operatorname{mult}_{f(a)} D - d| \leq D \cdot_f \mathbb{P}^1.$$

(2.11.2.2) [EKL95] studies a version of this result where the singularities of $f(\mathbb{P}^1)$ are also taken into account.

(2.11.2.3) It would be interesting to study (2.11) for $f: C \to X$ (where C is a proper curve), assuming only some positivity properties of f^*T_X/T_C.

2.12 Second Proof of (2.2). Pick a general $x \in X$ and fix $\epsilon > 0$. By (VI.2.15.7) there is a $k > 0$ and a divisor $D_k \in |kL|$ such that $\operatorname{mult}_x D_k \geq k\sqrt[n]{(L^n)} - k\epsilon$. Pick a general point $x' \notin \operatorname{Supp} D_k$. By (2.6) there is a connected chain of free rational curves $C = \cup C_i$ joining x and x'. By construction

$$\operatorname{mult}_x D_k \geq k\sqrt[n]{(L^n)} - k\epsilon \quad \text{and} \quad \operatorname{mult}_{x'} D_k = 0.$$

By (2.11) the largest possible multiplicity jump along C_i is at most $D_k \cdot C_i$. Thus

$$\operatorname{mult}_x D_k = |\operatorname{mult}_x D_k - \operatorname{mult}_{x'} D_k| \leq \sum_i (D_k \cdot C_i) \leq kn(n+1).$$

Compare with the previous inequality and let ϵ go to zero to obtain (2.2). □

Similar results hold if rank Pic > 1 but with larger bounds.

2.13 Theorem [Campana92; KoMiMo92c]. *Let X be a smooth Fano variety of dimension n over an algebraically closed field of any characteristic. Then X is rationally chain connected.*

2.14 Corollary [KoMiMo92c]. *Let X be a smooth Fano variety of dimension n over an algebraically closed field. Then:*

(2.14.1) *Any two points of X can be joined by a connected chain of rational curves of anticanonical degree at most $(n+1)2^n$.*

(2.14.2) *In characteristic zero there is a number $d(n)$ (depending only on n) such that any two points of X can be joined by an irreducible rational curve of anticanonical degree at most $d(\dim X)$.*

Proof. Let $C \subset X$ be a rational curve and $x_1, x_2 \in C$ points. By (II.5.6.2) there is a connected chain of rational curves $\sum C_i$ joining x_1, x_2 such that $-K_X \cdot C_i \leq n+1$ for every i.

Let $V_i \leftarrow U_i = V_i \times \mathbb{P}^1 \to X$, $i = 1, \dots, m$ be all the families of connected 1–cycles C with rational components such that $-K_X \cdot C \leq n+1$. By the above remark any two points of X can be joined by a $\langle U_1, \dots, U_m\rangle$-chain. Thus by (IV.4.17) any two points of X can be joined by a $\langle U_1, \dots, U_m\rangle$-chain of length at most 2^n.

The second assertion follows from the first one using (IV.3.10.1). □

By (2.5.1) and (2.9) this implies the boundedness of Fano varieties in any given dimension:

2.15 Corollary. *Let k be an algebraically closed field of characteristic zero. For every $n > 0$ there are only finitely many deformation types of n-dimensional smooth Fano varieties over k.* □

2.16 Proof of (2.13). By (IV.4.17) there is an open subset $X^0 \subset X$ and a proper morphism $\pi : X^0 \to Z^0$ such that the fibers of π are rationally connected and if $z \in Z^0$ is very general, then every rational curve which intersects $\pi^{-1}(z)$ is contained in it.

We are done if Z^0 is a point, thus assume that $\dim Z^0 > 0$.

Let $0 \in C$ be a pointed, irreducible, smooth and proper curve and $f : C \to X$ a morphism such that $x := f(0) \in \pi^{-1}(z)$ and $\pi \circ f : C \to Z$ is nonconstant. The bend-and-break technique of (II.5) produces rational curves on X through x. We follow the method and prove that under suitable additional conditions we can guarantee that the resulting rational curve is not contained in $\pi^{-1}(z)$. If $C \cong \mathbb{P}^1$, then we are done, so assume in the sequel that C is not rational.

Assume that there is a rational curve $h : \mathbb{P}^1 \to X$ which intersects $\pi^{-1}(z)$ but is not contained in it. Choose $p_1, p_2 \in \mathbb{P}^1$ such that $h(p_1) \in \pi^{-1}(z)$ and $h(p_2) \notin \pi^{-1}(z)$. By (II.5.6.2) we can deform $\operatorname{im} h$ to a connected 1-cycle of rational curves $\sum C_i$ such that $h(p_1), h(p_2) \in \operatorname{Supp} \sum C_i$ and $-K_X \cdot C_i \le n+1$ for every i. Thus there is an irreducible component C_j which intersects $\pi^{-1}(z)$ but is contained in it and in addition $-K_X \cdot C_j \le n+1$. Therefore, using (II.5.10) we see that it is sufficient to prove (2.13) in positive characteristic.

Let $Z^0 \subset Z$ be a compactification and $g : X' \to X$ a proper birational morphism such that g is an isomorphism over X^0 and π extends to a morphism $\pi' : X' \to Z$.

Let $V \subset \operatorname{Hom}(C, X, 0 \mapsto x)$ be an irreducible component containing $[f]$ and $U \subset (\mathbb{P}^1 \times V) \times_X X'$ the irreducible component which dominates $\mathbb{P}^1 \times V$. Then $g_U : U \to \mathbb{P}^1 \times V$ is birational, hence there is a dense open subset $V' \subset V$ such that g_U is an isomorphism over $\mathbb{P}^1 \times V'$. Therefore the universal morphism u factors as

$$u : C \times V' \xrightarrow{u'} X' \to X.$$

Thus we obtain morphisms

$$\rho : V' \to \operatorname{Hom}(C, X', 0 \mapsto x) \xrightarrow{\pi'_*} \operatorname{Hom}(C, Z, 0 \mapsto z).$$

2.16.1 Claim (Notation as above). *Assume that* $\dim_{[f]} \operatorname{Hom}(C, X, 0 \mapsto x) > 0$.

(2.16.1.1) *If the image of ρ is positive dimensional, then there is a rational curve $h : \mathbb{P}^1 \to X$ such that $h(\mathbb{P}^1)$ intersects $\pi^{-1}(z)$ but is not contained in it.*

(2.16.1.2) *Let H be ample on Z. If the image of ρ is zero dimensional, then there is a morphism $f' : C \to X$ such that $f'(0) \in \pi^{-1}(z)$ and*

$$-K_X \cdot_f C > -K_X \cdot_{f'} C \quad \textit{and} \quad H \cdot_{\pi\circ f} C = H \cdot_{\pi\circ f'} C.$$

Proof. Choose a smooth, irreducible curve and a nonconstant morphism $B^0 \to V'$. Let

$$h_X^0 : C \times B^0 \to X' \times B^0 \quad \text{and} \quad h_Z^0 : C \times B^0 \to Z \times B^0$$

be the cycle morphisms. By (II.5.5) there are normal compactifications S_X (resp. S_Z) of $C \times B^0$ such that h_X^0 (resp. h_Z^0) extend to finite morphisms

$$h_X : S_X \to X' \times B \quad \text{and} \quad h_Z : S_Z \to Z \times B.$$

$\pi' \circ h_X : S_X \to Z \times B$ is proper and agrees with $h_Z : S_Z \to Z \times B$ over a dense open set, thus there is a factorization

$$\pi' \circ h_X : S_X \xrightarrow{\pi_B} S_Z \to Z \times B.$$

Assume first that the image of $B^0 \to V' \to \operatorname{Hom}(C, Z, 0 \mapsto z)$ is one dimensional. By (II.5.5.1) there is a point $b \in B - B^0$ and a rational curve $C_Z \in p_Z^{-1}(b)$ such that $h_Z(C_Z)$ is a rational curve in Z passing through z.

Since $S_X \to S_Z$ is birational, the birational transform $C_X \subset S_X$ of C_Z is a rational curve. Moreover, $h_X(C_X) \subset X$ is a rational curve which maps onto $h_Z(C_Z)$. Thus $h_X(C_X)$ intersects $\pi^{-1}(z)$ but is not contained in it.

Assume next that the image of $B^0 \to V' \to \operatorname{Hom}(C, Z, 0 \mapsto z)$ is zero dimensional. Then $S_Z \cong C \times B$ but S_X has a contractible flat section. Therefore $S_X \to S_Z$ is not an isomorphism. Thus there is a point $b \in B - B^0$ such that $p_X^{-1}(b)$ is the union of a curve $C_b \cong C$ and of some rational curves E which are π_B-exceptional. Let $f' := h_X|C_b$. Then

$$-K_X \cdot_f C = -K_X \cdot_{f'} C_b - K_X \cdot_{h_X} E > -K_X \cdot_{f'} C_b.$$

By construction $\pi \circ f' = \pi \circ f$, hence $H \cdot_{\pi\circ f} C = H \cdot_{\pi\circ f'} C$. □

In the next step we use the following obvious

2.16.2 Claim. *Let $\pi : X \dashrightarrow Z$ be a map between schemes, H_X (resp. H_Z) ample divisors. There is an $\epsilon > 0$ such that if D is a smooth, irreducible and proper curve and $g : D \to X$ any morphism such that $\pi \circ g : D \to Z$ is defined, then*

$$H_X \cdot_g D \geq \epsilon(H_Z \cdot_{\pi\circ g} D).$$ □

Let H be an ample divisor on Z. Choose $\epsilon > 0$ such that $-K_X \cdot_g D \geq \epsilon(H \cdot_{\pi\circ g} D)$ for every $g : D \to X$ as in (2.16.2).

Choose any $f : C \to X$ such that $f(0) = x$ and $\pi \circ f : C \to Z$ is nonconstant. Composing f with a sufficiently high power of the Frobenius $C \to C$ we obtain $f_0 : C \to X$ such that

$$\epsilon H \cdot_{\pi \circ f_0} C - ng(C) > 0.$$

By the choice of ϵ this implies that

$$-K_X \cdot_{f_0} C - ng(C) > 0.$$

By (II.1.7) this implies that $\dim_{[f]} \operatorname{Hom}(C, X, 0 \mapsto x) > 0$ and (2.16.1) applies.

If (2.16.1.1) holds, then we have the sought after rational curve. If (2.16.2) holds, then we obtain $f_2 := f_1' : C \to X$ such that

$$\epsilon H \cdot_{\pi \circ f_1} C - ng(C) = \epsilon H \cdot_{\pi \circ f_2} C - ng(C) > 0 \quad \text{and} \quad -K_X \cdot_{f_1} C > -K_X \cdot_{f_2} C.$$

The same argument can be applied to f_2. Thus we either get the desired rational curve or a sequence of morphisms $f_i : C \to X$ such that

$$-K_X \cdot_{f_1} C > -K_X \cdot_{f_2} C > -K_X \cdot_{f_3} C > \cdots.$$

All of these are nonegative integers, thus the sequence stops with some morphism f_i. Applying (2.16.1) to f_i we obtain the required rational curve. □

V.3 Characterizations of $\mathbb{P}^n$

The aim of this section is to give several characterizations of projective spaces. Among algebraic curves $\mathbb{P}^1$ is characterized by the property that its tangent bundle admits a metric with positive curvature, or equivalently that its tangent budle has positive degree. The deepest results characterize $\mathbb{P}^n$ by some "positivity" property of $T_{\mathbb{P}^n}$.

3.1 Definition. For many purposes it is desirable to have a notion of "positivity" of vector bundles. Unfortunately, there are several different notions of "positivity" depending on the property of positive line bundles that we want to generalize. In complex differential geometry the usual notion is "positivity of curvature" (see e.g. [Griffiths-Harris78, 0.5]). In algebraic geometry the usual notion is called "ampleness" (see (3.9) or [Hartshorne70]) though in positive characteristic some other variants are also used. The precise definitions are not important for our purposes. We use only three basic properties which are shared by all variants:

(3.1.1) If E is "positive", then $\det E$ is an ample line bundle.

(3.1.2) If E is "positive" on X and $f : C \to X$ is a finite morphism from a curve C, then f^*E is "positive".

(3.1.3) $\sum \mathcal{O}_{\mathbb{P}^1}(a_i)$ is "positive" on $\mathbb{P}^1$ iff $a_i > 0$ for every i.

The main result of this section asserts that $\mathbb{P}^n$ is the only smooth projective variety whose tangent bundle satisfies certain weakened versions of the above three "positivity" properties.

3.2 Theorem [Mori79]. *Let X be a smooth projective variety of dimension n. Assume that*

(3.2.1) *K_X^{-1} is ample; and*

(3.2.2) *there is an $x \in X$ such that for every nonconstant morphism $f : (0, \mathbb{P}^1) \to (x, X)$ the pull back f^*T_X is the sum of line bundles of positive degree.*

Then $X \cong \mathbb{P}^n$.

Before giving the proof, we derive three corollaries. The first two answer a conjecture raised by [Frankel61] in the differential geometric context and generalized by [Hartshorne70] to the algebraic case.

3.3 Corollary [Mori79]. *Let X be a smooth projective variety of dimension n over an algebraically closed field of arbitrary characteristic. T_X is ample iff $X \cong \mathbb{P}^n$.*

3.4 Corollary [Mori79; Siu-Yau80]. *Let X be a compact complex manifold of dimension n. Then X admits a Kähler metric whose holomorphic bisectional curvature is everywhere positive iff $X \cong \mathbb{P}^n$.*

Proof. Let us show first that $T_{\mathbb{P}^n}$ has the required positivity properties. The usual presentation of $T_{\mathbb{P}^n}$ is given by

$$0 \to \mathcal{O}_{\mathbb{P}^n} \to \sum_1^{n+1} \mathcal{O}_{\mathbb{P}^n}(1) \to T_{\mathbb{P}^n} \to 0.$$

Elementary properties of ample vector bundles [Hartshorne70] imply that $T_{\mathbb{P}^n}$ is ample. The Fubini-Study metric on $\mathbb{P}^n$ has positive holomorphic bisectional curvature.

In any case we see that $\det T_{\mathbb{P}^n} = K_{\mathbb{P}^n}^{-1} = \mathcal{O}_{\mathbb{P}^n}(n+1)$ is ample. Also, setting $m = \deg f^*\mathcal{O}_{\mathbb{P}^n}(1)$ we obtain a surjection

$$\sum \mathcal{O}_{\mathbb{P}^1}(m) \to f^*T_{\mathbb{P}^n} \to 0.$$

By (II.3.8–9) this implies that $f^*T_{\mathbb{P}^n}$ is the sum of line bundles of positive degree.

Conversely, (3.1.1) imples that K_X^{-1} is ample. In the complex analytic case a Hermitian metric on T_X induces a metric on K_X^{-1} which has positive curvature. Therefore X is biholomorphic to an algebraic variety and K_X^{-1} is ample (Kodaira's Theorem, see e.g. [Griffiths-Harris78, I.4]).

(3.1.2–3) imply (3.2.2). □

Every projective algebraic variety of dimension n admits a finite morphism to $\mathbb{P}^n$. [Remmert-V.d.Ven61] conjectured that $\mathbb{P}^n$ is the only smooth variety with this property. The following result shows that this is indeed the case, at least in characteristic zero.

3.5 Corollary [Lazarsfeld84]. *Let X be a smooth projective variety of dimension n. Assume that there is a surjective and separable morphism $p : \mathbb{P}^n \to X$. Then $X \cong \mathbb{P}^n$.*

Proof. By the adjunction formula

$$f^*(-K_X) = -K_{\mathbb{P}^n} + (\text{ramification divisor}),$$

thus some multiple of $-K_X$ is effective. By (II.4.4) $\dim N_1(X) = 1$, hence $-K_X$ is ample. Pick a point $x \in X$ such that f is étale above x. Let $f : (0 \in \mathbb{P}^1) \to (x \in X)$ be a morphism and C the normalization of an irreducible component of $\mathbb{P}^n \times_X \mathbb{P}^1$. We have a diagram

$$\begin{array}{ccc} C & \xrightarrow{h} & \mathbb{P}^n \\ {\scriptstyle q}\downarrow & & \downarrow{\scriptstyle p} \\ \mathbb{P}^1 & \xrightarrow{f} & X. \end{array}$$

The natural map

$$r : h^*T_{\mathbb{P}^n} \to h^*(p^*T_X) = q^*(f^*T_X)$$

is a local isomorphism at $q^{-1}(0)$ since f is étale above x. Write $f^*T_X \cong \sum \mathcal{O}_{\mathbb{P}^1}(a_i)$. We need to prove that $a_i > 0$ for every i. For any j we have a map

$$q : \sum h^*\mathcal{O}_{\mathbb{P}^n}(1) \to h^*T_{\mathbb{P}^n} \to \sum q^*\mathcal{O}_{\mathbb{P}^1}(a_i) \to q^*\mathcal{O}_{\mathbb{P}^1}(a_j),$$

which is surjective over an open set $U \subset C$. Thus $q^*\mathcal{O}_{\mathbb{P}^1}(a_j)$ has a nonzero section which vanishes at some point. Therefore $a_i > 0$ for every i and so (3.2) implies (3.5). □

The proof of (3.2) uses the following lemma, which is a singular version of the bend-and-break results of (II.5).

3.6 Lemma. *Let X be a smooth projective variety and $x \in X$ a point. Let $B \subset \mathrm{Chow}_1(X)$ be a proper curve with the following properties:*

(3.6.1) *For every $b \in B$ the corresponding 1-cycle $C_b \subset X$ is an irreducible and reduced rational curve.*

(3.6.2) *$x \in C_b$ for every $b \in B$.*

(3.6.3) *C_b is singular for general $b \in B$.*

Then there is a $b \in B$ such that the normalization $f_b : \mathbb{P}^1 \to C_b \subset X$ is not an immersion.

Proof. Let $p : S \to B$ be the universal family and $u : S \to X$ the cycle morphism. We can normalize S and B to obtain $n : S_1 \to S$, $p_1 : S_1 \to B_1$ and $u_1 : S_1 \to X$. S_1 is a minimal ruled surface over B_1 and $f_b \cong u_1|p_1^{-1}(b)$. Set $E_1 = u_1^{-1}(x)$ (with reduced scheme structure).

Let $N \subset S_1$ be the set of points where n is not a local isomorphism. Pick $b \in B_1$. If $f_b : p_1^{-1}(b) \to C_b \subset X$ is not an immersion then we are done. Otherwise C_b has at least one singular point where two branches come together, thus N intersects the general fiber of p_1 in at least 2 distinct points.

After a suitable base change $B_2 \to B_1$ we obtain $p_2 : S_2 \to B_2$ and $u_2 : S_2 \to X$ with the following properties.

(3.6.4) $p_2 : S_2 \to B_2$ is a minimal ruled surface.

(3.6.5) There is a section $E \subset S_2$ such that $u_2(E) = \{x\}$.

(3.6.6) There are two sections $\sigma_i : B_2 \to S_2$ such that $u_2(\sigma_1(b)) = u_2(\sigma_2(b))$ for every $b \in B$. Equivalently, set $N_i := \operatorname{im}\sigma_i \subset S_2$, then there is a $g : B_2 \to X$ and a factorization

$$(u_2|N_1 \cup N_2) : N_1 \cup N_2 \xrightarrow{p_2} B_2 \xrightarrow{g} X.$$

3.6.7 Claim. *With the above assumptions,* $(N_1 \cdot N_2) > 0$.

Proof. Since E is contractible, (E^2) is negative. Let F be a fiber of p_2. Then $N_i \overset{n}{\approx} E + a_i F$ for some $a_i \geq 0$. At least one of the N_i is different from E, say $N_2 \neq E$. Then

$$(N_1 \cdot N_2) = (E \cdot N_2) + a_1(F \cdot N_2) \geq a_1.$$

We are done, unless $a_1 = 0$, that is $E = N_1$. Then $u_2(N_1) = \{x\}$, so by (3.6.6) $u_2(N_2) = \{x\}$. By (II.5.4.2) a minimal ruled surface can not have two contractible sections. □

Let $b \in B_2$ be a point such that N_1 and N_2 intersect in the fiber over b. Then $(N_1 \cup N_2) \cap p_2^{-1}(b)$ is an irreducibe subscheme of length 2 which is mapped to $g(b)$ by u_2. Thus $u_2|p_2^{-1}(b)$ is not an immersion. □

3.7 Proof of (3.2). Let us start with a general outline of the approach.

3.7.1 Idea of the Proof. We try to approach $\mathbb{P}^n$ via the family of lines on it. The family of all lines is too large, so we fix a point $x \in \mathbb{P}^n$ and consider the lines through x. These lines are naturally parametrized by $\mathbb{P}(T_x\mathbb{P}^n)$. Lines are characterized by the property that their anticanonical degree is $n+1$, which is the smallest possible. By (1.6) we are able to find morphisms $f : \mathbb{P}^1 \to X$ whose anticanonical degree is at most (n+1). First we prove that every such f is an embedding (3.7.3). Then we show that the lines through x make B_xX into a $\mathbb{P}^1$-bundle over the exceptional divisor $\mathbb{P}^{n-1} \subset B_xX$ (3.7.7). This easily implies that $X \cong \mathbb{P}^n$ (3.7.8).

3.7.2 Lemma. *Let $f : \mathbb{P}^1 \to X$ be a morphism such that $f^*T_X \cong \sum \mathcal{O}(a_i)$ and $a_i > 0$ for every i. Then $\deg f^*T_X \geq n+1$. If equality holds, then f is an immersion and*

$$f^*T_X \cong \mathcal{O}(2) + \mathcal{O}(1)^{n-1}.$$

Proof. Assume that $a_1 \geq a_2 \geq \ldots \geq a_n$. Then $a_i \geq 1$ by (3.2.2) and $a_1 \geq 2$ by (II.3.13). Thus $\deg f^*T_X = \sum a_i \geq n+1$ and if equality holds, then $a_1 = 2$ and $a_i = 1$ for $i > 1$. Therefore f is an immersion by (IV.2.11). □

3.7.3 Lemma (Notation and assumptions as in (3.2)). *Then*

(3.7.3.1) *There is a morphism $f : (0, \mathbb{P}^1) \to (x, X)$ such that* $\deg f^*T_X = n+1$.

(3.7.3.2) *Any morphism $f : (0, \mathbb{P}^1) \to (x, X)$ such that* $\deg f^*T_X = n+1$ *is an embedding.*

Proof. By (1.6.1) there is a morphism $f : (0, \mathbb{P}^1) \to (x, X)$ such that $\deg f^*T_X \leq n+1$. Thus by (3.7.2) in fact $\deg f^*T_X = n+1$ and any such morphism is an immersion.

Assume that f is not an embedding for some f. Let $C \subset \mathbb{P}^2$ be a nodal cubic. Then there are morphisms $g : C \to X$ and and $n : \mathbb{P}^1 \to C$ such that $f = g \circ n$. By (II.1.2)

$$\dim_{[g]} \operatorname{Hom}(C, X) \geq \deg g^*T_X + n\chi(\mathcal{O}_C) = n+1.$$

Passing through $x \in X$ is $(n-1)$-conditions, thus there is a 2-dimensional subscheme $[g] \in Z \subset \operatorname{Hom}(C, X)$ such that $x \in \operatorname{im} h$ for every $[h] \in Z$. $\operatorname{Aut}(C)$ is 1-dimensional, thus there is a curve $[g] \in B^0 \subset Z$ such that the induced morphism $t : B^0 \to \operatorname{Chow}_1(X)$ is nonconstant.

By (3.6) there is a point in the closure of $t(B^0)$ which corresponds to a rational curve $C_1 \subset X$ such that $x \in C_1$, $(-K_X \cdot C_1) = n+1$ and the normalization morphism $f_1 : \mathbb{P}^1 \to C_1 \subset X$ is not an immersion. This is impossibe by (3.7.2). □

3.7.4 Construction. Let $\pi : B_xX \to X$ be the blow up of $x \in X$ and $\mathbb{P}^{n-1} \cong E \subset B_xX$ the exceptional divisor. Let $C \subset B_xX$ be a 1-cycle with rational components such that $C \cdot E = 1$ and $C \cdot \pi^*(-K_X) = n+1$. Let $C' \subset C$ be an irreducible component which intersects E. $\pi(C')$ is a rational curve in X containing x such that $-K_X \cdot \pi(C') \leq n+1$. Thus C' is the only component of C not containded in E. By (3.7.3) $C' \cdot E = 1$, hence in fact $C' = C$.

Let $V \subset \operatorname{RatCurves}(B_xX)$ be the open subset parametrizing rational curves $C \subset B_xX$ such that $C \cdot E = 1$ and $C \cdot \pi^*(-K_X) = n+1$. By the above considerations V is proper. Let $f : \mathbb{P}^1 \to X$ be as in (3.7.3.1). The birational transform of $\operatorname{im} f$ is in V, thus V is not empty. Let

$$V \overset{s}{\longleftarrow} U \overset{u}{\longrightarrow} B_xX$$

be the universal family. $D := u^{-1}(E) \subset U$ is a divisor which intersects every fiber of s in a single point, thus $s : D \to V$ is an isomorphism.

It is a priori possible that V is reducible. We change the notation and replace V with the normalization of one of its irreducible components. $U \to V$ is a $\mathbb{P}^1$-bundle by (II.2.8).

The following result is the second step of the proof.

3.7.5 Proposition [Miyaoka92]. *Let X be a proper variety and $x \in X$ a smooth point. Let $V \subset \mathrm{RatCurves}(x, X)$ be a proper subvariety, $s : U \to V$ the universal family and $u : U \to X$ the cycle map. Then*

(3.7.5.1) *either u is generically one-to-one onto its image (hence birational in characteristic zero),*

(3.7.5.2) *or there is a point $v \in V$ such that $u(U_v)$ is singular at x.*

Proof. Assume that u is not generically one-to-one onto its image and let $u(U_v) \subset X$ be the image of a general fiber of s. By assumption there is an irreducible component $C' \subset u^{-1}(u(U_v))$ such that $C' \neq U_v$ and u maps C' onto $u(U_v)$. C' is not a fiber of s since two different fibers have different images in X. Let $C \to C'$ be the normalization. By base change we obtain a minimal ruled surface $S := U \times_V C \to C$, and two sections $C_1 := D \times_V C$ and $C_2 := C' \times_V C$ such that the assumptions of (3.7.6) are satisfied. Thus $u(U_{v'})$ is singular at x for some $v' \in V$. □

3.7.6 Lemma [Miyaoka92]. *Let $s : S \to C$ be a minimal ruled surface with typical fiber F and two sections $C_1, C_2 \subset S$. Let $u : S \to \mathbb{P}$ be a morphism to some projective space such that*

(3.7.6.1) $\mathrm{im}\, u$ *has dimension two,*

(3.7.6.2) *$u(C_1)$ is a single point $x \in \mathbb{P}$,*

(3.7.6.3) *$u(C_2)$ (with reduced scheme structure) is numerically equivalent to $u(F)$,*

(3.7.6.4) $x \in u(C_2)$.

Then there is a point $c \in C$ such that $\mathrm{length}(u^{-1}(x) \cap S_c) \geq 2$.

Proof. C_1 is contractible, thus $(C_1^2) < 0$ by (II.5.3.2). We can write $C_2 \overset{n}{\approx} C_1 + bF$ for some $b > 0$. We obtain that

$$\begin{aligned} \deg(C_2/u(C_2)) \cdot (F \cdot u^*H) &= \deg(C_2/u(C_2)) \cdot (u(C_2) \cdot H) \\ &= C_2 \cdot u^*H = b(F \cdot u^*H), \end{aligned}$$

hence $\deg(C_2/u(C_2)) = b$. In particular, there is a subscheme $Z \subset C_2$ of length b such that $u(Z) = \{x\}$ (as schemes).

If du has rank zero at some point $c' \in C_1$, then set $c = s(c')$. $u : S_c \to X$ is ramified at c' and we are done. Otherwise du has rank one everywhere along C_1 and $u^{-1}(x) = C_1 \cup T$ where T is disjoint from C_1. $C_2 \cdot C_1 = b + (C_1^2) < b$, thus Z is not a subscheme of $C_2 \cap C_1$. Let $z \in Z - C_1$ be a point and set $c = s(z)$ and $z' = S_c \cap C_1$. Then $u(z) = u(z')$ and $z \neq z' \in S_c$, thus we are again done. □

Applying this to our situation we obtain:

3.7.7 Lemma (Notation as in (3.7.4)). *Then $u : U \to B_xX$ is an isomorphism.*

Proof. The projection $B_xX \to X$ gives a natural map $\phi : V \to \mathrm{RatCurves}(x, X)$ which is injective on closed points. We can apply (3.7.5) to $\phi(V)$ to conclude that $\pi \circ u : U \to X$ is generically one-to-one. Thus $\pi \circ u$ and u are birational if char $= 0$.

In positive characteristic we need an additional observation. Let $f : \mathbb{P}^1 \to X$ be as in (3.7.3.1). f factors through $f' : \mathbb{P}^1 \to B_xX$. An easy computation using (3.7.2) gives that

$$f'^*T_{B_xX} \cong \mathcal{O}(2) + \mathcal{O}^{n-1}.$$

Thus by (II.2.16, II.3.5.3, I.7.4.3) $u : U \to B_xX$ is smooth along $[f'] \times \mathbb{P}^1$. It is also generically one-to-one, hence birational.

By (II.2.14.3) u is quasi-finite over $X - x$. Therefore the only possible exceptional divisor of u is D, but D is mapped to E. Thus there are no exceptional divisors and by (VI.1.4) u is an isomorphism. □

The following lemma completes the proof of (3.2):

3.7.8 Lemma. *Let $s : U \to \mathbb{P}^m$ be a $\mathbb{P}^1$-bundle with a section $D \subset U$. Let $u : U \to X$ be a birational morphism onto a smooth (or normal and factorial) variety X such that $u(D) =$ point and $u|U - D$ is an isomorphism.*

Then $X \cong \mathbb{P}^{m+1}$.

Proof. Let $\mathbb{P}^{m-1} \subset \mathbb{P}^m$ be a hyperplane and $H = s^*\mathbb{P}^{m-1} \subset U$. Since X is smooth, $u(H)$ is a Cartier divisor, hence

$$u^*u_*H = H + kD$$

for some natural number k. Let $L \subset D \cong \mathbb{P}^m$ be a line and $\mathcal{O}(D)|D \cong \mathcal{O}_D(d)$. Then

$$0 = L \cdot u^*u_*H = L \cdot H + kL \cdot D = 1 + kd.$$

Thus $k = 1$ and $d = -1$. Pushing down the sequence

$$0 \to \mathcal{O}_U \to \mathcal{O}_U(D) \to \mathcal{O}_D(d) \to 0$$

via s we obtain an exact sequence

$$0 \to \mathcal{O}_{\mathbb{P}^m} \to E \to \mathcal{O}_{\mathbb{P}^m}(-1) \to 0,$$

which splits since $\mathrm{Ext}^1_{\mathbb{P}^m}(\mathcal{O}(-1), \mathcal{O}) = 0$. Thus

$$U = \mathbb{P}(\mathcal{O}_{\mathbb{P}^m} + \mathcal{O}_{\mathbb{P}^m}(-1)).$$

Let $0 \in \mathbb{P}^{m+1}$ be a point. Projection from 0 gives a morphism $B_0\mathbb{P}^{m+1} \to \mathbb{P}^m$ and

$$B_0\mathbb{P}^{m+1} \cong \mathbb{P}(\mathcal{O}_{\mathbb{P}^m} + \mathcal{O}_{\mathbb{P}^m}(-1)).$$

Thus $U \cong B_0\mathbb{P}^{m+1}$ and under this isomorphism D corresponds to the exceptional divisor $E \subset B_0\mathbb{P}^{m+1}$. The pair $D \subset U$ determines X up to isomorphism, hence $X \cong \mathbb{P}^{m+1}$. □

3.7.9 Exercise [Ballico93]. Let $\operatorname{char} k = 2$ and denote homogeneous coordinates on $\mathbb{P}^{2k-1}$ by $(y_1, \ldots, y_k, z_1, \ldots, z_k)$. Choose $d > 0$ and homogeneous polynomials f, g of degrees $d-2$ resp. d. Set

$$X := \left(\left(\sum y_i z_i\right)^2 f^2 + g^2 = 0\right) \subset \mathbb{P}^{2k+1}.$$

Show that $T_X := \operatorname{Hom}(\Omega^1_X, \mathcal{O}_X)$ is locally free and it is generated by global sections. T_X is ample if X does not contain any lines. In particular, T_X is ample for general choices of f, g for $d \gg 1$.

3.7.10 Exercise (Mori). Let X be a smooth projective variety over a field of characteristic zero and $x \in X$ a closed point. Assume that $B_x X$ is Fano. Show that $X \cong \mathbb{P}^n$.

Several attempts have been made to characterize varieties X such that T_X is only "semipositive". The following provides the definitive generalization of (3.4).

3.8 Theorem [Mok88]. *Let X be a compact complex manifold. Assume that X admits a Kähler metric h whose holomorphic bisectional curvature is everywhere semi positive. Let $(\tilde{X}, \tilde{h})$ be the universal cover. Then*

$$(\tilde{X}, \tilde{h}) \cong (\mathbb{C}^m, e) \times \prod (\mathbb{P}^{n_i}, h_i) \times \prod (M_i, g_i)$$

where e is the flat metric on $\mathbb{C}^m$, h_i is some semipositive metric on $\mathbb{P}^{n_i}$ and M_i is a compact Hermitian symmetric space of rank at least two with its canonical metric g_i. (Here $\cong$ means an isometric biholomorphism.)

3.8.1 Example. Let X be a variety and assume that X is homogeneous under an algebraic group. (In positive characteristic we also assume that the stabilizer of a closed point is reduced.) By (1.5) T_X is generated by global sections, thus it can be written as the quotient of a trivial vector bundle. This implies that T_X is "semi positive" (for every definition that I know of). For instance, if X is over $\mathbb{C}$, then T_X carries a Hermitian metric whose curvature is semi positive, see e.g. [Griffiths-Harris78, 0.5].

In general the induced metric on X is not Kähler and most homogeneous spaces do not satisfy the conclusion of (3.8).

The generalization of (3.3) to the "semi positive" case is still mostly open. Some partial results are contained in [Campana-Peternell91; DPS94].

3.9 Definition. Let X be a variety and E a vector bundle on X. Let $\mathcal{O}(1)$ be the tautological line bundle on $\mathbb{P}_X(E)$. We say that E is *ample* resp. *nef* iff $\mathcal{O}(1)$ is ample resp. nef on $\mathbb{P}_X(E)$.

3.10 Conjecture [Campana-Peternell91]. (3.10.1) Let X be a smooth Fano variety. Then T_X is nef iff X is homogeneous.

(3.10.2) Let X be a smooth projective variety. Assume that T_X is nef. Then there is a finite étale cover $\tilde{X} \to X$ such that $\tilde{X}$ is a locally trivial (in the étale topology) fiber bundle over an Abelian variety whose fiber is a homogeneous Fano variety.

The conjecture is true if $\dim X \leq 3$ by [Campana-Peternell91].

3.10.3 Examples [DPS94]. (3.10.3.1) Let E be an elliptic curve with a 2-torsion point $\tau \in E$. Let $G < \operatorname{Aut}(E \times E \times E)$ be the subgroup generated by the two involutions

$$g_1(z_1, z_2, z_3) = (z_1 + \tau, -z_2, -z_3) \quad \text{and} \quad g_2(z_1, z_2, z_3) = (-z_1, z_2 + \tau, z_3 + \tau).$$

Show that $X := (E \times E \times E)/G$ is smooth, T_X is nef and $\dim \operatorname{Aut}(X) = 0$.

(3.10.3.2) Let E be an elliptic curve and F the rank two vector bundle obtained as the nonsplit extension of $\mathcal{O}_E$ and $\mathcal{O}_E$. Let $S = \mathbb{P}_E(F)$. Show that T_S is nef but no finite étale cover of S is homogeneous. By [DPS94, 1.7] there is no Hermitian metric on T_S whose curvature is everywhere nonnegative.

One expects that a "general" variety of dimension n does not admit any morphism onto smooth varieties except onto $\mathbb{P}^n$ and itself. Very few such results are proved. One possible generalization of (3.4) is the following:

3.11 Problem[4] [Lazarsfeld84]. Let X be an n-dimensional smooth projective variety, homogeneous under a linear algebraic group. Assume that $\dim N^1(X) = 1$. Let $f : X \to Y$ be a nonconstant morphism onto a smooth variety. Is it true that f is either an isomorphism or $Y \cong \mathbb{P}^n$.

An affirmative answer is known in some cases:

3.12 Theorem [Tsai93]. *Let X be a compact, irreducible Hermitian symmetric space of dimension n. Let $f : X \to Y$ be a nonconstant morphism onto a smooth variety. Then either f is an isomorphism or $Y \cong \mathbb{P}^n$.*

V.4 Lines on Fano Hypersurfaces

The aim of this section is to use and illustrate previous results by studying the family of lines on low degree hypersurfaces. We prove that lines generate the group of 1-cycles modulo rational equivalence and that the group of 1-cycles modulo algebraic equivalence is one dimensional. For smooth hypersurfaces this follows from (IV.3.14). The concrete description using lines allows us to extend the results to singular hypersurfaces as well.

[4] This is solved in [Hwang-Mok99]

4.1 Theorem. *Let $X \subset \mathbb{P}^n$ be a hypersurface of degree $d \leq n$ (with arbitrary singularities), $n \geq 4$. Then $B_1(X)_\mathbb{Q} \cong \mathbb{Q}$.*

4.1.1 Comments. By [Clemens83], if $X_5 \subset \mathbb{CP}^4$ is a very general quintic threefold, then $B_1(X_5)_\mathbb{Q}$ is infinite dimensional. Thus the result is sharp in this sense. In higher dimensions the situation is less clear.

It is conjectured that if $d \leq n/2$, then even $A_1(X)_\mathbb{Q} \cong \mathbb{Q}$ [Schoen93; Paranjape94]. I do not know how to obtain this stronger result. A result of this type with worse bound is, however, not hard to get. It is also a special case of [Paranjape94].

4.2 Theorem [Paranjape94]. *Let $X \subset \mathbb{P}^n$ be a hypersurface of degree d (with arbitrary singularities). If $\binom{d+1}{2} \leq n$, then $A_1(X)_\mathbb{Q} \cong \mathbb{Q}$.*

By the results of Sect. 1, the behavior of $A_1(X)$ of a Fano variety X is governed by low degree rational curves on X. In the case of hypersurfaces the simplest rational curves are lines. Theorems (4.1–2) are proved by studying the family of all lines on a hypersurface.

The Hilbert scheme of lines on a hypersurface X has been investigated classically. It is frequently called the *Fano variety of lines* of X. We denote it by $F(X)$. For special hypersurfaces it can be rather complicated, but for general hypersurfaces it is as nice as possible. The first general result in this direction was obtained by [Barth-V.d.Ven78]. The crucial codimension estimate (4.3.9) is sharpened here, which yields the optimal result in characteristic zero and also works in positive characteristic.

4.3 Theorem. *Let $X_d \subset \mathbb{P}^n$ be a hypersurface of degree d and $F(X_d)$ the Hilbert scheme of lines on X_d. Then*

(4.3.1) *$F(X_d)$ is empty for general X_d if $d > 2n-3$.*

(4.3.2) *$F(X_d)$ is smooth of dimension $2n-3-d$ for general X_d if $d \leq 2n-3$.*

(4.3.3) *$F(X_d)$ is connected for any X_d if $d \leq 2n-4$, except when $X_2 \subset \mathbb{P}^3$ is a smooth quadric.*

Proof. Fix n, d and let $H = H_{n,d}$ be the projective space parametrizing hypersurfaces of degree d in $\mathbb{P}^n$ and $G = \mathrm{Grass}(2, n+1)$ the Grassmann variety of lines in $\mathbb{P}^n$. Let $I \subset G \times H$ be the set of pairs $\{(l, X) : l \subset X\}$ and $p_G : I \to G$, $p_H : I \to H$ the projections. Let $S(I) \subset I$ be the set of pairs $\{(l, X) : l \subset X$ and X is singular at some point of $l\}$. By construction

(4.3.4) $$F(X) \cong p_H^{-1}([X]).$$

(4.3.5) Given $[l] \in G$, $p_G^{-1}(l)$ is the set of all hypersurfaces containing l. $H^0(\mathbb{P}^n, \mathcal{O}(d)) \to H^0(l, \mathcal{O}(d))$ is surjective and the kernel corresponds to $p_G^{-1}(l)$. Thus $p_G : I \to G$ is smooth and the fibers are linear subspaces of H of codimension $d+1$. Therefore $I \subset G \times H$ is irreducible, smooth and has codimension $d+1$.

4.3.6 Notation. Let $l \subset \mathbb{P}^n$ be a line. One can choose coordinates (x_i) on $\mathbb{P}^n$ such that $l = (x_0 = \ldots = x_{n-2} = 0)$. If a hypersurface X contains l, then its equation can be written as

$$\sum_0^{n-2} x_i f_i \quad \text{where} \quad \deg f_i = \deg X - 1.$$

4.3.7 Lemma (Notation as above).

(4.3.7.1) *X is singular at $p \in l$ iff $f_0(p) = \ldots = f_{n-2}(p) = 0$.*

(4.3.7.2) *If X is smooth along l, then p_H is smooth at (l, X) iff*

$$H^0(l, \mathcal{O}(d)) = \sum_i f_i H^0(l, \mathcal{O}(1)).$$

Proof. The first part is clear. Next assume that X is smooth along l and let $N_{l|X}$ be the normal bundle of l in X. We have an exact sequence

$$(4.3.7.3) \qquad 0 \to N_{l|X} \to \mathcal{O}_l(1)^{n-1} \xrightarrow{(f_0, \ldots, f_{n-2})} \mathcal{O}_l(d) \to 0.$$

From this we obtain that

$$H^1(l, N_{l|X}) = 0 \quad \text{iff} \quad H^0(l, \mathcal{O}(d)) = \sum_i f_i H^0(l, \mathcal{O}(1)).$$

$p_H : I \to H$ is a morphism between smooth varieties. Thus p_H is smooth at a point (l, X) iff the kernel of $dp_H(l, X)$ has the expected dimension at (l, X). By (4.3.4) the kernel of $dp_H(l, X)$ is the Zariski tangent space to $F(X)$ at l. Thus by (I.2.15) p_H is smooth at (l, X) iff $h^0(l, N_{l|X}) = 2n - 2 - (d + 1)$.

On the other hand from (4.3.7.3) we obtain that

$$h^0(l, N_{l|X}) = n - 1 - d + n - 2 - h^1(l, N_{l|X}) = 2n - 2 - (d+1) - h^1(l, N_{l|X}). \quad \square$$

4.3.8 Corollary. *$S(I) \subset I$ has codimension $n - 2$.*

Proof. For fixed $p \in l$ (4.3.7.1) gives $n - 1$ independent conditions. Varying p shows that $S(I) \cap p_G^{-1}(l)$ has codimension $n - 2$ in $p_G^{-1}(l)$. $\square$

4.3.9 Proposition. *Let $I^0 \subset I$ be the open subset consisting of those pairs (l, X) such that X is smooth along l. Let $Z^0 \subset I^0$ be the closed subset of those pairs such that p_H is not smooth at (l, X). Then* $\operatorname{codim}(Z^0, I^0) \geq 2n - 2 - d$.

Proof. It is sufficient to prove that $\operatorname{codim}(Z^0 \cap p_G^{-1}(l), p_G^{-1}(l)) = 2n - 2 - d$ for every $l \in G$ Choose coordinates as in (4.3.6). Then

$$I^0 \cap p_G^{-1}(l) \cong \{(f_0, \ldots, f_{n-2}) | f_i \in H^0(l, \mathcal{O}(d-1)) \text{ have no common zeros}\},$$

and

$$Z^0 \cap p_G^{-1}(l) \cong \{(f_0, \ldots, f_{n-2}) | \sum f_i H^0(l, \mathcal{O}(1)) \subsetneq H^0(l, \mathcal{O}(d))\}.$$

For a hyperplane $V \subset H^0(l, \mathcal{O}(d))$ set

$$Z_V^0 := \{(f_0, \dots, f_{n-2}) | \sum f_i H^0(l, \mathcal{O}(1)) \subset V\}.$$

Clearly $Z^0 \cap p_G^{-1}(l) = \cup_V Z_V^0$. Thus it is sufficient to prove that

$$\operatorname{codim}(Z_V^0, I^0 \cap p_G^{-1}(l)) \geq 2n - 2.$$

This is a straigtforward consequence of (4.3.11). □

4.3.10 Definition. Let

$$m : H^0(\mathbb{P}^1, \mathcal{O}(1)) \times H^0(\mathbb{P}^1, \mathcal{O}(d-1)) \to H^0(\mathbb{P}^1, \mathcal{O}(d))$$

be the multiplication map. If $V \subset H^0(\mathbb{P}^1, \mathcal{O}(d))$ is a subspace, then set

$$m^{-1}(V) := \{f \in H^0(\mathbb{P}^1, \mathcal{O}(d-1)) : m(H^0(\mathbb{P}^1, \mathcal{O}(1)) \times \{f\}) \subset V\}.$$

4.3.11 Lemma. *Let $V \subset H^0(\mathbb{P}^1, \mathcal{O}(d))$ be a hyperplane. Then either*

(4.3.1) $V = H^0(\mathbb{P}^1, \mathcal{O}(d)(-p))$ *for some* $p \in \mathbb{P}^1$ *and* $m^{-1}(V) = H^0(\mathbb{P}^1, \mathcal{O}(d-1)(-p))$*; or*

(4.3.2) *there is no such* p *and* $m^{-1}(V) \subset H^0(\mathbb{P}^1, \mathcal{O}(d-1))$ *has codimension* 2.

Proof. We identify $H^0(\mathbb{P}^1, \mathcal{O}(k))$ with the vectorspace of polynomials of degree k. Let $\sum_0^d u_i x^i$ denote a general polynomial of degree d and $\sum_0^{d-1} v_i x^i$ a general polynomial of degree $d-1$. If V is given by a linear equation $\sum c_i u_i = 0$, then $m^{-1}(V)$ is given by the two equations

$$\sum_0^{d-1} c_i v_i = 0 \quad \text{and} \quad \sum_0^{d-1} c_{i+1} v_i = 0.$$

If these two equations are linearly dependent, then there is a point $p = (s : t) \in \mathbb{P}^1$ such that $sc_i = tc_{i+1}$ for $0 \leq i \leq d-1$. Equivalently,

$$(c_0, \dots, c_d) = \text{const} \cdot (t^d, t^{d-1}s, \dots, s^d).$$

This means that $V = H^0(\mathbb{P}^1, \mathcal{O}(d)(-p))$. □

We are now ready to prove (4.3). If $d > 2n-3$, then $\dim I < \dim H$, thus on a general hypersurface there are no lines. If $d \leq 2n-3$, then by (4.3.9) p_H is smooth at a general point, in particular it is surjective.

Let $H^0 \subset H$ be the open subset parametrizing smooth hypersurfaces. If $[X] \in H^0$, then $p_H^{-1}(X)$ is smooth iff $[X] \notin p_H(Z^0)$. By (4.3.9) the codimension of Z^0 is greater than the generic fiber dimension of p_H. This shows that $F(X)$ is smooth for general X. (In characteristic zero this also follows from generic smoothness.)

In order to prove (4.3.3) let $L \subset H$ be a general line. Then by (4.3.8–9)

$$\operatorname{codim}(Z^0 \cap p_H^{-1}(L), p_H^{-1}(L)) \geq 2n - 2 - d$$

and

$$\operatorname{codim}(S(I) \cap p_H^{-1}(L), p_H^{-1}(L)) \geq n - 1.$$

Thus if $2n - 3 - d \geq 2$ and $n \geq 3$, then $p : p_H^{-1}(L) \to L$ is smooth outside a codimension two set. Let

$$p : p_H^{-1}(L) \xrightarrow{p_1} C \xrightarrow{p_2} L$$

be the Stein factorization. If p_2 is ramified at some point $c \in C$, then p is not smooth along $p_1^{-1}(c)$. $p_1^{-1}(c)$ has codimension one in $p_H^{-1}(L)$, a contradiction. Thus $p_2 : C \to L$ is étale. The Hurwitz formula gives that $2g(C) - 2 = -2 \deg p_2$. Therefore p_2 is an isomorphism and p has connected fibers over L. Since L is a general line, p_H has connected fibers. □

4.4 Exercises. Let $X_d \subset \mathbb{P}^n$ be a hypersurface of degree d and $l \subset X$ a line. Assume that X_d is smooth along l. Let $N_{l|X_d} = \sum \mathcal{O}(a_i)$ be the normal bundle of l in X. Prove that:

(4.4.1) $1 \geq a_i \geq 2 - d$ for every i. Therefore if X_3 is a smooth cubic, then $F(X_3)$ is smooth.

(4.4.2) Assume that X_d and $l \subset X_d$ are general. Then

$$N_{l|X} \cong \mathcal{O}^{d-1} + \mathcal{O}(1)^{n-1-d} \quad \text{if } d \leq n-1, \text{ and}$$
$$N_{l|X} \cong \mathcal{O}^{2n-3-d} + \mathcal{O}(-1)^{d-n+1} \quad \text{if } d \geq n-1.$$

(4.4.3) Conclude that a general hypersurface $X_d \subset \mathbb{P}^n$ is separably uniruled by lines if $d \leq n - 1$.

(4.4.4) By analyzing the family of smooth conics, conclude that a general hypersurface $X_n \subset \mathbb{P}^n$ is separably uniruled by conics.

I do not know whether every smooth hypersurface $X_d \subset \mathbb{P}^n$ is separably uniruled for $d \leq n$.

4.5 Exercises. Let $X_d \subset \mathbb{P}^n$ be a general hypersurface. Prove the following:

(4.5.1) If $n \geq 4$ and $d \leq 2n-3$, then X_d contains a pair of nonintersecting lines.

(4.5.2) If $n \geq 4$ and $(3n-3)/2 < d \leq 2n-3$, then any two lines in X_d are disjoint.

(4.5.3) If $n \geq 4$ and $(3n-5)/2 < d \leq 2n-3$ and $l \subset X_d$ is any line then

$$N_{l|X} \cong \mathcal{O}^{2n-3-d} + \mathcal{O}(-1)^{d-n+1}.$$

(4.5.4) If $n \geq 4$ and $(3n-3)/2 < d \leq 2n-3$, then the union of all lines is a smooth subvariety of X_d.

4.6 Exercises. (4.6.1) Let $X_d \subset \mathbb{P}^n$ be a hypersurface of degree d. Let $F(X_d, x)$ denote the Hilbert scheme of lines in X_d passing through x. Give explicit equations for $F(X_d, x)$ in terms of the equation of X_d. Show that $\dim F(X_d, x) \geq n - 1 - d$ for every X_d and x.

(4.6.2) If X_d is smooth and char $= 0$, then $\dim F(X_d, x) = n - 1 - d$ for general $x \in X_d$. Prove the same conclusion if char > 0 and X_d general.

(4.6.3) [Collino79] Let char $= p$, $d = p^r + 1$ and $X_d = (\sum x_i^d = 0) \subset \mathbb{P}^n$. Show that $\dim F(X_d, x) \geq n - 4$ for every $x \in X_d$.

(4.6.4) Let $L \subset \mathbb{P}^n$ be a general linear subspace of dimension d. Using the family of lines on X_d intersecting L show that if $d \leq n - 1$ then X_d is degree $d!$ uniruled (IV.1.1.3) if $\dim F(X_d, x) = n - 1 - d$ for general $x \in X_d$.

4.7 Exercise (cf. [Altman-Kleiman77]). Let Q be the rank two tautological quotient bundle on Grass$(2, n+1)$. Show that every $X \subset \mathbb{P}^n$ of degree d gives a section $s(X) \in H^0(\text{Grass}(2, n+1), S^d Q)$ and that $F(X) = (s(X) = 0)$. Using the Plücker embedding this gives a canonical embedding

$$F(X) \subset \text{Grass}(2, n+1) \subset \mathbb{P}^N \quad \text{where} \quad N = \binom{n+1}{2} - 1.$$

Assume that $F(X)$ is smooth of dimension $2n - 3 - d$. Show that

$$K_{F(X)} \cong \mathcal{O}_{F(X)}\left(\binom{d+1}{2} - n - 1\right).$$

Therefore $F(X_d)$ is a smooth Fano variety if $\binom{d+1}{2} \leq n$ and X_d is general.

4.8 Proof of (4.1). Assume first that $d \leq n - 1$. In this case through every point of X there is a line in X. First we prove the following:

4.8.1 Lemma. *Let $X \subset \mathbb{P}^n$ ($n \geq 4$) be a hypersurface of degree $d \leq n - 1$ and $x_1, x_2 \in X$ two closed points. Then there is a connected curve $C \subset X$ of degree at most $n - 1$ containing x_1 and x_2 such that every irreducible component of C is a line.*

Proof. Let $x_1, x_2 \in H \subset \mathbb{P}^n$ be a general hypersurface of degree d. If a curve $C_H \subset H$ can be found with the above requirements, then we can specialize it to a curve $C \subset X$.

H is smooth hence by Lefschetz theory $\rho(X) = 1$ (see e.g. [Grothendieck68]). Since H is general, it is separably uniruled by lines (4.4.3). The proof of (2.6) used only separably uniruledness, thus we obtain that two general points of H can be connected by a chain of lines of length at most $n - 1$. By specialization, any two points of X can be connected by a chain of lines whose length is at most $n - 1$. □

Let $Y \subset \text{Chow}(X)$ be the closed subset parametrizing connected curves $C \subset X$ of degree at most $n - 1$ such that every irreducible component of C

is a line. Applying (4.8.1) to X_K where K is an uncountable algebraically closed field, we see that Y satisfies the assumptions of (IV.3.13.3). Thus $A_0(F(X))_{\mathbb{Q}} \to A_1(X)_{\mathbb{Q}}$ is surjective.

By (4.3) $F(X)$ is connected, hence $B_1(X)_{\mathbb{Q}} = \mathbb{Q}$.

If $d = n$, then lines do not cover X and we use the family of conics in a similar manner. Since we have not proved the connectedness of the family of conics on a hypersurface, we proceed in a slightly more roundabout way.

Let T be the spectrum of a DVR and X_T a family of hypersurfaces over T such that $X_0 = X$ and the generic fiber X_g is smooth. Let $Q_T \to T$ be the family of conics on X_T. By suitable choice of T we may assume that there is a subfamily $Q_T^0 \subset Q_T$ such that

(4.8.2.1) the geometric generic fiber of $Q_T^0 \to T$ is irreducible, and

(4.8.2.2) $u : U_T^0 \to X_T$ is dominant where $U_T^0 \to Q_T^0$ is the universal family and u the cycle map.

Let $Y_T \subset \mathrm{Chow}(X_T)$ be the closed subset parametrizing connected curves $C \subset X_T$ of degree at most $2n-2$ such that every irreducible component of C is a subset of a conic in Q_T^0. Let $U_T \to Y_T$ be the universal family. As before we obtain that

$$u^{(2)} : U_T \times_{Y_T} U_T \to X_T \times_T X_T$$

is dominant on the generic fiber, hence also on the special fiber. Thus $A_1(X)_{\mathbb{Q}}$ is generated by the irreducible components of the curves in Q_0^0, where Q_0^0 is the fiber of Q^0 over the closed point of T. By construction Q_0^0 is connected.

If every conic in Q_0^0 is smooth, then we are done. Otherwise every conic in Q_0^0 is algebraicaly equivalent to a pair of lines on X. Since the family of lines is connected, we are done again. □

4.9 Proof of (4.2). We prove that $F(X)$ is rationally chain connected if $\binom{d+1}{2} \leq n$. If X is general, then $F(X)$ is a smooth Fano variety by (4.7) hence rationally chain connected by (2.13).

For general X pick two lines $l_1, l_2 \in X$. One can get a family of general hypersurfaces X_t and a family of pairs of nonintersecting lines $l_{1,t}, l_{2,t} \subset X_t$ such that $X_0 = X, l_{i,0} = l_1$ and $l_{2,0} = l_2$. We already showed that $[l_{1,t}], [l_{2,t}] \in F(X_t)$ can be connected by a chain of rational curves. By degeneration we obtain a chain of rational curves $C \subset F(X_0)$ connecting $[l_1]$ and $[l_2]$.

By (IV.3.13) $A_0(F(X)) \cong \mathbb{Z}$. We already noted that $A_0(F(X))_{\mathbb{Q}} \to A_1(X)_{\mathbb{Q}}$ is surjective. Thus $A_1(X)_{\mathbb{Q}} \cong \mathbb{Q}$.

The next two exercises check the existence of lines on most Fano weighted complete intersections.

4.10 Exercise: Lines on Complete Intersections.

(4.10.1) Let $X = X_{d_1,\dots,d_k} \subset \mathbb{P}^n$ be a complete intersection and $l \subset X$ a line such that X is smooth along l. Show that

$$\dim_{[l]} \mathrm{Hilb}(X) \geq 2n - 2 - \sum (d_i + 1).$$

(4.10.2) If equality holds for one pair $l \subset X$, then every complete intersection with the same $d_1, \ldots, d_k$ contains a line.

(4.10.3) Fix a line $l \subset \mathbb{P}^n$ and let $l \subset X$ be a general complete intersection as above. Assume that $2n - 2 - \sum(d_i + 1) \geq 0$. Show that $\operatorname{Hilb}(X)$ is smooth at $[l]$ of the expected dimension. Thus X contains a line if $\sum(d_i + 1) \leq 2n - 2$.

(4.10.4) Show that if $\sum(d_i + 1) > 2n - 2$ and X is general, then X does not contain a line.

(4.10.5) Show that if $\sum d_i \leq n - 1$, then there is a line through every point of X. If $\sum d_i > n - 1$ and X is general, then X does not contain a line through every point.

4.11 Exercise: Lines on Weighted Complete Intersections.

(4.11.1) Let $X = X_{d_1,\ldots,d_k} \subset \mathbb{P}(a_0, \ldots, a_n)$ be a weighted complete intersection. Use the method of (IV.6.7) to show that there is a complete intersection

$$Y = Y_{d_1,\ldots,d_k} \subset \mathbb{P}^{\sum a_i - 1}$$

and a dominant map $Y \dashrightarrow X$.

(4.11.2) If $\sum d_i \leq \sum a_i - 2$, then through every point $x \in X$ there is a rational curve $C \subset X$ such that $\mathcal{O}_X(1) \cdot C \leq 1$. Thus if X is smooth, then X is covered by lines.

(4.11.3) Assume that $\sum(d_i + 1) \leq \sum(a_i + 1) - 4$. It is quite likely that X contains a rational curve C such that $\mathcal{O}_X(1) \cdot C \leq 1$. I have not checked this in all cases.

V.5 Nonrational Fano Varieties

As we saw in (2.13), Fano varieties are rationally connected, thus they are very similar to $\mathbb{P}^n$. In dimension two, all Fano (=Del Pezzo) surfaces are rational (III.2.4). One of the early problems of higher dimensional algebraic geometry was to understand whether Fano varieties, and especially low degree hypersurfaces in $\mathbb{P}^{n+1}$ are rational or not.

The rationality question for threefolds of degrees 3 and 4 in $\mathbb{P}^4$ has been open for a long time. In the early seventies two different approaches were discovered. Together they settled the question completely.

[Iskovskikh-Manin71] developed the Noether–Fano method which enabled them to prove that any birational selfmap of any smooth quartic is an isomorphism. This in particular imples that they are not rational. The same method shows that they are not birational to conic bundles or to a family of Del Pezzo surfaces. This approach was further developed and applied to many other threefolds by Iskovskikh and his students, see e.g. [Iskovskikh80a,b; Sarkisov81,82]. This method in principle works in any dimension, but the technical difficulties have been overcome in a few cases only. [Pukhlikov87] proves that any birational selfmap of a smooth quintic in $\mathbb{P}^5$ is an isomorphism. Some additional cases are treated in [Pukhlikov89].

[Clemens-Griffiths72] initiated the study of intermediate Jacobians and used it to show that smooth cubic threefolds are not rational. This method has been very succesful in dimension three, see e.g. [Beauville77]. Unfortunately it does not work in higher dimensions, at least not in the usual formulation.

[Artin-Mumford72] observed that the torsion subgroup of $H^3(X,\mathbb{Z})$ is a birational invariant and used this to find examples of nonrational but unirational varieties in any dimension.

Recently, [Kollár95a] observed that there are Fano varieties in positive characteristic which are not separably uniruled, in particlular, they are not ruled. Using the degeneration techniques of Matsusaka (IV.1.6–8) these give examples of Fano varieties over $\mathbb{C}$ which are not ruled.

The basis of the construction is the following simple remark:

5.1 Lemma. *Let X be a smooth proper variety and M a big line bundle on X. Assume that there is an injection $M \to \wedge^i \Omega^1_X$ for some $i > 0$. Then X is not separably uniruled. If X is over a field of* char p, *then $p|\bar{u}(X)$ (IV.1.7.3).*

Proof. Since M is big, there is an open set $U \subset X$ such that sections of M^k separate points of U for $k \gg 1$. In particular, if $f : C \to X$ is a morphism from a smooth proper curve to X whose image intersects U, then $\deg f^*M > 0$. By shrinking U we may also assume that $M \to \wedge^i \Omega^1_X$ has rank one at every point of U.

Assume that X is separably uniruled. By (IV.1.9) there is a free morphism $f : \mathbb{P}^1 \to X$ whose image intersects U. $f^*T_X \cong \sum \mathcal{O}_{\mathbb{P}^1}(a_j)$ is semi positive, thus $a_j \geq 0$. The injection $M \to \wedge^i \Omega^1_X$ pulls back to an injection

$$f^*M \hookrightarrow \wedge^i f^*\Omega^1_X \cong \wedge^i \left(\sum \mathcal{O}_{\mathbb{P}^1}(-a_j)\right).$$

f^*M has positive degree, which is a contradiction.

Let $f : Y \times \mathbb{P}^1 \dashrightarrow X$ be a degree d uniruling of X. Then f is not separable, thus $p|\deg f$. Therefore $p|\bar{u}(X)$. □

5.1.1 Remark. By the Bogomolov–Sommese vanishing theorem (see, for instance, [Esnault-Viehweg92, 6.9]) the assumptions of (5.1) can be satisfied in characteristic zero only if $i = \dim X$, when X is of general type.

The construction of the requisite examples is based on cyclic covers (II.6.1.5). Here we use another variant of the same construction, which is better suited for our purposes.

5.2 Notation. Let X be a scheme and L a line bundle on X. Set $U := \operatorname{Spec}_X \sum_{m\geq 0} L^{-m}$ and let $\pi : U \to Z$ denote the projection. Fix k and a section $s \in H^0(X, L^k)$.

$\pi_*(\pi^*L) = \sum_{m\geq -1} L^{-m}$, hence π^*L has a canonical section corresponding to $1 \in H^0(X, L^0 \cong \mathcal{O}_X)$. Denote this section by y_L. Since

$\pi_*(\pi^* L^k) = \sum_{m \geq -k} L^{-m}$, both y_L^k and s can be viewed as sections of $\pi^* L^k$. Let $Y := (y_L^k - s = 0) \subset U$ denote the zero set. The restriction of π to Y is also denoted by π. Y is also denoted by $X[\sqrt[k]{s}]$.

Choose local coordinates x_i at a point $x \in X$. On U we can use $y := y_L$ and the x_i as local coordinates. Locally Y is given by the equation $y^k - s(x_1, \dots, x_n) = 0$.

In order to use (5.1) we need to study Ω_Y^1 and find a big line bundle in a suitable wedge power of it. It turns out that in the cases when such a line bundle exists, Y is singular. Therefore we also need to study the singularities of Y and their resolution.

5.3 Lemma (Notation as above).

(5.3.1) *There is an exact sequence* $0 \to \pi^* \Omega_X^1 \to \Omega_U^1 \to \pi^* L^{-1} \to 0$.

(5.3.2) $\mathcal{O}_U(-Y) \cong \pi^* L^{-k}$.

(5.3.3) *There is an exact sequence* $\pi^* L^{-k} \xrightarrow{d_Y} \Omega_U^1 \to \Omega_Y^1 \to 0$. *In the above local coordinates the image of* d_Y *is given by*

$$\left(-\frac{\partial s}{\partial x_1} dx_1, \dots, -\frac{\partial s}{\partial x_n} dx_n, k y^{k-1} dy \right).$$

(5.3.4) *Assume that the characteristic* p *of the base field divides* k. *Then we obtain an exact sequence*

$$0 \to \operatorname{coker}\left[\pi^* L^{-k} \xrightarrow{d_Y} \pi^* \Omega_X^1 \right] \to \Omega_Y^1 \to \pi^* L^{-1} \to 0.$$

Proof. (5.3.1–3) are immediate from the construction (5.2). If p divides k, then $k y^{k-1} dy = 0$. Thus $\operatorname{im} d_Y \subset \pi^* \Omega_X^1$ and we get (5.3.4). □

The crucial observation is (5.3.4), where the unexpected subsheaf emerges. In order to understand it better, we need to write it as the pull back of a sheaf from X.

5.4 Definition–Lemma. Let X be a smooth variety over a field of char p and L a line bundle on X. If $p | k$, then there is a natural differential

$$d : L^k \to L^k \otimes \Omega_X^1, \tag{5.4.1}$$

constructed as follows. Let τ be a local generator of L, $s = f\tau^k$ a local section of L^k and the x_i local coordinates. Set

$$d(s) := \sum \frac{\partial f}{\partial x_i} \tau^k dx_i.$$

This is independent of the choices made and thus defines d.

For a fixed $s \in H^0(X, L^k)$ we can view $d(s)$ as a sheaf homomorphism $d(s) : \mathcal{O}_X \to L^k \otimes \Omega_X^1$. Tensoring with L^{-k} we obtain

$$ds : L^{-k} \to \Omega_X^1. \tag{5.4.2}$$

Comparing the definitions of d_Y in (5.3.3) and of ds in (5.4.2) we obtain at once:

5.5 Lemma (Notation as above). *Then $d_Y = -\pi^* ds$, thus there is an exact sequence*

$$0 \to \pi^* \operatorname{coker}\left[L^{-k} \xrightarrow{ds} \Omega^1_X\right] \to \Omega^1_Y \to \pi^* L^{-1} \to 0. \qquad \square$$

The singularities of Y are related to the critical points of s. Usually the notion of critical point does not make sense for sections of line bundles. In positive characteristic there are many exceptions:

5.6 Definition. Let X be a smooth variety over a field of char p and L a line bundle on X. Let k be an integer divisible by p.

(5.6.1) We say that a local section s of L^k has a *critical point* at $x \in X$ if $d(s) \in \Gamma(L^k \otimes \Omega^1_X)$ vanishes at x.

(5.6.2) Pick local coordinates x_i near x and a local generator τ of L at x. Write $s = f\tau^k$. The matrix

$$H(s) := \left(\frac{\partial^2 f}{\partial x_i \partial x_j}\right)$$

is called the *Hessian* of s with respect to the coordinates x_i and the generator τ.

(5.6.3) The critical point of s at x is called *nondegenerate* if $\operatorname{rank} H(s)(x) = \dim X$.

5.6.4 Lemma. *The rank of $H(s)$ at a point $x \in X$ is independent of the choices of x_i and τ.*

Proof. If we replace τ by $\tau = h\tau'$, then $s = (fh^k)\tau'^k$ and $h(x) \neq 0$. The new Hessian is

$$H'(s) = \left(\frac{\partial^2 (fh^k)}{\partial x_i \partial x_j}\right) = \left(\frac{\partial^2 f}{\partial x_i \partial x_j}\right) h^k.$$

Thus $H'(s)(x)$ is a constant multiple of $H(x)$. If we change the local coordinate system, the Hessian is transformed by the Jacobian as usual. $\square$

5.6.5 Remark. If $L = \mathcal{O}_X$, then we recover the classical notion of critical points and Hessians of functions.

5.6.6 Exercise. (5.6.6.1) Show that the hypersurface $(y^k - f(x_1, \dots, x_n) = 0)$ is singular at the point $(y, x_1, \dots, x_n)$ iff $(x_1, \dots, x_n)$ is a critical point of f (assuming $p|k$).

(5.6.6.2) Show that f has a nondegenerate critical point at x iff $\partial f/\partial x_1, \dots, \partial f/\partial x_n$ generate the maximal ideal of the local ring $\mathcal{O}_{x,X}$.

(5.6.6.3) Show that f has a nondegenerate critical point at x, iff in suitable local coordinates f can be written as

$$f = c + \begin{cases} x_1x_2 + x_3x_4 + \ldots + x_{n-1}x_n + f_3, & \text{if } n \text{ is even,} \\ x_1^2 + x_2x_3 + \ldots + x_{n-1}x_n + f_3, & \text{if } n \text{ is odd,} \end{cases} \quad \text{where } f_3 \in m_x^3.$$

The usual Morse lemma can be generalized to positive characteristic:

5.7 Exercise. Let X be a smooth variety over a field of char p and L a line bundle on X. Let k be an integer divisible by p and $W \subset H^0(X, L^k)$ a finite dimensional subvectorspace. Let m_x denote the ideal sheaf of $x \in X$.

(5.7.1) Assume that $p \neq 2$ or $p = 2$ and $\dim X$ is even. If for every closed point $x \in X$ the restriction map $W \to (\mathcal{O}_X/m_x^3) \otimes L^k$ is surjective, then a general section $f \in W$ has only nondegenerate critical points.

(5.7.2) If $p = 2$ and $\dim X$ is odd, then every critical point is degenerate.

(5.7.3) Assume that $p = 2$, $\dim X$ is odd and for every closed point $x \in X$ the restriction map $W \to (\mathcal{O}_X/m_x^4) \otimes L^k$ is surjective. Let $f \in W$ be a general section and $x \in X$ any critical point of f. Then length $\mathcal{O}_{x,X}/(\partial f/\partial x_1, \ldots, \partial f/\partial x_n) = 2$. Such a critical point (assuming $p = 2$, $\dim X$ odd) will be called *almost nondegenerate*.

(5.7.4) Show that f has an almost nondegenerate critical point at x iff in suitable local coordinates f can be written as

$$f = c + ax_1^2 + x_2x_3 + \ldots + x_{n-1}x_n + bx_1^3 + f_3$$

where $b \neq 0$, $f_3 \in m_x^3$ and the coefficient of x_1^3 in f_3 is zero.

The coherent sheaf coker $\left[L^{-k} \xrightarrow{ds} \Omega_X^1\right]$ has rank $n-1$ and it is almost never locally free. Also, it turns out that in most cases it does not lift to a resolution of Y. To overcome these difficulties, we concentrate at its determinant, which gives a subline bundle of Ω_Y^{n-1}.

5.8 Definition (Notation as above). Set

$$Q(L, s) := \left(\det \operatorname{coker} \left[L^{-k} \xrightarrow{ds} \Omega_X^1\right]\right)^{**}, \tag{5.8.1}$$

where ** denotes double dual. $Q(L, s)$ is a line bundle on X. By construction there are natural maps

$$q : \wedge^{n-1}\Omega_X^1 \to Q(L, s), \quad \text{and} \tag{5.8.2}$$

$$\pi^*Q(L, s) \hookrightarrow (\Omega_Y^{n-1})^{**}. \tag{5.8.3}$$

Let $r : Y' \to Y$ be a resolution of singularities. Local sections of Ω_Y^{n-1} lift to rational sections of $\Omega_{Y'}^{n-1}$, which may have poles along the exceptional divisors. We would like to investigate when the sections coming from $\pi^*Q(L, s)$ lift to regular sections of $\Omega_{Y'}^{n-1}$. This question is local above the singular points of Y and will be settled by an explicit computation in various cases. Before doing this, we need to study $Q(L,s)$.

5.9 Lemma (Notation as above). *Let $(s=0)=\sum a_j D_j$ and assume that $(p,a_j)=1$ for every j. Assume furthermore that the set of critical points of s has codimension at least 2 in $X-\sum D_j$. Then:*

(5.9.1) $Q(L,s)\cong K_X\otimes L^k(-\sum(a_j-1)D_j)$.

(5.9.2) *Choose local coordinates x_i at $x\in X$ and a local generator τ of L near x. Write $s=f\tau^k$ and $f=\prod_j f_j^{a_j}$. Let*

$$\eta_i := \Big(\prod_j f_j^{a_j-1}\Big)\frac{dx_1\wedge\ldots\wedge\widehat{dx_i}\wedge\ldots\wedge dx_n}{\partial f/\partial x_i}.$$

(η_i is undefined if $\partial f/\partial x_i$ is identically zero.)

Then $q(\eta_{i_1})=\pm q(\eta_{i_2})$ and they give a local generator η_x of $\operatorname{im}[\pi^*Q(L,s)\to(\Omega_Y^{n-1})^{**}]$ above x.

Proof. The image of ds is generated by $\sum_i(\partial f/\partial x_i)dx_i$. If $a_j>1$ the all these partials vanish along $(f_j=0)$ with multiplicity a_j-1. Thus ds can be extended to a map

$$ds^*: L^{-k}\Big(\sum_j(a_j-1)D_j\Big)\to\Omega_X^1.$$

This implies that $q(\eta_{i_1})=\pm q(\eta_{i_2})$ and that η_i is a local generator of $Q(L,s)$ at $x\in X$ if $(\prod_j f_j^{1-a_j})(\partial f/\partial x_i)$ is nonzero at x. On $X-\sum D_j$ this holds outside a codimension two set by assumption. ds^* has rank one at every smooth point of $\sum D_j$. Thus $q(\eta_i)$ is a local generator outside a codimension two set of X, hence everywhere. □

The explicit form of η_x makes it possible to compute $r^*\eta_x$. I do now know any general result which describes when $r^*\eta_x$ is regular. The following three cases are sufficient for many examples.

5.10 Proposition. *Let $Y=X[\sqrt[k]{s}]$ and $r:Y'\to Y$ a resolution of singularities. (The existence of a resolution is established in the course of the proof.) Let η_x be the local generator of $\operatorname{im}[\pi^*Q(L,s)\to(\Omega_Y^{n-1})^{**}]$ constructed in (5.9.2).*

(5.10.1) *If s has a nondegenerate critical point at x, then $r^*\eta_x$ is a regular section of $\Omega_{Y'}^{n-1}$ over a neighborhood of x for $n\ge3$.*

(5.10.2) *If s has an almost nondegenerate critical point at x and $k=2$ then $r^*\eta_x$ is a regular section of $\Omega_{Y'}^{n-1}$ over a neighborhood of x for $n\ge3$.*

(5.10.3) *Assume that locally at x we can write $f=x_1^a g$ where $(x_1=0)$ and $(g=0)$ intersect transversally. Then $r^*(x_1\eta_x)$ is a regular section of $\Omega_{Y'}^{n-1}$ over a neighborhood of x.*

Proof. In the third case choose local coordinates such that $\partial g/\partial x_n$ is nonzero at x. Then

$$\eta_n = x_1^{a-1}\frac{dx_1\wedge\ldots\wedge dx_{n-1}}{x_1^a\partial g/\partial x_n} = \frac{dx_1\wedge\ldots\wedge dx_{n-1}}{x_1\partial g/\partial x_n}.$$

Let $\zeta_n := (\partial g/\partial x_n)^{-1}dx_1\wedge\ldots\wedge dx_{n-1}$ as a local section of Ω_X^{n-1}. $\pi\circ r : Y'\to X$ is a morphism of smooth varieties, thus $(\pi\circ r)^*\zeta_n$ is a regular section of $\Omega_{Y'}^{n-1}$ over a neighborhood of x. By construction $(\pi\circ r)^*\zeta_n = r^*(x_1\eta_x)$. This shows (5.10.3). The resolution can be constructed by blowing up $(y = x_1 = 0)$.

I do the first two cases only for $p = k = 2$. This is enough for many of the applications. The general case can be done by the same method and is left as an exercise; see [Kollár95a, 20-22] for details. Thus assume that $p = k = 2$. Then locally above x the equation of Y is

$$y^2 = c + \begin{cases} x_1x_2 + x_3x_4 + \ldots + x_{n-1}x_n + f_3, & \text{if } n \text{ is even,}\\ ax_1^2 + x_2x_3 + \ldots + x_{n-1}x_n + bx_1^3 + f_3, & \text{if } n \text{ is odd.}\end{cases}$$

where a, b, c are constants, $b\neq 0$ and $f_3\in m_x^3$. Replace y by $y-\sqrt{c}-\sqrt{a}x_1$ to get rid of c and ax_1^2. The only singularity is at the origin $y = x_1 = \ldots = x_n = 0$. Blowing up the origin we obtain a smooth variety $r : Y'\to Y$ (this is the point where we need that $b\neq 0$). One chart of Y' is given by the coordinate changes

$$y' = y, x_i' = x_i/y,\ i = 1,\ldots,n.$$

In this chart we compute that

$$\begin{aligned} r^*\eta_n = r^*\frac{dx_1\wedge\ldots\wedge dx_{n-1}}{x_{n-1}+\partial f_3/\partial x_n} &= \frac{d(y'x_1')\wedge\ldots\wedge d(y'x_{n-1}')}{y'x_{n-1}'+\partial f_3/\partial x_n}\\ &= y'^{n-2}\frac{dx_1'\wedge\ldots\wedge dx_{n-1}'}{x_{n-1}'+y'h}\\ &\quad + \sum_i \pm y'^{n-3}\frac{dx_1'\wedge\ldots\wedge \widehat{dx_i'}\wedge\ldots\wedge dx_{n-1}'\wedge dy'}{x_{n-1}'+y'h},\end{aligned}$$

where $h = y'^{-2}(\partial f_3/\partial x_n)(y'x_1',\ldots,y'x_n')$ is regular since $f_3\in m_x^3$.

We only have to check that $r^*\eta_n$ does not have a pole along the exceptional divisor E of r. In our case E is given by the equation $y' = 0$ and $x_{n-1}' + y'h$ is nonzero generically along E. Thus $r^*\eta_n$ is regular as claimed. □

5.10.4 Remark. Let $p = 2$ and $s\in H^0(\mathbb{P}^2, \mathcal{O}(4))$ a general section. Then $\mathbb{P}^2[\sqrt[2]{s}]$ is a Del Pezzo surface and rational. Thus the restriction $n\geq 3$ is necessary.

The following is a general nonrationality result in positive characteristic:

5.11 Theorem. *Let X be a smooth projective variety of dimension $n\geq 3$ over a field of* char p *and L a line bundle on X. Assume that:*

(5.11.1) *For every closed point* $x \in X$ *the restriction map*

$$H^0(X, L^p) \to (\mathcal{O}_X/m_x^4) \otimes L^p \quad \textit{is surjective.}$$

(5.11.2) $K_X \otimes L^p$ *is ample.*

Then, for general $s \in H^0(X, L^p)$*, the corresponding* p*-fold cover* $Y = X[\sqrt[p]{s}]$ *is not separably uniruled. In particluar,* p *divides* $\bar{u}(X)$.

Proof. By (5.7) s has only (almost) nondegenerate critical points. Thus $(s = 0)$ is reduced and by (5.9.1) $Q(L, s) \cong K_X \otimes L^p$ and hence ample. By (5.10) there is a resolution of singularities $r : Y' \to Y$ and an injection $r^*\pi^*Q(L, s) \to \Omega_{Y'}^{n-1}$. Thus by (5.1) Y' is not separably uniruled.

(If $p > 2$ or $\dim X$ is even, then $\mathcal{O}_X/m_x^4$ in (5.11.1) can be replaced by $\mathcal{O}_X/m_x^3$ by (5.7).) □

The method of (IV.1.6–8) will be used to get characteristic zero examples of nonruled Fano varieties by lifting positive characteristic examples. The proof is immediate from (IV.1.6–8).

5.12 Proposition. *Let* $f : X \to S$ *be a proper and flat morphism with irreducible and reduced fibers,* S *irreducible. Let* T *be the spectrum of a DVR with generic point* g*, closed point* 0*, and* $g : Z \to T$ *a proper and flat morphism with reduced fibers. Let* Z_0 *be an irreducible component of* $g^{-1}(0)$ *and* Z_g *the generic fiber. Assume that* Z_g *is birational to a fiber of* f*. Then*

(5.12.1) *If* Z_0 *is not geometrically ruled, then* X_s *is not geometrically ruled for very general* $s \in S$.

(5.12.2) $\bar{u}(Z_0) | \bar{u}(X_s)$ *for very general* $s \in S$. □

The simplest application is to cyclic covers of projective spaces:

5.13 Theorem. *Let* p *be a prime and* $s \in H^0(\mathbb{CP}^n, \mathcal{O}(pd))$ *a very general section. Then the corresponding cyclic cover* $Y = \mathbb{CP}^n[\sqrt[p]{s}]$ *is a smooth Fano variety which is not ruled provided*

$$\frac{n+1}{p-1} > d > \frac{n+1}{p} \quad \textit{and} \quad n \geq 3.$$

Proof. $K_Y = \pi^*(K_{\mathbb{P}^n} + (1 - 1/p)(s = 0))$, thus Y is Fano iff $(1 - 1/p)pd < n+1$. This gives the upper bound for d.

The cyclic covers $\mathbb{CP}^n[\sqrt[p]{s}]$ form an irreducible family parametrized by (the projectivization of) $H^0(\mathbb{CP}^n, \mathcal{O}(pd))$. Any cyclic cover $\mathbb{P}^n[\sqrt[p]{s}]$ in characteristic p can be lifted to a cyclic cover in characteristic zero. By (5.11) the characteristic p cover is not ruled if $-n-1+pd > 0$. Thus by (5.12) the same holds in characteristic zero. □

5.13.1 Remarks. (5.13) is ineffective in the sense that it does not give any particular cyclic cover which is not ruled.

The method implies that a similar assertion holds in all but finitely many charateristics, but gives no method to decide which (if any) charateristics should be excluded.

Similar techniques can be applied to hypersurfaces as well. In this case sometimes one can use several different characteristics simultaneously. ($\ulcorner x \urcorner$ denotes the smallest integer $\geq x$.)

5.14 Theorem [Kollár95a]. *Let $X_d \subset \mathbb{CP}^{n+1}$ be a very general hypersurface of degree d.*

(5.14.1) *If*

$$d \geq 2\ulcorner (n+3)/3 \urcorner$$

then X_d is not ruled.

(5.14.2) *If*

$$d \geq 3\ulcorner (n+3)/4 \urcorner$$

then X_d is not birational to a conic bundle.

(5.14.3) *Let $d = n+1$, Y a variety of dimension $n-1$ and $\phi : Y \times \mathbb{P}^1 \dashrightarrow X_d$ a dominant map. Then $\deg \phi$ is divisible by every prime $p \leq \sqrt{n+2} - 1$.*

Proof. As a first step we need to lift positive characteristic cyclic covers to hypersurfaces in characteristic zero. This is accomplished by the following:

5.14.4 Construction [Mori75, 4.3]. Let $T = \operatorname{Spec} R$ be the spectrum of a DVR with local parameter t, quotient field K and residue field k. Let $f, g \in R[x_0, \dots, x_r]$ be homogeneous polynomials of degrees cd and d respectively such that $g^c - f$ is not identically zero in $k[x_0, \dots, x_r]$. Let $\mathbb{P}_T(x_0, \dots, x_r, y)$ be the weighted projective space with weights $(1^{r+1}, d)$. The scheme

$$Z = (y^c - f = ty - g = 0) \subset \mathbb{P}_T(x_0, \dots, x_r, y)$$

defines a family of weighted complete intersections over T. The general fiber is isomorphic to the hypersurface

$$(g^c - t^c f = 0) \subset \mathbb{P}^{r+1}_K.$$

The closed fiber is isomorphic to a degree c cover of the hypersurface $(g(s) = 0) \subset \mathbb{P}^{r+1}_k$ ramified along $(f(s) = 0)$.

As a corollary we obtain a special case of (5.14):

5.14.5 Lemma. *Let $X_{pe} \subset \mathbb{CP}^{n+1}$ be a very general hypersurface such that $(p+1)e \geq n+3$. Then $p | \bar{u}(X)$.*

Proof. Let $Y \subset \mathbb{P}^n$ be a smooth hypersurface of degree e over a field of characteristic p and $s \in H^0(Y, \mathcal{O}_Y(pe))$ a general section. Set $Z_0 = Y[\sqrt[p]{s}]$. Then $K_Y \otimes L^p \cong \mathcal{O}_Y(e - 2 - n + pe)$ is ample iff $(p+1)e \geq n+3$. By (5.14.4) Z_0 can be lifted to a hypersurface of degree pe in $\mathbb{CP}^{n+1}$. Thus (5.12) implies (5.14.5). □

Let X_d be a hypersurface of degree d. Pick a prime p. If $d = pe$, then (5.14.5) applies, but not in general. To reduce to this case, we need one more degeneration.

If $d > d'$, then construct a flat family of hypersurfaces whose general fiber is a smooth hypersurface of degree d and whose special fiber is the union of a hypersurface X' of degree d' and $(d - d')$ hyperplanes. If $p | \bar{u}(X')$ then by (5.12) the same holds for the general fiber.

5.15.6 Lemma. *Let $X_d \subset \mathbb{CP}^{n+1}$ be a very general hypersurface of degree d. If*

$$d \geq p \ulcorner (n+3)/(p+1) \urcorner \quad \textit{then } p | \bar{u}(X_d).$$

Proof. We apply the above degeneration with $e = \llcorner d/p \lrcorner$, $d' = pe$. The rest follows from (5.14.5) once we note that $d \geq p \ulcorner (n+3)/(p+1) \urcorner$ is equivalent to $(p+1) \llcorner d/p \lrcorner \geq n+3$. □

If X_d is ruled, then $\bar{u}(X_d) = 1$. Applying (5.15.6) with $p = 2$ gives (5.14.1).

Assume that X_d is birational to a conic bundle $X_1 \to Y_1$. A conic defined over a field F always has a point over a degree 2 field extension of F. Thus there is a subvariety $Y_2 \subset X_1$ such that $Y_2 \to Y_1$ has degree 2. $X_1 \times_{Y_1} Y_2 \to Y_2$ is a conic bundle with a section, thus it is birational to $Y_2 \times \mathbb{P}^1$. Therefore

$$Y_2 \times \mathbb{P}^1 \dashrightarrow X_1 \times_{Y_1} Y_2 \dashrightarrow X_1 \dashrightarrow X$$

is a degree 2 uniruling of X. Applying (5.15.6) with $p = 3$ gives (5.14.2).

Finally let $d = n + 1$. If $p \leq \sqrt{n+2} - 1$, then easy computation shows that

$$n + 1 \geq p(n + 3 + p)/(p + 1) \geq p \ulcorner (n+3)/(p+1) \urcorner.$$

By (5.15.6) this implies (5.14.3). □

It is also possible to get explicit examples of Fano varieties defined over $\mathbb{Z}$ which are not ruled. It is clear from the example that one can write down many similar ones with some work. The hard part is to find functions whose critical points are easy to understand.

5.16 Theorem. *Choose a prime p and integers n, d such that $2n < pd$. Let c be an integer and $F \in \mathbb{Z}[x_1, \ldots, x_{2n}]$ any polynomial of degree at most $p(d+1)$. Then (projective models of) the affine hypersurface over $\mathbb{C}$*

$$y^p = c + x_1x_2 + x_3x_4 + \ldots + x_{2n-1}x_{2n} + x_1^{pd+1} + \ldots + x_{2n}^{pd+1} + pF$$

are not ruled.

Proof. The equation is set up in such a way that it already defines a family of p-fold covers over $\operatorname{Spec} \mathbb{Z}$. If we prove that the fiber in characteristic p is not ruled, then the same holds for the generic fiber which is the above hypersurface over $\mathbb{C}$. In characteristic p we can replace y by $y - \sqrt[p]{c}$ to get the simpler equation

$$y^p = x_1x_2 + x_3x_4 + \ldots + x_{2n-1}x_{2n} + x_1^{pd+1} + \ldots + x_{2n}^{pd+1}.$$

This corresponds to the p-fold cyclic cover where $L = \mathcal{O}(d+1)$ and

$$s = x_0^{p-1}g \in H^0(\mathbb{P}^{2n}, \mathcal{O}(p(d+1))), \quad \text{where}$$
$$g = x_0^{pd-1}x_1x_2 + \ldots + x_0^{pd-1}x_{2n-1}x_{2n} + x_1^{pd+1} + \ldots + x_{2n}^{pd+1}.$$

The critical points are easy to describe:

5.16.1 Claim. *The critical points of s lying in the affine chart $x_0 \neq 0$ are all nondegenerate. $(g = 0)$ intersects $(x_0 = 0)$ transversally.*

Proof. $\partial^2 x_1^{pd+1}/\partial x_i \partial x_j = 0$ for any i, j. Thus the Hessian of f is the same as the Hessian of $x_1x_2 + x_3x_4 + \ldots + x_{2n-1}x_{2n}$, hence everywhere nondegenerate.

The second part follows since $x_1^{pd+1} + \ldots + x_{2n}^{pd+1} = 0$ defines a smooth hypersurface inside $(x_0 = 0)$. □

By (5.9) we obtain that $Q(L, s) \cong \mathcal{O}(-2n - 1 + (d+1)p - (p-2)) = \mathcal{O}(dp + 1 - 2n)$. Let $r : Y' \to Y$ be a desingularization. By (5.10) we have an injection $r^*\pi^*Q(L,s)(-1) \to \Omega_{Y'}^{2n-1}$. Thus Y' is not ruled if $dp > 2n$. □

5.16.3 Exercise. Assume that $n < d < 2n$. Let $|c| > 7n$ be an integer. Show that the double cover of $\mathbb{CP}^{2n}$ given by the affine equation

$$y^2 = c + x_1x_2 + \ldots + x_{2n-1}x_{2n} + \sum_i x_i^{2d+1} + 2\sum_i x_i^{2d+2}$$

is a smooth Fano variety which is not ruled.

Hint. Show that if $(x_1, \ldots, x_{2n})$ is a critical point of the function

$$x_1x_2 + \ldots + x_{2n-1}x_{2n} + \sum_i x_i^{2d+1} + 2\sum_i x_i^{2d+2},$$

then $|x_i| \leq 1$ for every i. A more careful analysis of the critical points may give the same result for $c \neq 0$.

The above results show that there are many Fano varieties which are not rational and not even ruled. For three dimensional Fano varieties the rationality question is mostly settled, see [Iskovskikh80a,b] for a survey. Very little is known in higher dimensions. In particular, I do not know a single example of a smooth hypersurface of degree at least 4 which is rational.

5.17 Exercise. Let $X_2 \in \mathbb{P}^n$ be a smooth quadric. Show that X_2 is rational. Let $X_3 \subset \mathbb{P}^{2n+1}$ be a smooth cubic which contains a pair of skew n-dimensional linear spaces. Show that X_3 is rational.

Unirationality of hypersurfaces has been investigated intensively, mainly by Morin and Segre; see the survey article [Segre50] for a summary of results and for references. Modern treatment of some of the results is given in [Ramero90; Paranjape-Srinivas92]. The aim of the next exercise is to study the unirationality of hypersurfaces of degree at most four.

5.18 Exercise. Let $X_d \subset \mathbb{P}^{n+1}$ be a hypersurface of degree d. We always assume that X is irreducible.

(5.18.1) Let $P(X) \to X$ be the projectivized tangent sheaf. As a point set, $P(X)$ consists of pairs (x, l) where $x \in X$ and $x \in l$ is a line which is tangent to X at x. Inside P there are subschemes $P_k(X) \subset P(X)$ parametrizing pairs (x, l) where l is at least k-fold tangent at x. For $x \in X$ choose local linear coordinates $x_1, \dots, x_{n+1}$ and expand the equation of X as a sum of homogeneous terms $f = f_0 + f_1 + \dots + f_d$. The fiber of $P_k(X) \to X$ over x is given by $f_1 = \dots = f_k = 0$. If $Z \subset X$ is a subscheme, let $P_k(Z, X)$ denote the preimage of Z in $P_k(X)$.

(5.18.2) Let $(x, l) \in P_{d-1}(X)$. If $l \not\subset X$, then there is a unique d^{th} intersection point of X and l; denote it by $\phi(x, l)$. This gives a rational map $\phi : P_{d-1}(X) \dashrightarrow X$. Usually $P_{d-1}(X)$ is irreducible and ϕ is defined on an open set, but it may happen that it is reducible and ϕ need not be defined on every irreducible component.

(5.18.3) Let $d = 3$ and $Z = L \subset X$ a line such that X is smooth along L. Show that $P_2(L, X)$ is irreducible, rational and $\phi : P_2(L, X) \dashrightarrow X$ is dominant of degree two.

(5.18.4) Conclude that if X is a cubic defined over a field k which contains a line L defined over k such that X is smooth along L, then X is unirational over k. ([Manin72, II.2] gives a unirationality criterion which assumes only the existence of a k-point.) In particular, if k is algebraically closed then any smooth cubic of dimension at least two is unirational.

(5.18.5) Let $d = 4$ and $Z = H \subset X$ a 2-plane such that X is smooth along H. Show that $P_3(H, X)$ is irreducible, rational and $\phi : P_3(L, X) \dashrightarrow X$ is domimant of degree six.

(5.18.6) Conclude that if X is a quartic defined over a field k which contains a 2-plane H defined over k such that X is smooth along H, then X is unirational over k. In particular, if k is algebraically closed then any smooth quartic of dimension at least six is unirational.

The method of inseparable cyclic covers gives further interesting examples:

5.19 Exercise (Notation as in (5.2–8)).

(5.19.1) Assume that $Q(L, s)$ is nef and there is an injection $r^*\pi^*Q(L, s) \to \Omega_{Y'}^{n-1}$. Show that Y' is not separably rationally connected.

(5.19.2) Set $X = \mathbb{P}^n$, p a prime divisor of $n+1$ and $L = \mathcal{O}(n+1/p)$. Let $s \in H^0(\mathbb{P}^n, L^p \cong \mathcal{O}(n+1))$ a general section. Let $\ell \subset \mathbb{P}^n$ be a general line. Show that

$$\operatorname{coker}\left[L^{-p} \xrightarrow{ds} \Omega^1_{\mathbb{P}^n}\right]|\ell \cong \mathcal{O}_\ell^{n-1}.$$

Hint. View s as a homogeneous polynomial of degree $n+1$. ds lifts to a map

$$\sigma : \mathcal{O}(-n-1) \to \mathcal{O}(-1)^{n+1}, \quad \text{given by} \quad \left(\frac{\partial s}{\partial x_0}, \dots, \frac{\partial s}{\partial x_n}\right).$$

Show that $\operatorname{coker}\sigma|\ell \cong \mathcal{O}_\ell^n$ iff $H^0(\ell, \mathcal{O}_\ell(n))$ is generated by the polynomials $(\partial s/\partial x_i)|\ell$. The latter is a open condition, thus it is enough to write down one example. For instance

$$s = x_0 x_n^n + x_0^n x_n + \sum_{i=1}^{n-1} x_i x_0^i x_n^{n-i} \quad \text{and} \quad \ell = (x_1 = \dots = x_{n-1} = 0).$$

(5.19.3) Let $Y = \mathbb{P}^n[\sqrt[p]{s}]$ as above and $r : Y' \to Y$ a resolution. Show that Y' is separably uniruled, rationally connected but not separably rationally connected.

(5.19.4) Let $C \subset Y'$ be the preimage of a general line in $\mathbb{P}^n$, $h : \mathbb{P}^1 \to Y'$ its normalization. Show that C is a free rational curve with $n+1$ singular points of the form $y^p = x^2$ if $p \neq 2$ and $y^2 = x^3$ if $p = 2$. Also,

$$h^* T_{Y'} \cong \mathcal{O}_{\mathbb{P}^1}(n+1) + \mathcal{O}_{\mathbb{P}^1}^{n-1}.$$

Thus preimages of lines account for all deformations of h; cf. (II.3.14).

5.20 Exercise (Canonical maps of Fano varieties).

(5.20.1) Let X be a smooth proper variety of dimension n. The usual pluricanonical maps are defined using global sections of $(\Omega_X^n)^{\otimes m}$. Define maps $\phi(X, i, m) : X \dashrightarrow G$ using sections of $S^m(\Omega_X^i)$ where G is a suitable Grassmannian.

(5.20.2) Show that the closure of the image of $\phi(X, i, m)$ is a birational invariant of X for every $i, m > 0$.

(5.20.3) Let X be a smooth projective variety of dimension $n \geq 3$ over a field of characteristic p and L an ample line bundle on X. Let $s \in H^0(X, L^p)$ be a section with only (almost) nondegenerate critical points. Let $r : Y' \to X[\sqrt[p]{s}]$ be a resolution of singularities. Assume that $Q(L, s)$ is ample on X.

Show that $X[\sqrt[p]{s}]$ is a birational invariant of Y'.

(5.20.4) Assume in addition that $Q(L, s)^{-1} \otimes L$ is ample. Show that the morphism $\pi : X[\sqrt[p]{s}] \to X$ is a birational invariant of Y'.

(5.20.5) Let $X = \mathbb{P}^n$, $L = \mathcal{O}(d)$ and assume that $n + 1 < pd \leq n + d$. Let $s_i \in H^0(X, L^p)$ be sections with only (almost) nondegenerate critical points and $\pi : Y_i := \mathbb{P}^n[\sqrt[p]{s_i}] \to \mathbb{P}^n$ the corresponding cyclic covers. Show that Y_1 and Y_2 are birational iff the hypersurfaces $(s_1 = 0)$ and $(s_2 = 0)$ are projectively equivalent.

(5.20.6) Use this to construct many examples of smooth Fano varieties in characteristic zero which are not birational to each other.

Chapter VI. Appendix

The aim of this chapter is to discuss some auxiliary results that do not fit anywhere else. Section 1 considers rational curves on exceptional loci of maps. The main result (1.2) is a theorem of [Abhyankar56] which says that if $f : X \to Y$ is a birational morphism and Y is smooth, then every exceptional divisor of f is ruled. This is clear if resolution of indeterminacies holds for f^{-1}, but the result can be proved using only some partial resolution results. The relevant assertion about resolutions (1.3) is contained in [Zariski39] and it is useful in many instances.

Exceptional sets of birational maps are considered at the end of the section. Theorem (1.10) can be applied to conclude that the exceptional sets of flips and flops are uniruled in most cases.

Section 2 is an introduction to the intersection theory of divisors. It is much easier to intersect Cartier divisors than higher codimension subvarieties. Also, the case of divisors has special importance in understanding Riemann–Roch and the cone of curves of a variety. The definition and basic properties of the intersection numbers are given in (2.1–14). The presentation follows the unpublished manuscript [Kleiman74]. The first application is the asymptotic Riemann–Roch theorem for nef line bundles (2.15):

Let X be a proper scheme of dimension m and L a nef line bundle on X. Then

$$h^0(X, L^n) = \frac{(L^m)}{m!}n^m + O(n^{m-1}).$$

Previous proofs of this result relied either on Kodaira type vanishing theorems or on some ingenious application of Frobenius morphisms [Fujita83]. It is nice to have a proof which relies only on basic properties of cohomology.

Kleiman's characterization of nef divisors has been very important in earlier chapters; it is finally proved in (2.17). The proof of the basic ampleness criteria are given in (2.18–19). These also have been used in connection with the cone of curves.

VI.1 Rational Curves on Exceptional Loci

The aim of this section is to discuss some situations where rational curves appear on exceptional loci of maps.

1.1 Definition. Let $f : X \dashrightarrow Y$ be a birational map. Assume that Y is normal. The closed subset

$$E(f) = \{x \in X |\ f \text{ is not a local isomorphism at } x\} \subset X$$

is called the *exceptional set* of f.

The most important theorem about rational curves on exceptional loci is the following:

1.2 Theorem [Abhyankar56]. *Let X, Y be normal, excellent schemes and $f : Y \to X$ a birational morphism of finite type. Let $E \subset Y$ be the exceptional set of f. Assume that X is regular. Then E has ruled components over X.*

The proof relies on two results which are of considerable interest themselves.

1.3 Theorem. *Let X, Y be integral, excellent schemes and $f : Y \to X$ a dominant morphism of finite type. Let $D \subset Y$ be an irreducible divisor and $y \in D$ the generic point. Asume that Y is normal at y. We define a sequence of schemes and maps as follows:*

$X_0 = X, f_0 = f.$

If $f_i : Y \dashrightarrow X_i$ is already defined, then let $Z_i \subset X_i$ be the closure of $f_i(y)$. By shrinking X_i we may assume that Z_i is nonsingular. Let $X_{i+1} = B_{Z_i}X_i$, $f_{i+1} : Y \dashrightarrow X_{i+1}$ the induced map and $F_{i+1} \subset X_{i+1}$ the exceptional divisor of the projection $p_i : X_{i+1} \to X_i$.

Then $\dim Z_n \geq \dim X - 1$ and X_n is normal at the generic point of Z_n for some $n \geq 0$.

Before we get to the proof, some definitions are needed.

1.3.1 Definition. Let X be an integral scheme with field of rational functions F. $k(x)$ denotes the residue field of a point $x \in X$. A *prime divisor of F centered at x* is a DVR (R, m_R) with quotient field F such that

$$m_R \cap \mathcal{O}_{x,X} = m_{x,X} \quad \text{and} \quad \operatorname{tr deg}(R/m_R : k(x)) \geq \dim \mathcal{O}_{x,X} - 1.$$

If $x \in X$ is a codimension one point, then $R = \mathcal{O}_{x,X}$ is a prime divisor centered at x iff X is regular at x. A prime divisor defines a valuation v_R of F by setting

$$v_R(f) := \max\{s \in \mathbb{Z} | f \in m_R^s\} \quad \text{for } 0 \neq f \in F.$$

1.3.2 Construction. The following is the ring theoretic version of the sequence of blow ups defined in (1.3). We use the notation of (1.3.1). Set $\mathcal{O}_1 := \mathcal{O}_{x,X}$ with closed point $x_1 = x$. Assume that $\mathcal{O}_n, x_n$ are already defined. Let $z_1, \dots, z_r$ be generators of the maximal ideal of $\mathcal{O}_n$. Assume that $v_R(z_1) \le v_R(z_i)$ for every i. Let $\mathcal{O}'_n = \mathcal{O}_n[z_i/z_1 : i = 2, \dots, r]$, $\mathcal{O}_{n+1}$ the localization of $\mathcal{O}'_n$ at $\mathcal{O}'_n \cap m_R$ and x_{n+1} the closed point of $\operatorname{Spec} \mathcal{O}_{n+1}$.

The key to proving (1.3) is the following characterization of prime divisors. The statement and the proof closely follows [Artin86, 5.1].

1.4 Theorem [Zariski39]. *Let X be an excellent integral scheme with field of rational functions F and $x \in X$ a point. Let (R, m_R) be a DVR whose quotient field is F such that $m_R \cap \mathcal{O}_{x,X} = m_{x,X}$. The following are equivalent:*

(1.4.1) *R is the localization of an $\mathcal{O}_{x,X}$ algebra of finite type;*
(1.4.2) $\operatorname{tr}\deg(R/m_R : k(x)) = \dim \mathcal{O}_{x,X} - 1$;
(1.4.3) $\operatorname{tr}\deg(R/m_R : k(x)) \ge \dim \mathcal{O}_{x,X} - 1$;
(1.4.4) *$R = \mathcal{O}_n$ for $n \gg 1$, where $\mathcal{O}_n$ is defined in (1.3.2).*

Proof. (1.4.2) ⇒ (1.4.3) and (1.4.4) ⇒ (1.4.1) are clear. Let A be an $\mathcal{O}_{x,X}$ algebra of finite type whose localization at a prime p is R. Since A and $\mathcal{O}_{x,X}$ have the same quotient field, they have the same dimension. p is a codimension one prime, thus A/p has dimension $\dim \mathcal{O}_{x,X} - 1$. Therefore its quotient field R/m has transcendence degree $\dim \mathcal{O}_{x,X} - 1$ (see e.g. [Matsumura80, 14.C]). This shows that (1.4.1) ⇒ (1.4.2).

It remains to show that (1.4.3) ⇒ (1.4.4). Assume that there is $u \in R - m$ such that $u \in R/m$ is transcendental over $k(x)$. Write $u = y_1/y_2$ where $y_i \in \mathcal{O}_1$ and $v_R(y_2)$ is the smallest possible. We can write $y_i = \sum_j z_j y_{ij}$ where $y_{ij} \in \mathcal{O}_1$ and the z_i are generators of the maximal ideal as in (1.3.2). Let $y'_i := \sum_j (z_j/z_1) y_{ij} \in \mathcal{O}_2$. Then $y_i = z_1 y'_i$ and $u = y'_1/y'_2$. By construction $v_R(y'_2) < v_R(y_2)$. Iterating this procedure eventually we obtain that $u = y_1^{(s)}/y_2^{(s)}$ and $v_R(y_2^{(s)}) = 0$. This implies that $u \in \mathcal{O}_s$.

$\mathcal{O}_k$ is the localization of an $\mathcal{O}_{x,X}$-algebra of finite type by construction. Therefore

$$\operatorname{tr}\deg(k(x_s) : k(x_1)) = \dim \mathcal{O}_s - \dim \mathcal{O}_1.$$

This implies that

$$\operatorname{tr}\deg(R/m_R : k(x_s)) \ge \dim \mathcal{O}_s - 1.$$

Repeating the above procedure starting with $\mathcal{O}_s$ we eventually stop with $\mathcal{O}_r$ such that R/m_R is an algebraic extension of $k(x_r)$ and $\dim \mathcal{O}_r = 1$. R is, then the normalization of $\mathcal{O}_r$, thus $R/\mathcal{O}_r$ has finite length over $\mathcal{O}_r$.

$\mathcal{O}_r \subset \mathcal{O}_{r+1} \subset \dots \subset R$ is an increasing chain of rings, thus there is an index n such that $\mathcal{O}_n = \mathcal{O}_{n+1}$. Let z_i be generators of the maximal ideal m_n of $\mathcal{O}_n$. Then $z_i = z_1(z_i/z_1)$, thus z_1 alone generates m_n. Therefore $\mathcal{O}_n$ is a DVR and $R = \mathcal{O}_n$. □

1.4.5 Exercise. (1.4.5.1) Let $F = \mathbb{C}(x,y)$. Show that there is an injection $i : F \to \mathbb{C}((t))$ given by $x \mapsto t, y \mapsto e^t - 1$. Let $R \subset \mathbb{C}(x,y)$ be the intersection of $i(F)$ and $\mathbb{C}[[t]]$. Show that $R \subset F$ is a DVR with quotient field F but it is not a prime divisor for any scheme of finite type $X/\mathbb{C}$.

(1.4.5.2) Let $E \subset F$ be a field extension, (R_F, m_F) a DVR with quotient field F. Set $R_E = R_F \cap E$ and $m_E = m_F \cap E$. Show that either $E = R_E$ or (R_E, m_E) is a DVR and

$$\operatorname{tr}\deg(R_F/m_F : R_E/m_E) \le \operatorname{tr}\deg(F : E).$$

(1.4.5.3) Let $f : X \to Y$ be a finite type dominant morphism of integral schemes. Let $E \subset F$ be the induced injection on quotient fields. Let (R_F, m_F) be a prime divisor of X. Show that either $E \subset R_F$ or (R_E, m_E) is a prime divisor of Y.

1.4.6 Proof of (1.3). Let E (resp. F) be the field of rational functions on X (resp. Y). Let $R_F \subset F$ be the DVR associated to D. If $D \to Y$ is dominant, then $Z_0 = X$ and we are done. Otherwise $R_E := R_F \cap E$ is a DVR which is a prime divisor on Y by (1.4.5.3).

(1.4) shows that after a sequence of blow ups we obtain $f_n : Y \dashrightarrow X_n$ such that $Z_n = \overline{f(Y)}$ is a divisor and X_n is normal at its generic point. □

1.4.7 Another Proof of (1.3). Here is another proof following [Johnston89] which assumes that X is smooth and f is separable but gives an explicit upper bound for the number of necessary steps.

The problem concerns the codimension one point $y \in D$. We can remove the singular set of Y, hence we may assume that Y is also smooth. Since $p_i : X_{i+1} \to X_i$ is proper, f_{i+1} is defined at y and $f_{i+1}(y) \subset F_{i+1}$.

f induces a map $f^*\omega_X \to \wedge^{\dim X} \Omega^1_Y$ which is an injection since f is separable. Let

$$f^*\omega_X(mD) \subset \wedge^{\dim X} \Omega^1_Y$$

be the saturation of the image. Let $k_i = \operatorname{codim}(Z_{i-1}, X_{i-1}) - 1$. Then

$$\omega_{X_i} = p_i^* \omega_{X_{i-1}}(k_i F_i).$$

Repeatedly applying the above identity we obtain

$$\begin{aligned} \wedge^{\dim X} \Omega^1_Y \supset f_n^* \omega_{X_n} &= f_{n-1}^* \omega_{X_{n-1}} \otimes \mathcal{O}_Y(k_n f_n^{-1}(F_n)) = \cdots \\ &= f^*\omega_X \otimes \mathcal{O}_Y\Big(\sum_{1 \le i \le n} k_i f_i^{-1}(F_i)\Big) \\ &\supset f^*\omega_X \otimes \mathcal{O}_Y\Big(\Big(\sum_{1 \le i \le n} k_i\Big) D\Big). \end{aligned}$$

Therefore $k_n = 0$ for some $n \le m$. □

1.4.7.1 Exercise. Find examples where the above process takes m steps.

1.5 Theorem. *Let Y be a normal $\mathbb{Q}$-factorial scheme, X a normal scheme and $f : X \to Y$ a birational morphism. Then every irreducible component of the exceptional set E of f has codimension one in X.*

Proof. Let $x \in E$ be a point and set $y = f(x)$. Let t_i be local parameters of X at x. Since f^{-1} is not regular at y, we may assume that t_1 is not regular at y. By assumption the divisor class group of $\mathcal{O}_{y,Y}$ is torsion, thus we can write

$$t_1^m = \frac{u}{v} \quad \text{for some relatively prime} \quad u, v \in \mathcal{O}_{y,Y}.$$

Observe that $v(y) = 0$ since t_1 is not regular at y. Thus $(u = v = 0)$ has codimension two in Y and

$$(f^*u = f^*v = 0) = (t_1^m f^*v = f^*v = 0) \supset (f^*v = 0) \ni x$$

has codimension one at x. □

1.5.1 Proof of (1.2). Let $E_k \subset E$ be an irreducible component. By (1.4) E_k has codimension one in X. Using the notation of (1.3) we obtain that E_k is birational to an F_j. F_j is ruled by construction and we are done. □

1.5.2 Exercise. Let $f : X \dashrightarrow Y$ be a birational map between normal and proper varieties. Assume that X is $\mathbb{Q}$-factorial and f is not a morphism. Show that

$$\dim E(f) + \dim E(f^{-1}) \geq \dim X - 1.$$

Show that both assumptions are necessary.

Keeping (1.2) in mind, it is useful to introduce a definition:

1.6 Definition. Let Y be a normal scheme. We say that Y has *ruled* (resp. *uniruled*) *modifications* if the following condition is satisfied:

If X is normal and $f : X \to Y$ birational, then every irreducible component of the exceptional set of f is ruled (resp. uniruled) over Y.

1.6.1 Exercise. Find a flat family of surface singularities $f_t(x, y, z) = 0$ such that for $t = 0$ the singularity at the origin has ruled modifications but for $t \neq 0$ it does not.

The following are some easy properties:

1.6.2 Lemma. (1.6.2.1) *If Y is smooth, then it has ruled modifications.*

(1.6.2.2) *Assume that Y has uniruled modifications. If $f : X \to Y$ is birational, then every irreducible component of the exceptional set of f is uniruled over Y.*

(1.6.2.3) *Let $g : Z \to Y$ be finite and dominant. Assume that Z has uniruled modifications. Then Y has uniruled modifications as well.*

Proof. (1.6.2.1) follows from (1.2). If $\bar{f} : \bar{X} \to Y$ is the normalization, then the exceptional set of $\bar{f}$ finitely dominates the exceptional set of f, which shows (1.6.2.2).

Finally assume that X is normal and $f : X \to Y$ is birational. Let Z' be the birational transform of Z in $Z \times_Y X$. The exceptional set of $Z' \to Z$ finitely dominates the exceptional set of f. □

1.7 Lemma. *Let Y be normal, $\mathbb{Q}$-factorial and $g : Z \to Y$ a proper and birational morphism. Assume that*

(1.7.1) *Z has ruled (resp. uniruled) modifications;*

(1.7.2) *The exceptional set of g is ruled (resp. uniruled) over Y.*

Then Y has ruled (resp. uniruled) modifications.

Proof. By (1.5) the exceptional set of f has pure codimension one. Let E be an irreducible component with generic point e. Let $h : X \dashrightarrow Z$ be the induced map. It is defined at e. If h is a local isomorphism at e, then E is birational to a component of the exceptional set of g, thus E is ruled (resp. uniruled). Otherwise E is h-exceptional and E is ruled (resp. uniruled) by (1.7.1). □

1.7.3 Exercise. Find an example showing that the $\mathbb{Q}$-factoriality assumption is necessary.

1.8 Remark. It is conjectured that if (X, D) is $\mathbb{Q}$-factorial and canonical or even dlt (see [Kollár et al.92, 2.13] for the definition), then X has uniruled modifications. Using (IV.1.9) it is easy to see that the log-minimal model program implies this conjecture. In particular, it is true if $\dim X \leq 3$. In dimension three this can also be proved using the explicit partial resolution of canonical singularities given in [Reid80].

1.9 Theorem. *Let X, Y be normal schemes. Assume that X has uniruled modifications. Let $f : X \dashrightarrow Y$ be a map, $\Gamma \subset X \times Y$ the closure of the graph of f and $E \subset \Gamma$ the exceptional set of $p_1 : \Gamma \to X$. Then*

(1.9.1) *$p_2(E) \subset Y$ has uniruled components.*

(1.9.2) *If X has ruled modifications, then every codimension one irreducible component of $p_2(E)$ is ruled.*

(1.9.3) *If in addition Y is proper and it does not contain rational curves, then f extends to a morphism $f : X \to Y$.*

Proof. E is uniruled over X by assumption. Let $y \in p_2(E)$ and take an arbitrary $z \in E$ such that $p_2(z) = y$. Since E is uniruled over X, there is a rational curve $z \in C \subset E$ such that $p_1(C) = $ point (IV.1.5). $C \subset X \times Y$ and therefore $p_2 : C \to Y$ is injective. Thus $p_2(E) \subset Y$ is uniruled by (IV.1.3).

Let $V \subset p_2(E)$ be a codimension one irreducible component. $\Gamma \to Y$ is birational, therefore a local isomorphism over the generic point of V. Thus

V is birational to an irreducible component of E, hence ruled if E is ruled over X.

If Y does not contain any rational curves, then $E = \emptyset$ hence $\Gamma \to X$ is quasifinite. If in addition Y is proper, then $p_1 : \Gamma \to X$ is proper, hence an isomorphism. This implies the last claim. □

1.9.4 Exercise. Show that the following proper schemes do not contain rational curves:

(1.9.4.1) Products of curves of genus at least one.

(1.9.4.2) Schemes which admit a finite morphism to an Abelian scheme.

(1.9.4.3) (Clemens, Voisin) A very general hypersurface of degree at least $2n-2$ in $\mathbb{P}^n$ for $n \geq 4$. (This is quite hard, see [CKM88, #21; Voisin96].)

The natural examples in connection with the following result are flips and flops.

1.10 Theorem. *Let $f : X \dashrightarrow Y$ be a birational map between normal schemes, Y proper. Assume that X is $\mathbb{Q}$-factorial and Y has uniruled modifications. Let $U_X \subset X$ be the largest open subset such that $f|U_X$ is an isomorphism and set $U_Y = f(U_X)$. Assume that* $\operatorname{codim}(Y - U_Y, Y) \geq 2$.

Then $X - U_X$ is uniruled.

Proof. Let $\Gamma \subset X \times Y$ be the closure of the graph of f, $E \subset \Gamma$ the exceptional set of p_Y. It is sufficient to show that $p_X(E) = X - U_X$.

Let $x \in X - U_X$ and $x' \in p_X^{-1}(x)$ arbitrary. If $x' \notin E$, then p_Y is a local isomorphism at x'. Since $x \notin U_X$, this implies that p_X can not be a local isomorphism at x'. Therefore by (1.5) there is a divisor $x' \in V \subset \Gamma$ which is p_X-exceptional. Since $p_Y(V) \subset Y - U_Y$ has codimension at least two, $x' \in V \subset E$. This is a contradiction. □

1.10.1 Exercise. Find an example showing that the assumption about the $\mathbb{Q}$-factoriality of X is necessary.

1.10.2 Exercise. Let X, Y be smooth proper varieties such that K_X and K_Y are numerically trivial. Let $f : X \dashrightarrow Y$ be a birational map. Prove that $E(f)$ and $E(f^{-1})$ have codimension at least 2. Thus every irreducible component of $E(f)$ and $E(f^{-1})$ is uniruled.

More generally, the same is true if K_X and K_Y are both nef, cf. [Kollár89].

VI.2 Intersection Numbers of Divisors and Asymptotic Riemann-Roch

The aim of this appendix is to give a quick definition of intersection numbers of Cartier divisors and to derive their basic properties needed elsewhere. This approach was developed in the articles [Snapper60; Cartier60; Kleiman66]. The presentation in (2.1–14) closely follows the manuscript [Kleiman74]. I would like to thank him for allowing me to use his unpublished notes. For a more detailed exposition of the general theory see [Fulton84].

We work with schemes but all the proofs apply without change for algebraic spaces.

2.1 Definition. Let X/S be a Noetherian scheme. Let $K(X)$ denote the abelian group generated by the symbols $[F]$ where F is coherent sheaf on X whose support is proper over a 0-dimensional subscheme of S, modulo the relations $[F_2] = [F_1] + [F_3]$ for every short exact sequence

$$0 \to F_1 \to F_2 \to F_3 \to 0.$$

$K(X)$ is called the *Grothendieck group* of X. By abusing notation, we use F to denote its class in $K(X)$. For each r, let $K_r(X) \subset K(X)$ denote the subgroup generated those F whose support has dimension at most r.

The method of dévissage was invented by Grothendieck to facilitate inductive proofs for $K(X)$. It can be formulated in several forms. For us the following is the most convenient:

2.2 Lemma (Dévissage of Grothendieck). *Let X/S be a Noetherian scheme and $\mathcal{F}$ a class of coherent sheaves on X, closed under isomorphisms. Assume the following:*

(2.2.1) *If $Z \subset X$ is a closed, integral subscheme whose support is proper over a 0-dimensional subscheme of S, then $\mathcal{O}_Z \in \mathcal{F}$;*

(2.2.2) *If $0 \to F_1 \to F_2 \to F_3 \to 0$ is exact and $F_1, F_3 \in \mathcal{F}$, then $F_2 \in \mathcal{F}$.*

(2.2.3) *If $Z \subset X$ is an integral subscheme whose support is proper over a 0-dimensional subscheme of S, $0 \to F_1 \to \mathcal{O}_Z^r \to F_3 \to 0$ is exact,* $\dim \operatorname{Supp} F_3 < \dim Z$ *and $F_3 \in \mathcal{F}$, then $F_1 \in \mathcal{S}$.*

Then $\mathcal{F}$ is the class of all coherent sheaves on X whose support is proper over a 0-dimensional subscheme of S.

Proof. We use induction on $\dim \operatorname{Supp} F$.

Every coherent sheaf F on X has a finite filtration $F = F_1 \supset F_2 \supset \dots$ such that the quotients F_i/F_{i+1} are torsion free sheaves over an integral subscheme $X_i \subset X$. Thus it is sufficient to consider the case when X is integral and F is torsion free.

By [Matsumura80, 22.A] there is an open set $U \subset X$ such that $F|U$ is locally free of rank r. This gives an injection $i : F \to k(X)^r$ where $k(X)$ is the (quasi coherent) sheaf of rational functions. Let

$$G := i(F) \cap \mathcal{O}_X^r \subset k(X)^r.$$

The natural injections $G \to F$ and $G \to \mathcal{O}_X^r$ are isomorphisms over U. We get exact sequences

$$\begin{array}{ccccccccc} 0 & \longrightarrow & G & \longrightarrow & F & \longrightarrow & Q_1 & \longrightarrow & 0, \quad \text{and} \\ 0 & \longrightarrow & G & \longrightarrow & \mathcal{O}_X^r & \longrightarrow & Q_2 & \longrightarrow & 0, \end{array} \tag{2.2.4}$$

where $\dim \operatorname{Supp} Q_i < \dim X$. By induction $Q_i \in \mathcal{F}$. Thus by (2.2.3) $G \in \mathcal{F}$ which in turn implies that $F \in \mathcal{F}$. □

Frequently the following simpler version is all that one needs:

2.2.5 Corollary. *Let X/S be a Noetherian scheme and $T_1, T_2 : K_r(X) \to A$ additive maps to an Abelian group A. Assume that if $Z \subset X$ is a closed, integral subscheme whose support is proper over a 0-dimensional subscheme of S, then $T_1(\mathcal{O}_Z) = T_2(\mathcal{O}_Z)$.*

Then $T_1 = T_2$ on $K_r(X)$. □

2.3 Corollary. *Let F be a coherent sheaf on X whose support is proper over a 0-dimensional subscheme of S and has dimension at most r. Let $Y_1, \dots, Y_n$ be the r-dimensional irreducible components of* $\operatorname{red}\operatorname{Supp} F$ *with generic points $y_i \in Y_i$. Then*

$$F \equiv \sum \operatorname{length}_{y_i} F \cdot \mathcal{O}_{Y_i} \bmod K_{r-1}(X). \tag{2.3.1}$$

Proof. We can view both sides of (2.3.1) as additive maps $T_i : K_r(X) \to K_r(X)/K_{r-1}(X)$. They clearly agree if $F = \mathcal{O}_Z$ and Z is integral. Thus (2.2.5) implies (2.3). □

2.4 Definition. Let L be an ivertible sheaf on X. $F \mapsto L \otimes F$ is an automorphism of the Abelian group $K(X)$. We define an endomorphism of $K(X)$ by the formula:

$$c_1(L) \cdot F := F - L^{-1} \otimes F.$$

Basic properties of these operators are described next:

2.5 Proposition (Notation as above).

(2.5.1) *$F \mapsto c_1(L) \cdot F$ is a well defined additive endomorphism of $K(X)$.*

(2.5.2) *$c_1(L)$ has degree -1, that is $c_1(L) \cdot K_r(X) \subset K_{r-1}(X)$ for every r.*

(2.5.3) *$c_1(L_1)$ and $c_1(L_2)$ commute for any two line bundles L_i.*

(2.5.4) *If $Z \subset X$ is integral of dimension r and $L|Z \cong \mathcal{O}_Z(D)$ where D is an effective Cartier divisor, then $c_1(L) \cdot \mathcal{O}_Z = \mathcal{O}_D$.*

(2.5.5) *If $Z \subset X$ is integral of dimension r and $L|Z \cong \mathcal{O}_Z(Z_1 - Z_2)$ where the Z_i are effective Weil divisors, then*

$$c_1(L) \cdot \mathcal{O}_Z \equiv \mathcal{O}_{Z_1} - \mathcal{O}_{Z_2} \bmod K_{r-2}(X).$$

(2.5.6) $c_1(L_1 \otimes L_2) \cdot F \equiv c_1(L_1) \cdot F + c_1(L_2) \cdot F \bmod K_{r-2}(X)$.

Proof. If $F_2 = F_1 + F_3$, then $L^{-1} \otimes F_2 = L^{-1} \otimes F_1 + L^{-1} \otimes F_3$ since tensoring with an invertible sheaf is an exact functor. This shows (2.5.1). Let $y_i \in \operatorname{Supp} F$ be as in (2.3). Then $\operatorname{length}_{y_i} F = \operatorname{length}_{y_i} L^{-1} \otimes F$. Thus, computing modulo $K_{r-1}(X)$,

$$F \equiv \sum \operatorname{length}_{y_i} F \cdot \mathcal{O}_{Y_i} = \sum \operatorname{length}_{y_i} (L^{-1} \otimes F) \cdot \mathcal{O}_{Y_i} \equiv L^{-1} \otimes F.$$

This shows (2.5.2) and (2.5.3) is clear from the definition.

(2.5.4) follows from the exact sequence

$$0 \to \mathcal{O}_Z(-D) \to \mathcal{O}_Z \to \mathcal{O}_D \to 0.$$

In order to see (2.5.5), consider the exact sequences

$$0 \to \mathcal{O}_Z(-Z_1) \to \mathcal{O}_Z \to \mathcal{O}_{Z_1} \to 0, \quad \text{and}$$
$$0 \to \mathcal{O}_Z(-Z_1) \to L^{-1} \to L^{-1} \otimes \mathcal{O}_{Z_2} \to 0.$$

These show that $c_1(L) \cdot \mathcal{O}_Z = \mathcal{O}_{Z_1} - L^{-1} \otimes \mathcal{O}_{Z_2}$. $\mathcal{O}_{Z_2} \equiv L^{-1} \otimes \mathcal{O}_{Z_2} \bmod K_{r-2}(X)$ by (2.5.2). Putting these together gives (2.5.5).

Finally, we have a formal equality

$$c_1(L_1 \otimes L_2) \cdot F = c_1(L_1) \cdot F + c_1(L_1) \cdot F - c_1(L_2) \cdot c_1(L_2) \cdot F.$$

By (2.5.2) $c_1(L_2) \cdot c_1(L_2) \cdot F \in K_{r-2}$ which shows (2.5.6). □

2.6 Definition. Let X/S be a Noetherian scheme and F a coherent sheaf on X whose support is proper over a 0-dimensional subscheme of S. Assume that $m \geq r = \dim \operatorname{Supp} F$. The *intersection number* of $L_1, \ldots, L_m$ on F is defined by

(2.6.1) $$(L_1 \cdots L_m \cdot F) := \chi(X, c_1(L_1) \cdots c_1(L_m) \cdot F).$$

From (2.4) we obtain the alternate expression:

(2.6.2) $$(L_1 \cdots L_m \cdot F) = \sum_{\sigma:\{1,\ldots,m\} \to \{0,1\}} (-1)^{|\sigma^{-1}(1)|} \chi(X, F \otimes \prod_i L^{-\sigma(i)}).$$

Frequently the parentheses are omitted from the notation.

If $L = L_1 = \ldots = L_m$, then we write $(L^m \cdot F)$ instead of $(L_1 \cdots L_m \cdot F)$. If $Y \subset X$ is a closed subscheme, then we write $(L^m \cdot Y)$ instead of $(L^m \cdot \mathcal{O}_Y)$.

We also use $L_1 \cdots L_m$ to denote $L_1 \cdots L_m \cdot X$.

If $L_i = \mathcal{O}_X(D_i)$ for some Cartier divisors D_i, then $(D_1 \cdots D_m \cdot F)$ stands for $(L_1 \cdots L_m \cdot F)$.

Sometimes there is a possibility of confusion since L^m can mean tensor power, direct sum of m copies or selfintersection number. In such cases I try to be systematic and use (L^m) to denote the selfintersection number.

The intersection numbers depend very little on F. The main reason of defining it in this generality is ease of induction.

2.7 Proposition (Notation as above).

(2.7.1) $(L_1 \cdots L_m \cdot F) = 0$ *if* $m > r = \dim \operatorname{Supp} F$. *Thus for fixed* $L_1, \ldots, L_m$ *we can view the intersection product as an additive function* $K_m(X)/K_{m-1}(X) \to \mathbb{Z}$.

(2.7.2) $(L_1 \cdots L_m \cdot F)$ *is a symmetric* m*-linear function on* $\operatorname{Pic}(X)$ *for fixed* F.

(2.7.3) *Let* Y_i *be the* m*-dimensional irreducible components of* $\operatorname{Supp} F$ *and* y_i *the generic point of* Y_i. *Then*

$$(L_1 \cdots L_m \cdot F) = \sum \operatorname{length}_{y_i} F \cdot (L_1 \cdots L_m \cdot Y_i).$$

Proof. If $F \in K_r(X)$, then by (2.4.2) $c_1(L_1) \cdots c_1(L_m) \cdot F \in K_{r-m}(X)$, hence zero if $m > r$. This shows the first part of (2.7.1). The intersection product is a linear map on $K_m(X)$ since the operators $c_1(L_i)$ are additive (2.5.1) and the Euler characteristic is also additive.

The symmetric nature of the intersection product follows from (2.5.3); additivity from (2.5.6). Finally (2.3) and (2.6.1) imply (2.7.3). □

2.7.4 Definition – Corollary. The intersection product extends by linearity to the vector spaces $\operatorname{Pic}(X) \otimes \mathbb{Q}$ and $\operatorname{Pic}(X) \otimes \mathbb{R}$. We keep the same notation for the extensions.

The following result connects our definition of intersection numbers with the more classical ones.

2.8 Theorem. *Let* S *be a zero dimensional scheme. Let* X/S *be a scheme and* F *a coherent sheaf on* X *such that* $\operatorname{Supp} F$ *is proper and has dimension at most* m. *Let* L_i *be line bundles on* X *such that* $L_i = \mathcal{O}_X(D_i)$ *for some effective Cartier divisors* D_i. *Assume that for every* $x \in \cap D_i$ *the sheaf* F *is locally free at* x *and* $D_1, \ldots, D_m$ *(or rather their local equations* $f_{i,x}$*) form a regular sequence in* $\mathcal{O}_{x,X}$. *Then*

$$(L_1 \cdots L_m \cdot F) = \sum_{x \in \cap D_i} \operatorname{rank}_x F \cdot \operatorname{length}(\mathcal{O}_{x,X}/(f_{1,x}, \ldots, f_{m,x})).$$

Proof. Both sides are additive an short exact sequences, hence by (2.2.5) we may assume that $F = \mathcal{O}_X$ and X is integral. $m \geq \dim X$ and for every $x \in \cap D_i$ the ring $\mathcal{O}_{x,X}$ has a regular sequence of length m. Therefore $m = \dim_x X = \dim X$ for every $x \in \cap D_i$. The theorem is obviously true for $\dim X = 0$.

By (2.5.4), $c_1(L) \cdot \mathcal{O}_X = \mathcal{O}_{D_m}$. Using this in the first line and, then induction on dim Supp F we obtain that

$$\begin{aligned}(L_1 \cdots L_m \cdot \mathcal{O}_X) &= (L_1 \cdots L_{m-1} \cdot \mathcal{O}_{D_m}) \\ &= \sum_{x \in \cap D_i} \operatorname{length}(\mathcal{O}_{x,D_m}/(f_{1,x}, \dots, f_{m-1,x})) \\ &= \sum_{x \in \cap D_i} \operatorname{length}(\mathcal{O}_{x,X}/(f_{1,x}, \dots, f_{m,x})). \qquad \square\end{aligned}$$

2.9 Proposition. *Let $f : X \to S$ be a morphism, L_i line bundles on X and F a sheaf on X, flat over S, such that* Supp F *is proper over S. Then the function*

$$s \mapsto (L_1 \cdots L_m \cdot (F \otimes \mathcal{O}_{X_s}))$$

is locally constant on S

Proof. This can be seen from the definition (2.6.2). F is flat over S and so are the sheaves $M \otimes F$ where M is a line bundle. The Euler characteristic is locally constant in flat families, hence

$$s \mapsto \chi(X, (F \otimes \mathcal{O}_{X_s}) \otimes \prod_i L^{-\sigma(i)}) \quad \text{is locally constant.}$$

Thus their alternating sum is also locally constant. □

2.9.1 Corollary. *Let L_i and L'_i be line bundles on X. Assume that L_i and L'_i are algebraically equivalent for every i. Then*

$$L_1 \cdots L_m \cdot F = L'_1 \cdots L'_m \cdot F. \qquad \square$$

2.10 Proposition. *Let $f : X \to S$ be a proper morphism and L_i line bundles on X. Assume that f is a well defined family of proper algebraic cycles. Then the function*

$$s \mapsto (L_1 \cdots L_m \cdot f^{[-1]}(s))$$

is locally constant on S

Proof. This can be proved as (I.3.12). □

2.11 Proposition (Projection formula). *Let $f : Y/S \to X/S$ be a morphism between schemes, L_i line bundles on X and F a sheaf on Y whose support is proper over a 0-dimensional subscheme of S. Let $m \geq$* dim Supp F. *Then*

$$f^*L_1 \cdots f^*L_m \cdot F = L_1 \cdots L_m \cdot f_*F.$$

Proof. $Rf_* : F \mapsto \sum_i (-1)^i R^i f_* F$ is an additive map $K_m(Y) \to K_m(X)$. If M is a line bundle on X, then $R^i f_*(f^*M \otimes F) = M \otimes R^i f_* F$. This shows that

$$Rf_*(c_1(f^*M) \cdot F) = c_1(M) \cdot Rf_* F.$$

If G has zero dimensional support, then $\chi(Y, G) = \chi(X, f_*G) = \chi(X, Rf_*G)$. Thus

$$\begin{aligned} f^*L_1 \cdots f^*L_m \cdot F &= \chi(Y, c_1(f^*L_1) \cdots c_1(f^*L_m) \cdot F) \\ &= \chi(X, Rf_*(c_1(f^*L_1) \cdots c_1(f^*L_m) \cdot F)) \\ &= \chi(X, c_1(L_1) \cdots c_1(L_m) \cdot Rf_* F). \end{aligned}$$

We still need to prove that

(2.11.1) $$L_1 \cdots L_m \cdot Rf_* F = L_1 \cdots L_m \cdot f_* F.$$

Both sides are additive, hence it is enough to check this for $F = \mathcal{O}_Z$ where Z is integral. If $\dim f(Z) < m$, then both sides are zero by (2.7.1). If $\dim f(Z) = m = \dim Z$, then $f : Z \to f(Z)$ is generically finite, thus $\dim \operatorname{Supp} R^i f_* \mathcal{O}_Z < m$ for $i > 0$. Thus (2.11.1) again holds by (2.7.1). □

2.12 Lemma. *Let L be an invertible sheaf, and F a class in $K_r(X)$. Then*

$$L^{\otimes m} \otimes F = \sum_{i=0}^{r} \binom{m+i-1}{i} c_1(L)^i F \quad \textit{for all } m \in \mathbb{Z}.$$

Proof. Setting $n := -m$ we have the formal identity,

$$(1+x)^n = \sum_{i \geq 0} \binom{n}{i} x^i.$$

Substitute $x = y^{-1} - 1$, and use that

$$\binom{-m}{i} = (-1)^i \binom{m+i-1}{i},$$

to obtain

(2.12.1) $$y^m = \sum_{i \geq 0} \binom{m+i-1}{i} (1 - y^{-1})^i.$$

Use this in case y stands for the operator $F \mapsto L \otimes F$. Then $1 - y^{-1} = c_1(L)$. Also, $c_1(L)^i F = 0$ for $i > r$ by (2.7.1). Thus (2.12.1) yields (2.12). □

As a corollary we obtain a precise form of [Snapper60]:

2.13 Theorem. *Let X/S be a scheme and F a coherent sheaf on X such that $\operatorname{Supp} F$ is proper and has dimension at most r. Let $L_1 \ldots, L_n$ be invertible sheaves on X and $m_1, \ldots, m_n$ arbitrary integers. Then*

$$\chi(X, L_1^{\otimes m_1} \otimes \cdots \otimes L_n^{\otimes m_n} \otimes F)$$
$$= \sum \chi(X, c_1(L_1)^{i_1} \cdots c_1(L_n)^{i_n} \cdot F) \cdot \binom{m_1 + i_1 - 1}{i_1} \cdots \binom{m_n + i_n - 1}{i_n},$$

where the sum ranges over all n-tuples of nonnegative integers i_j whose sum is at most r.

Proof. Let $y_1, \dots, y_n$ be commuting variables. Then (2.12.1) gives the identity

$$y_1^{m_1} \cdots y_n^{m_n}$$
$$= \sum_{i_1 \geq 0, \dots, i_n \geq 0} (1 - y_1^{-1})^{i_1} \cdots (1 - y_1^{-1})^{i_n} \cdot \binom{m_1 + i_1 - 1}{i_1} \cdots \binom{m_n + i_n - 1}{i_n}.$$

Using this for $y_i = L_i$ and taking Euler characteristic gives (2.13).

If $\sum_j i_j > r$, then $c_1(L_1)^{i_1} \cdots c_1(L_n)^{i_n} \cdot F = 0$, thus we can omit that term from the sum. □

2.13.1 Exercise. Use (2.13) to obtain the following forms of Riemann–Roch for curves and surfaces. Assume that S is Artinian.

(2.13.1.1) Let C be a proper curve over S and L an invertible sheaf on C. Then $\chi(C, L) = \deg(L) + \chi(\mathcal{O}_C)$.

(2.13.1.2) Let X be a smooth, proper surface over S with relative dualizing sheaf $\omega_{X/S}$ and L an invertible sheaf on X. Then

$$\chi(X, L) = \frac{(c_1(L) \cdot c_1(L \otimes \omega_{X/S}^{-1}))}{2} + \chi(\mathcal{O}_X).$$

2.14 Corollary (Asymptotic Riemann-Roch I.). *Let X be a proper scheme of dimension m, F a coherent sheaf on X and L a line bundle on X. Then*

$$\chi(X, L^n \otimes F) = \frac{(L^m \cdot F)}{m!} n^m + O(n^{m-1}),$$

where $O(n^{m-1})$ depends on X, F, L.

Proof. Use (2.13) for one line bundle $L = L_1$. The right hand side is a polynomial in $n = m_1$ of degree at most m. The degree m term is

$$\binom{n + m - 1}{m} \chi(X, c_1(L)^m \cdot F) = \frac{(L^m \cdot F)}{m!} n^m + O(n^{m-1}).$$ □

2.15 Theorem (Asymptotic Riemann-Roch II.). *Let X be a proper scheme of dimension m, L a nef line bundle on X and F a coherent sheaf on X. Then*

(2.15.1) $$h^i(X, L^n \otimes F) = O(n^{m-1}) \quad \textit{for } i > 0, \textit{ and}$$

(2.15.2) $$h^0(X, L^n \otimes F) = \frac{(L^m \cdot F)}{m!} n^m + O(n^{m-1}),$$

where $O(n^{m-1})$ *depends on* X, F, L.

2.15.3 Comment. In the proof we use the stronger definition of nef: $L^{\dim Y} \cdot Y \geq 0$ for every integral subscheme $Y \subset X$. Later (using (2.15)) we prove that this is equivalent to the weaker definition: $L \cdot C \geq 0$ for every integral curve $C \subset X$.

Proof. By (2.14), (2.15.1) implies (2.15.2). The condition (2.15.1) defines a class of sheaves on X; we need to prove that this is the class of all coherent sheaves. We check the conditions of (2.2). If $0 \to F_1 \to F_2 \to F_3 \to 0$ is exact, then

$$h^i(X, L^n \otimes F_2) \leq h^i(X, L^n \otimes F_1) + h^i(X, L^n \otimes F_3),$$

which shows (2.2.2). Also

$$h^i(X, L^n \otimes F_1) \leq h^i(X, L^n \otimes F_2) + h^{i-1}(X, L^n \otimes F_3).$$

If $\dim \operatorname{Supp} F_3 < m$, then $h^0(X, L^n \otimes F_3) = O(n^{m-1})$, thus (2.2.3) is also satisfied.

In order to prove (2.2.1) we use induction on the dimension. We have to prove that $F = \mathcal{O}_X$ satisfies (2.15.1). Let $k(X)$ denote the (quasi coherent) sheaf of rational functions. As in (2.2) choose an injection $L \subset k(X)$. Let

$$I_1 = L \cap \mathcal{O}_X, \quad I_2 = \mathcal{O}_X \cap L^{-1} \quad \text{and} \quad \mathcal{O}_{Z_i} = \mathcal{O}_X / I_i.$$

Observe that $I_2 \otimes L \cong I_1$ and $\dim Z_i < \dim X$. We obtain exact sequences

$$\begin{array}{ccccccccc} 0 & \longrightarrow & I_1 & \longrightarrow & \mathcal{O}_X & \longrightarrow & \mathcal{O}_{Z_1} & \longrightarrow & 0 \\ & & \cong\downarrow & & & & & & \\ 0 & \longrightarrow & I_2 \otimes & \longrightarrow & L & \longrightarrow & L \otimes \mathcal{O}_{Z_2} & \longrightarrow & 0 \end{array}$$

Tensoring with L^n and writing out the cohomology sequences we get:

(2.15.4) $$\begin{array}{rcl} & H^{i-1}(X, L^n) & \to H^{i-1}(Z_1, L^n \otimes \mathcal{O}_{Z_1}) \to \\ H^i(X, I_1 \otimes L^n) \to & H^i(X, L^n) & \to H^i(Z_1, L^n \otimes \mathcal{O}_{Z_1}), \quad \text{and} \\ & H^{i-1}(X, L^{n+1}) & \to H^{i-1}(Z_2, L^{n+1} \otimes \mathcal{O}_{Z_2}) \to \\ H^i(X, I_2 \otimes L^{n+1}) \to & H^i(X, L^{n+1}) & \to H^i(Z_2, L^{n+1} \otimes \mathcal{O}_{Z_2}). \end{array}$$

Since $H^i(X, I_1 \otimes L^n) = H^i(X, I_2 \otimes L^{n+1})$ we obtain the inequality

(2.15.5) $$\begin{aligned} |h^i(X, L^{n+1}) - h^i(X, L^n)| \leq{}& h^i(Z_2, L^{n+1} \otimes \mathcal{O}_{Z_2}) + h^{i-1}(Z_2, L^{n+1} \otimes \mathcal{O}_{Z_2}) \\ &+ h^i(Z_1, L^n \otimes \mathcal{O}_{Z_1}) + h^{i-1}(Z_1, L^n \otimes \mathcal{O}_{Z_1}). \end{aligned}$$

Therefore by induction

$$|h^i(X,L^{n+1}) - h^i(X,L^n)| = O(n^{m-2}) \quad \text{for } i \geq 2, \text{ and hence}$$
$$h^i(X,L^n) = O(n^{m-1}) \quad \text{for } i \geq 2.$$

Thus by (2.14)

$$h^0(X,L^n) - h^1(X,L^n) = \frac{(L^m)}{m!}n^m + O(n^{m-1}). \tag{2.15.6}$$

We distinguish two cases.

If $h^0(X,L^m) = 0$ for every $n > 0$, then the left hand side of (2.15.6) is nonpositive. $(L^m) \geq 0$ by assumption, thus in fact $(L^m) = 0$ and $h^1(X,L^n) = O(n^{m-1})$, as required.

We are left with the case when $L^k \cong \mathcal{O}(D)$ for an effective Cartier divisor D for some $k > 0$. From the sequence

$$0 \to \mathcal{O}(-D) \to \mathcal{O}_X \to \mathcal{O}_D \to 0$$

we obtain that

$$h^1(X,L^{n+k}) - h^1(X,L^n) \leq h^1(D,L^{n+k} \otimes \mathcal{O}_D) = O(n^{m-2}).$$

Therefore again

$$h^1(X,L^n) = O(n^{m-1}).$$

□

2.15.7 Corollary. *Let X be a proper and irreducible variety of dimension m, M a nef and big line bundle on X and $x \in X$ a smooth point. For every $\epsilon > 0$ there is a $\mathbb{Q}$-Cartier divisor $D = D(x,\epsilon)$ such that $M \equiv D$ and*

$$\operatorname{mult}_x D \geq \sqrt[m]{(M^m)} - \epsilon.$$

Proof. Fix $s,t > 0$ and let $m \subset \mathcal{O}_X$ be the ideal sheaf of x. From the sequence

$$0 \to m^s \otimes M^t \to M^t \to (\mathcal{O}_X/m^s) \otimes M^t \cong \mathcal{O}_X/m^s \to 0$$

we see that

$$h^0(X, m^s \otimes M^t) > 0 \quad \text{if} \quad H^0(X,M^t) > H^0(X,\mathcal{O}_X/m^s).$$

Since $x \in X$ is a smooth point,

$$H^0(X,\mathcal{O}_X/m^s) = \dim_k k[x_1,\dots,x_m]/(x_1,\dots,x_m)^s$$
$$= \binom{m+s-1}{m} = \frac{s^m}{m!} + O(s^{m-1}),$$

and by (2.15)

$$H^0(X,M^t) = \frac{(M^m)}{m!}t^m + O(t^{m-1}).$$

Choose $t \gg 1$ and $\sqrt[m]{(M^m)} > s/t > \sqrt[m]{(M^m)} - \epsilon$. Let $D(s,t,x)$ be the zero set of a nonzero section of $m^s \otimes M^t$ and $D(x,\epsilon) = D(s,t,x)/t$. By construction $\operatorname{mult}_x D(x,\epsilon) \geq s/t$, as required. □

The following exercise develops some effective upper bounds for $h^0(X,L)$.

2.15.8 Exercise (Matsusaka estimates).

(2.15.8.1) Recall the Hodge index theorem for surfaces: if S is a proper surface and A, B Cartier divisors on S, B ample, then $(A^2)(B^2) \leq (A \cdot B)^2$.

(2.15.8.2) Let A, B be ample divisors on an irreducible, proper variety X of dimension m. Let $b_j = \log(A^j \cdot B^{m-j})$. Show that $b_{j-1} + b_{j+1} \leq 2b_j$, thus the sequence b_j is concave.

(2.15.8.3) Let A, B be nef divisors on an irreducible, proper variety X of dimension m. Using (2.18.7.2) show that for every $0 \leq i \leq m$,

$$(A^m)^i(B^m)^{m-i} \leq (A^i \cdot B^{m-i})^m \quad \text{and} \quad (A^i \cdot B^{m-i})(B^m)^{i-1} \leq (A \cdot B^{m-1})^i.$$

(2.15.8.4) (Bertini) Let $Z \subset \mathbb{P}^n$ be an irreducible variety which spans $\mathbb{P}^n$. Show that $n + 1 \leq \deg Z + \dim Z$.

(2.15.8.5) Let A, B be nef divisors on an irreducible, proper variety X of dimension m. Assume that B is big and $|A|$ is base point free giving a morphism $q : X \to Z$. Then

$$h^0(X, \mathcal{O}(A)) \leq (A^{\dim Z} \cdot B^{m-\dim Z}) + \dim Z.$$

(2.15.8.6) [Matsusaka72] Let X be an irreducible, normal and proper variety, D a Cartier divisor on X. Assume that $|D|$ gives a map $f : X \dashrightarrow Z \subset \mathbb{P}^m$. Let H be a nef and big Cartier divisor on X. Then

$$h^0(X, \mathcal{O}(D)) \leq \frac{(D \cdot H^{m-1})^{\dim Z}}{(H^m)^{\dim Z - 1}} + \dim Z.$$

Hint. Let $\Gamma \subset X \times Z$ be the closure of the graph of f with projections $p : \Gamma \to X$ and $q : \Gamma \to Z$. Use (2.15.8.5) on Γ with $A = q^*\mathcal{O}(1)$ and $B = p^*H$.

(2.15.8.7) Notation as in (2.15.8.6). Then

$$h^0(X, \mathcal{O}(nH)) \leq (H^m)n^m + \dim X.$$

Asymptotically this is not as good as (2.15.2) but the effective upper bound is frequently very useful.

(2.15.8.8) By looking at the proof carefully, show that (2.15.8.7) is true if H is a $\mathbb{Q}$-Cartier Weil divisor and $|nH|$ gives a birational map. Study cones over rational curves to see that the latter assumption is necessary.

(2.15.8.9) [Kollár-Matsusaka83] Let X be a smooth projective variety of dimension m over a field of characteristic zero and H a nef and big divisor on X. Use (2.15.8.6) to get explicit upper bounds for $H^0(X, \mathcal{O}_X(K_X + nH))$ in terms of $m, (H^m), (K_X \cdot H^{m-1})$. Using the Kodaira vanishing theorem this gives a bound for $\chi(X, \mathcal{O}(tH))$ for negative values of t. Use Lagrange interpolation to show that all the coefficients of $\chi(X, \mathcal{O}(tH))$ can be bounded in terms of $m, (H^m), (K_X \cdot H^{m-1})$.

The strongest explicit bound for the space of global sections of nef and big line bundles is given by the following result. For the proof consult the original papers.

2.15.9 Theorem [Kollár-Matsusaka83; Luo89; Matsusaka91]. *There exists a function $Q(x, y, z)$ with the following property. Let X be a normal, proper variety of dimension m over a field of characteristic zero and H a nef and big Cartier divisor on X. Then*

$$\left| h^0(X, \mathcal{O}_X(nH)) - \frac{(H^m)}{m!} n^m + \frac{(K_X \cdot H^{m-1})}{2(m-1)!} n^{m-1} \right| \\ \leq n^{m-2} Q(m, (H^m), (K_X \cdot H^{m-1})).$$

From the proof one can write down $Q(x, y, z)$ explicitly, though it is rather large.

The following result, or sometimes only the equivalence of (2.16.2) and (2.16.3) is known as Kodaira's lemma. The equivalence with (2.16.1) is usually proved using Kodaira's vanishing theorem, which is applicable only in characteristic zero. The general case was proved by [Fujita83] using fairly delicate arguments.

2.16 Corollary (Kodaira-Fujita lemma). *Let X be a projective scheme of dimension m with an ample divisor H. Let M be a nef Cartier divisor on X. The following are equivalent:*

(2.16.1) $(M^m) > 0$;
(2.16.2) $h^0(X, \mathcal{O}(nM)) > const \cdot n^m$ *for* $n \gg 1$;
(2.16.3) *There is a* $k > 0$ *and an effective divisor* E *such that* $kM \sim H + E$.

Proof. (2.16.1) and (2.16.2) are equivalent by (2.15). Assume (2.16.2). Then

$$h^0(X, \mathcal{O}(nM - H)) \geq h^0(X, \mathcal{O}(nM)) - h^0(H, \mathcal{O}_H(nM)) \\ \geq const \cdot n^m - O(n^{m-1}) > 0.$$

This shows (2.16.3). Finally if (2.16.3) holds, then

$$h^0(X, \mathcal{O}(nkM)) \geq h^0(X, nH) \geq const \cdot n^m \quad \text{for } n \gg 1,$$

which shows (2.16.2). □

The next result shows the promised equivalence of the two definitions of nef.

2.17 Theorem [Kleiman66]. *Let X be a proper scheme and L a line bundle on X. The following are equivalent:*

(2.17.1) $L \cdot C \geq 0$ *for every integral curve* $C \subset X$;
(2.17.2) $(L^{\dim Y} \cdot Y) \geq 0$ *for every integral subscheme* $Y \subset X$.

Proof. Clearly (2.17.2) implies (2.17.1). To see the converse it is sufficient to consider the case when X is integral. By induction on the dimension we may assume that $L^k \cdot Y \geq 0$ if $Y \subsetneq X$.

Assume first in addition that there is line bundle H on X such that H is generated by global sections and $(H^m) > 0$. (For instance, H is very ample.)

2.17.3 Claim. *If $Y \subset X$ has dimension k, then $H^{k-i} \cdot L^i \cdot Y \geq 0$ for every $i < \dim X$.*

Proof. We may assume that Y is integral. If $k - i = 0$, then $\dim Y < \dim X$ and $L^i \cdot Y \geq 0$ by the inductive assumption. By assumption we can write $H|Y = \mathcal{O}(Z)$ where Z is an effective Cartier divisor on Y. If $k - i > 0$, then by (2.5.4) $H^{k-i} \cdot L^i \cdot Y = H^{k-i-1} \cdot L^i \cdot Z$, and we are done by induction on $\dim Y$. □

Consider $L + tH \in \operatorname{Pic}(X) \otimes \mathbb{R}$ for $t \geq 0$. If $C \subset X$ is an integral curve, then $(L+tH) \cdot C = L \cdot C + tH \cdot C \geq 0$. Thus by induction $(L+tH)^k \cdot Y \geq 0$ for $\dim Y < \dim X$. Furthermore,

$$((L+tH)^m) = (L^m) + \sum_{i=1}^{m-1} \binom{m}{i} t^i (H^{k-i} \cdot L^i) + t^m (H^m)$$

is a monotone increasing function of t for $t > 0$ which goes to infinity since $(H^m) > 0$. If $(L^m) < 0$, then there is a $t_0 \geq 0$ such that $((L + t_0 H)^m) = 0$ and $((L+tH)^m) > 0$ for every $t > t_0$.

Choose $t > t_0$ rational. Then $(L + tH)^k \cdot Y \geq 0$ for every Y and $((L + tH)^m) > 0$. Thus by (2.15) there is a positive integer k such that $k(L+tH)$ is linearly equivalent to an effective divisor D. Look at a different expansion:

$$((L+tH)^m) = t^m (H^m) + \sum_{i=1}^{m-1} t^{i-1} H^{i-1} \cdot L \cdot (L+tH)^{m-i}.$$

I claim that in the sum every term is nonnegative. Indeed,

$$H^{i-1} \cdot L \cdot (L+tH)^{m-i} = \frac{1}{k} H^{i-1} \cdot L \cdot (L+tH)^{m-i-1} \cdot D,$$

which can be written as the sum of terms of the form $H^j \cdot L^{m-1-j} \cdot D$, and these are nonnegative by (2.17.3). This gives that

$$0 = ((L+t_0 H)^m) = \lim_{t \to t_0} ((L+tH)^m) \geq \lim_{t \to t_0} t^m (H^m) = t_0^m (H^m).$$

Thus $t_0 = 0$ and $(L^m) = ((L + t_0H)^m) \geq 0$. This settles the case assuming that H exists.

In general by the Chow lemma there is a projective variety X' and a proper birational morphism $p : X' \to X$. By (2.11) $(L^m \cdot X) = (p^*L^m \cdot X')$. On X' we can choose a very ample line bundle H', thus by the already established case

$$(L^m \cdot X) = (p^*L^m \cdot X') \geq 0. \qquad \square$$

2.17.4 Exercise. (2.17.4.1) Let X be an irreducible, projective, m-dimensional variety, L a line bundle on X and H an ample line bundle on X. Use the Hodge index theorem for surfaces to show that

$$L \equiv 0 \Leftrightarrow (H^{m-1} \cdot L) = (H^{m-2} \cdot L^2) = 0.$$

(2.17.4.2) Let $H \subset X$ be an ample divisor, $\dim H \geq 2$. Show that $L \equiv 0 \Leftrightarrow L|H \equiv 0$. Thus the restriction map $\operatorname{Pic}(X) \to \operatorname{Pic}(H)$ is injective.

Using (2.15) it is not hard to prove the basic ampleness criteria. In increasing generality they can be found in [Nakai63, Kleiman66, Moishezon67, Hartshorne70].

2.18 Theorem. *Let X be a proper scheme and H a line bundle on X. The following are equivalent:*

(2.18.1) *H is ample on X;*

(2.18.2) (Nakai–Moishezon criterion) $(H^{\dim Y} \cdot Y) > 0$ *for every integral subscheme $Y \subset X$;*

(2.18.3) (Seshadri criterion) *For every closed point $x \in X$ there is a number $\epsilon(x) > 0$ such that if $x \in C \subset X$ is an integral curve, then $H \cdot C \geq \epsilon(x) \cdot \operatorname{mult}_x C$.*

Proof. Assume that H is ample and let H^k be very ample. We can view X as a subscheme of $\mathbb{P}^N$ such that $H^k = \mathcal{O}_X(1)$. Thus

$$(H^{\dim Y} \cdot Y) = k^{-\dim Y}(\mathcal{O}_X(1)^{\dim Y} \cdot Y) = k^{-\dim Y} \deg Y > 0.$$

Also, $\deg C \geq \operatorname{mult}_x C$ for every $x \in C$, thus we can take $\epsilon(x) = 1/k$ (independent of $x \in X$).

Next assume (2.18.2). H is ample on X iff H is ample on every irreducible component of $\operatorname{red} X$ (see e.g. [Hartshorne77, III.Ex.5.7]). Thus we may assume that X is integral.

By induction we may assume that $H|Z$ is ample for every subscheme $Z \subsetneq X$. By (2.15) H^n has a nontrivial section for some $n > 0$. This gives an exact sequence

$$0 \to H^{-n} \to \mathcal{O}_X \to \mathcal{O}_D \to 0.$$

$H|D$ is ample, thus $H^{nk}|D$ is generated by global sections and $H^1(D, H^{nk}|D) = 0$ for $k \gg 1$. This shows that $H^1(X, H^{nk-n}) \to H^1(X, H^{nk})$ is surjective for $k \gg 1$. Thus there is an $m \gg 1$ such that $H^1(X, H^{nm-n}) \to H^1(X, H^{nm})$ is an isomorphism. This implies that sections of $H^{nm}|D$ lift to sections of H^{nm}. In particular H^{nm} is generated by global sections. Let $f : X \to \mathbb{P}^N$ be the corresponding morphism. f is proper since X is proper. If $C \subset X$ is an integral curve which is contained in a fiber of f, then $(H^{nm} \cdot C) = 0$ which contradicts (2.18.2). Thus f is finite. Therefore $H^{nm} = f^*\mathcal{O}(1)$ is ample.

Finally we show that (2.18.3) implies (2.18.2). Let $y \in Y$ be a closed point. Let $y \in Y_0 \subset Y$ be an affine neighborhood and choose an embedding $Y_0 \subset V \cong \mathbb{A}^k$. Let $p : B_yY \to Y$ be the blow up of $y \in Y$ with exceptional divisor $E \subset B_yY$. There is a natural embedding $B_yY_0 \subset B_yV$. If $\mathbb{P}^{k-1} \cong E' \subset B_yV$ denotes the exceptional divisor, then $E = E' \cap B_yY_0$.

Choose a rational number $0 < \eta < \epsilon(y)$. I claim that $p^*H - \eta E$ is nef on B_yY. Let $D \subset B_yY$ be an integral curve. If $D \subset E$, then $(p^*H - \eta E) \cdot D = \eta \deg D$ where $\deg D$ is the degree of D in the projective space E'. If $D \not\subset E$, then

$$\begin{aligned}(p^*H - \eta E) \cdot D &= H \cdot p(D) - \eta \operatorname{mult}_y p(D) \\ &\geq \begin{cases} (\epsilon(y) - \eta) \operatorname{mult}_y p(D) & \text{if } y \in p(D); \\ H \cdot p(D) & \text{if } y \notin p(D). \end{cases}\end{aligned}$$

Thus $(p^*H - \eta E) \cdot D \geq 0$ always. By (2.17) this implies that

$$(p^*H - \eta E)^{\dim Y} \cdot B_yY \geq 0. \tag{2.18.4}$$

2.18.5 Claim. $(p^*H - \eta E)^{\dim Y} \cdot B_yY = (H^{\dim Y} \cdot Y) - \eta^{\dim Y} \deg E$.

Proof. By (2.12) $p^*H^i \cdot \mathcal{O}(E)^j \cdot B_yY = p^*H^i \cdot \mathcal{O}(E)^{j-1} \cdot E = (p^*H|E)^i \cdot (\mathcal{O}(E) \otimes \mathcal{O}_E)^{j-1} \cdot E$. $p^*H|E \cong \mathcal{O}_E$ and $\mathcal{O}(E) \otimes \mathcal{O}_E \cong \mathcal{O}_{B_yV}(E') \otimes \mathcal{O}_E \cong \mathcal{O}_E(-1)$. Thus

$$\begin{aligned}p^*H^i \cdot \mathcal{O}(E)^j \cdot B_yY &= \mathcal{O}_E^i \cdot \mathcal{O}_E(-1)^{j-1} \cdot E \\ &= \begin{cases} 0 & \text{if } i > 0, j > 0, \text{ and} \\ (-1)^{\dim Y - 1} \deg E & \text{if } i = 0,\ j = \dim Y. \end{cases}\end{aligned}$$

Thus

$$\begin{aligned}(p^*H - \eta E)^{\dim Y} \cdot B_yY &= (p^*H)^{\dim Y} \cdot B_yY + \eta^{\dim Y}(-1)^{2\dim Y - 1} \deg E \\ &= H^{\dim Y} \cdot Y - \eta^{\dim Y} \deg E. \qquad \square\end{aligned}$$

Combining (2.18.4–5) gives that $H^{\dim Y} \cdot Y \geq \eta^{\dim Y} \deg E > 0$. Thus (2.18.3) implies (2.18.2) and we are finished. □

2.18.6 Definition [Demailly92]. Let X be a proper scheme, M a nef line bundle on X and $x \in X$ a closed point. The number

$$\epsilon(M, x) := \inf_{x \in C} \frac{(M \cdot C)}{\operatorname{mult}_x C}$$

is called the *Seshadri constant* of M at x.

2.18.6.1 Exercise. Show that $\epsilon(M, x) > 0$ for some x iff $(M^{\dim X}) > 0$.

2.18.7 Exercises. Let X be a proper scheme of dimension m, H and L_i Cartier divisors on X.

(2.18.7.1) If H is ample and L is nef, then $H + L$ is ample.

(2.18.7.2) If X is projective and L is nef, then there are ample $\mathbb{Q}$-divisors H_i such that $L = \lim H_i$.

(2.18.7.3) If all the L_i are nef, then $(L_1 \cdots L_m \cdot X) \geq 0$.

(2.18.7.4) If $L_1 \equiv 0$ and $L_2, \ldots, L_m$ are arbitrary, then $(L_1 \cdots L_m \cdot X) = 0$.

2.18.8 Exercise. Let $f : X \to Y$ be a morphism between proper schemes.

(2.18.8.1) Let L be an f-ample divisor on X and H an ample divisor on Y. Show that $L + nf^*H$ is ample for $n \gg 1$.

(2.18.8.2) Let $0 \in C$ be a smooth, proper curve, $X = C \times C$ and $f : X \to C$ the second projection. Let $\Delta \subset X$ be the diagonal and $C_2 = \{0\} \times C$. Then $L = \Delta - C_2$ is f-nef but $L + f^*H$ is not nef for any divisor H on C for $g(C) > 0$. (It's square is $-2g(C)$!)

Finally we prove the Kleiman ampleness criterion. In contrast with the criteria in (2.18) it does not hold for an arbitrary proper scheme and it also uses the theorem on the base.

2.19 Theorem [Kleiman66]. *Let X be a proper scheme and H a line bundle on X. Assume that one of the following conditions is satisfied:*

(2.19.1) *X is projective;*

(2.19.2) *X is smooth, or more generally $\mathbb{Q}$-factorial.*

Then H is ample iff $H^\cap$ is a positive linear function on $\overline{NE}(X) - \{0\}$.

Proof. One direction is in (II.4.8). Conversely, assume that $H^\cap$ is a positive linear function on $\overline{NE}(X) - \{0\}$. We check that (2.18.2) is satisfied. Let $Y \subset X$ be an integral subscheme. By induction we may assume that if $Z \subsetneq Y$ is a subscheme, then $(H^{\dim Z} \cdot Z) > 0$.

As a first step, we find an effective Cartier divisor M_Y such that $c_1(M_Y) \cdot \mathcal{O}_Y = \mathcal{O}_Z$ where $Z \subset Y$ is a suitable nonempty subscheme. If X is projective, then a general very ample divisor will do by (2.5.4). If X is $\mathbb{Q}$-factorial, then pick any affine set $U \subset X$ which intersects Y. Take a general very ample divisor $M_U \subset U$ which intersects $Y \cap U$ but does not contain it. The closure of M_U in X is a Weil divisor, hence some multiple of it is Cartier. That is our M_Y.

By (II.4.8) $H - \eta M_Y$ is nonnegative on $\overline{NE}(X)$ for $1 \gg \eta > 0$, thus $H - \eta M_Y$ is nef. Thus by (2.18.7.3)

$$((H - \eta M_Y) \cdot H^{\dim Y - 1} \cdot Y) \geq 0.$$

Therefore

$$(H^{\dim Y} \cdot Y) \geq \eta(M_Y \cdot H^{\dim Y-1} \cdot Y) = \eta(H^{\dim Y-1} \cdot Z) > 0.$$

Thus H is ample by (2.18). □

2.19.3 Exercise. If X is a smooth, proper algebraic space then the above proof does not work since not all points have affine Zariski open neighborhoods. Let $X_1 \subset \mathbb{P}^4$ be a general quintic threefold with a single node. Blow up the node to obtain $X_2 \to X_1$ with exceptional divisor $E \cong \mathbb{P}^1 \times \mathbb{P}^1$. Show that there is a smooth algebraic space X and a morphism $X_2 \to X$ which is an isomorphism outside E and induces a coordinate projection on E. Let $\mathbb{P}^1 \cong C \subset X$ be the exceptional curve of $p : X \to X_1$. By [Ceresa-Collino83] C is numerically zero on X. $H = p^*\mathcal{O}_{X_1}(1)$ is positive on $\overline{NE}(X) - \{0\}$ but it is not ample.

2.20 Exercise. As a consequence of (2.19), prove the following conjecture of Chevalley, settled in [Kleiman66]:

Let X be a proper, $\mathbb{Q}$-factorial variety with Picard number $\rho(X) = r$. Assume that for any $r+1$ points $p_0, \ldots, p_r \in X$, there is an open affine subset $U \subset X$ containing $p_0, \ldots, p_r$. Then X is projective.

The proof is in several steps:

(2.20.1) If X is not projective then there are $0 \neq x, y \in \overline{NE}(X)$ such that $x + y = 0$.

(2.20.2) Choose $x^i \in NE(X)$ such that $\lim x^i = x$. Use (II.4.10.4) to find a sequence of irreducible curves C_{i0} and positive numbers a_{i0} such that $\lim a_{i0}[C_{i0}] = x_0 \neq 0$ and $x = x_0 + x_0'$ where $x_0' \in \overline{NE}(X)$.

(2.20.3) Apply (II.4.10.4) to conclude that there are irreducible curves C_{ij} and constants a_{ij} such that

$$\lim_{i\to\infty} \sum_{j=1}^{r} a_{ij}[C_{ij}] = x_0' + y.$$

(2.20.4) By repeatedly replacing C_{ij} with a subsequence, show that there are closed irreducible subsets $Z_0, \ldots, Z_r \subset X$ such that:

$C_{ij} \subset Z_j$ for every i, and
if $Z_j' \subsetneq Z_j$ then Z_j' contains only finitely many of the curves C_{ij}.

(2.20.5) By assumption there is an open affine subset $U \subset X$ which intersects each of the Z_j. Let $D = X - U$; it is a $\mathbb{Q}$-Cartier divisor.

(2.20.6) Show that if $C \subset X$ is an irreducible curve, $C \not\subset D$ and L is any line bundle on X then $(L(nD) \cdot C) \geq 0$ for $n \gg 1$.

(2.20.7) Conclude that if L is any line bundle on X then $(L(nD) \cdot x_0) = 0$ for $n \gg 1$, hence also $(L \cdot x_0) = 0$. This means that $x_0 = 0$, a contradiction.

References

[Abhyankar56] S. S. Abhyankar: On the valuations centered in a local domain. Amer. J. Math. **78** (1956) 321–348

[Altman-Kleiman77] A. Altman, S. Kleiman: Foundations of the theory of Fano schemes. Comp. Math. **34** (1977) 3–47

[Andreotti-Bombieri69] A. Andreotti, E. Bombieri: Sugli omeomorfismi delle varietá algebriche. Ann. Sci. Norm. Sup. Pisa **23** (1969) 430–450

[Andreotti-Noguet67] A. Andreotti, F. Noguet: La convexité holomorphe dans l'espace analytique des cycles d'une variété algébrique. Ann. Sci. Norm. Sup. Pisa **21** (1967) 31–82

[Angéniol81] B. Angéniol: Familles de Cycles Algébrique – Schéma de Chow. Lecture Notes in Mathematics, vol. 896. Springer, Berlin Heidelberg New York 1981

[Angehrn-Siu95] U. Angehrn, Y.-T. Siu: Effective Freeness and Point Separation for Adjoint Bundles. Invent. Math. **122** (1995) 291–308

[Artin62] M. Artin: Some numerical criteria for contractibility of curves on algebraic surfaces. Amer. J. Math. **84** (1962) 485–496

[Artin69a] M. Artin: Algebraic approximation of structures over complete local rings. Publ. Math. IHES **36** (1969) 23–58

[Artin69b] M. Artin: Algebraisation of formal moduli I. In: D. C. Spencer and S. Iyanaga (eds.), Global Analysis. Univ. Tokyo Press - Princeton Univ. Press 1969, pp. 21–72

[Artin76] M. Artin: Deformations of Singularities. Tata Lecture Notes, vol. 54, 1976

[Artin86] M. Artin: Néron Models. In: G. Cornell and J. Silverman (eds.), Arithmetic Geometry. Springer, Berlin Heidelberg New York 1986, pp. 213–230

[Artin-Mumford72] M. Artin, D. Mumford: Some elementary examples of uniruled varieties which are not rational. Proc. London. Math. Soc. **25** (1972) 75–95

[Ballico93] E. Ballico: On singular varieties with ample tangent bundle. Indag. Math. (1993) 1–10

[Barlet75] D. Barlet: Espace analytique réduit des cycles analytiques complexes de dimension finie. In: Séminaire Norguet, Lecture Notes in Mathematics, vol. 482. Springer, Berlin Heidelberg New York 1975, pp. 1–158

[Barth-Peters83] W. Barth, C. Peters: Automorphisms of Enriques surfaces. Inv. Math. **37** (1983) 383–411

[Barth-V.d.Ven78] W. Barth, A. Van de Ven: Fano-varieties of lines on hypersurfaces. Arch. Math. **31** (1978) 96–104

[BPV84] W. Barth, C. Peters, A. Van de Ven: Compact Complex Surfaces. Springer, Berlin Heidelberg New York 1984

[Batyrev82] V. V. Batyrev: Boundedness of the degree of multidimensional Fano varieties. Vestnik M. G. U. (1982) pp. 22–27 (in Russian)

[Beauville77] A. Beauville: Variété de Prym et Jacobiennes intermédiaire. Ann. Sci. E.N.S. **10** (1977) 309–391

[BDW96] A. Bertram, G. Daskalopoulos, R. Wentworth: Gromov invariants for holomorphic maps from Riemann surfaces to Grassmannians. J. AMS **9** (1996) 529–571

[BBK96] F. Bien, A. Borel, J. Kollár: Rationally connected homogeneous spaces. Invent. Math. **124** (1996) 103–127

[Bingener87] J. Bingener: Lokale Modulräume in der analytischen Geometrie I–II. Viehweg 1987

[Blanchard56] M. A. Blanchard: Sur les variétés analytiques complexes. Ann. Sci. Ec. Norm. Sup. **73** (1956) 157–202

[Blass-Levine82] P. Blass, M. Levine: Families of Zariski surfaces. Duke Math. J. **49** (1982) 129–136

[Bloch-Srinivas83] S. Bloch, V. Srinivas: Remarks on correspondences and algebraic cycles. Amer. J. Math. **105** (1983) 1235–1253

[Borel69] A. Borel: Linear algebraic groups. Benjamin 1969

[Borevich-Shafarevich64] Z. I. Borevich, I. R. Shafarevich: Number Theory. Nauka (1964) (in Russian)

[Bott57] R. Bott: Homogeneous vector bundles. Ann. Math. **66** (1957) 203–248

[Buchsbaum-Eisenbud74] D. Buchsbaum, D. Eisenbud: Some structure theorems for finite free resolutions. Adv. in Math. **12** (1974) 84–139

[Campana81] F. Campana: Coréduction algébrique d'un espace analytic faiblements Kählerien compact. Inv. Math. **63** (1981) 187–223

[Campana91] F. Campana: Une version géométrique généralisée du théorème du produit de Nadel. Bull. Soc. Math. France **119** (1991) 479–493

[Campana92] F. Campana: Connexité rationnelle des variétés de Fano. Ann. Sci. E.N.S. **25** (1992) 539–545

[Campana-Peternell91] F. Campana, T. Peternell: Projective manifolds whose tangent bundles are numerically effective. Math. Ann. **289** (1991) 169–187

[Campana-Peternell94] F. Campana, T. Peternell: Cycle spaces. In: H. Grauert, Th. Peternell, R. Remmert (eds.), Several Complex Variables VII. Enc. of Math. Sci., vol. 74. Springer, Berlin Heidelberg New York 1994, pp. 319–350

[Cartier60] P. Cartier: Sur un théorème de Snapper. Bull. Soc. Math. France **88** (1960) 333–343

[Catanese92] F. Catanese: Chow Varieties, Hilbert Scemes and Moduli Spaces of Surfaces of General Type. J. Alg. Geom. **1** (1992) 561–596

[Cayley1860] A. Cayley: A new analytic representation of curves in space. Quarterly J. Math. **3** (1860) 225–236

[Cayley1862] A. Cayley: A new analytic representation of curves in space. Quarterly J. Math. **5** (1862) 81–86

[Ceresa-Collino83] G. Ceresa, A. Collino: Some remarks on algebraic equivalence of cycles. Pacific J. Math. **105** (1983) 285–290

[Chevalley35] C. Chevalley: Démonstration d'une hypothèse de E. Artin. Abh. Math. Sem. Hansischen Univ. **11** (1935) 73

[Cho-Miyaoka98] K. Cho, Y. Miyaoka: A characterization of projective spaces in terms of the minimum degrees of rational curves (preprint) 1998

[Chow-v.d.Waerden37] W.-L. Chow, B. L. van der Waerden: Zur algebraischen Geometrie, IX. Math. Ann. **113** (1937) 692–704

[Clemens83] H. Clemens: Homological equivalence, modulo algebraic equivalence is not finitely generated. Publ. Math. IHES **58** (1983) 231–250

[Clemens-Griffiths72] H. Clemens, P. Griffiths: The intermediate Jacobian of the cubic threefold. Ann. Math. **95** (1972) 281–356

[CKM88] H. Clemens, J. Kollár, S. Mori: Higher Dimensional Complex Geometry. Astérisque **166** (1988)

[Collino79] A. Collino: Lines on quartic threefolds. J. London. Math. Soc. **19** (1979) 257–267

[Colliot-Thélène86] J.-L. Colliot-Thélène: Arithmétique des variétés rationnelles et problèmes birationnelles. Proc. Int. Congr. Math. 1986, pp. 641–653

[Corti95] A. Corti: Factoring birational maps of threefolds after Sarkisov. J. Alg. Geom. **4** (1995) 223–254

[Debarre97] O. Debarre: Variétés de Fano. Astérisque Exposé 827, Séminaire Bourbaki 1996/97 vol. 245, 1997, pp. 197–221

[Demailly92] J.-P. Demailly: Singular Hermitian metrics on positive line bundles. In: Complex Algebraic Varieties, Lecture Notes in Mathematics, vol. 1507. Springer, Berlin Heidelberg New York 1992, pp. 87–104

[Demailly93] J. P. Demailly: A numerical criterion for very ample line bundles. J. Diff. Geom. **37** (1993) 323–374

[DPS94] J. P. Demailly, T. Peternell, M. Schneider: Compact complex manifolds with numerically effective tangent bundles. J. Alg. Geom. **3** (1994) 295–345

[Demazure80] M. Demazure: Surfaces de del Pezzo I–V. In: Séminaire sur les Singularités des Surfaces, Lecture Notes in Mathematics, vol. 777. Springer, Berlin Heidelberg New York 1980, pp. 21–69

[Dolgachev82] I. Dolgachev: Weighted projective spaces. Group actions and vector fields. Lecture Notes in Mathematics, vol. 956. Springer, Berlin Heidelberg New York 1982

[Dolgachev-Hu98] I. Dolgachev, Y. Hu: Variation of geometric invariant theory quotient. Publ. Math. IHES **87** (1998) 5–56

[Douady66] A. Douady: Le problème des modules pour les sous espaces analytiques d'un espace analytique donné. Ann. Inst. Fourier **16** (1966) 1–95

[Douady74] A. Douady: Le problème des modules locoux pour les espaces C-analytiques compactes. Ann. Sci. Éc. Norm. Sup. **17** (1974) 569–602

[Drezet-Narasimhan89] J.-M. Drezet, M. S. Narasimhan: Groupe de Picard des variétés de fibrés semistables sur les courbes algébriques. Inv. Math. **97** (1989) 53–94

[EGA] A. Grothendieck, J. Dieudonné: Eléments de Gémétrie Algébrique. Publ. Math. IHES **4, 8, 11, 17, 20, 24, 28, 32** (1960–67)

[Ein86] L. Ein: Hilbert Scheme of Smooth Space Curves. Ann. Sci. E. N. S. **19** (1986) 469–478

[Ein88] L. Ein: Subvarieties of generic complete intersections. Inv. Math. **94** (1988) 163–170

[EKL95] L. Ein, O. Küchle, R. Lazarsfeld: Local positivity of ample line bundles. J. Diff. Geom. **42** (1995) 193–219

[Ekedahl88] T. Ekedahl: Canonical models of surfaces of general type in positive characteristic. Publ. Math. IHES **67** (1988) 97–144

[Esnault-Viehweg92] H. Esnault, E. Viehweg: Lectures on vanishing theorems. In: DMV Semnar, vol. 20, Birkhäuser 1992

[Fletcher89] A. Fletcher: Working with weighted complete intersections. MPI 1989 (preprint)

[Frankel61] T. Frankel: Manifolds with positive curvature. Pacific J. Math. **11** (1961) 165–174

[Fujiki81] A. Fujiki: Deformation of uni-ruled manifolds. Publ. RIMS Kyoto Univ. **17** (1981) 687–702

[Fujiki82] A. Fujiki: On the Douady space of a compact complex space in the category C. Nagoya Math. J. **85** (1982) 189–211

[Fujiki84] A. Fujiki: On the Douady space of a compact complex space in the category C, II. Publ. RIMS Kyoto Univ. **20** (1984) 461–489

[Fujita83] T. Fujita: Semipositive line bundles. J. Fac. Sci. Tokyo Univ. **30** (1983) 353–378

[Fujita90] T. Fujita: Classification theories of polarized varieties. Cambridge University Press, Cambridge 1990

[Fulton84] W. Fulton: Intersection Theory. Springer, Berlin Heidelberg New York 1984

[FMSS95] W. Fulton, R. MacPherson, F. Sottile, B. Sturmfels: Intersection theory on spherical varieties. J. Alg. Geom. **4** (1995) 181–194

[Green-Morrison86] M. Green, I. Morrison: The equations defining Chow varieties. Duke Math. J. **53** (1986) 733–747

[Griffiths-Harris78] P. Griffiths, J. Harris: Principles of Algebraic Geometry. John Wiley, New York 1978

[Grothendieck62] A. Grothendieck: Fondéments de la Géométrie Algébrique. Sec. Math. Paris (1962)

[Grothendieck68] A. Grothendieck: Cohomologie Locale des Faisceaux Cohérent et Théorèmes de Lefschetz Locaux et Globaux - SGA 2. North Holland, Amsterdam 1968

[Hartshorne70] R. Hartshorne: Ample subvarieties of Algebraic Varieties. Lecture Notes in Mathematics, vol. 156. Springer, Berlin Heidelberg New York 1970

[Hartshorne77] R. Hartshorne: Algebraic Geometry. Springer, Berlin Heidelberg New York 1977

[Hironaka60] H. Hironaka: On the theory of birational blowing-up. Thesis, 1960 (unpublished)

[Hironaka64] H. Hironaka: Resolution of singularities of an algebraic variety over a field of characteristic zero. Ann. of Math **79** (1964) 109–326

[Hodge-Pedoe52] W. Hodge, D. Pedoe: Methods of Algebraic Geometry. Cambridge University Press, Cambridge 1952

[Hwang-Mok99] J.-M. Hwang, N. Mok: Holomorphic Maps from Rational Homogeneous Spaces. Invent. Math. (to appear) 1999

[Illusie71] L. Illusie: Complex cotangent et déformations, I–II. Lecture Notes in Mathematics, vols. 239 and 283. Springer, Berlin Heidelberg New York 1971 and 1972

[Ionescu86] P. Ionescu: Generalized adjunction and applications. Math. Proc. Cambridge Phil. Soc. **99** (1986) 457–442

[Iskovskikh77] V. A. Iskovskikh: Fano 3-folds, I. Math. USSR Izv. **11** (1977) 485–527

[Iskovskikh78] V. A. Iskovskikh: Fano 3-folds, II. Math. USSR Izv. **12** (1978) 469–506

[Iskovskikh80a] V. A. Iskovskikh: Anticanonical models of three-dimensional algebraic varieties. J. Soviet Math **13** (1980) 745–814

[Iskovskikh80b] V. A. Iskovskikh: Birational automorphisms of three-dimensional algebraic varieties. J. Soviet Math **13** (1980) 815–868

[Iskovskikh80c] V. A. Iskovskikh: Minimal models of rational surfaces over arbitrary fields. Math. USSR Izv. **14** (1980) 17–39

[Iskovskikh83] V. A. Iskovskikh: Algebraic Threefolds with Special Regard to the Problem of Rationality. Proc. ICM, Warszawa, 1983, pp. 733–746

[Iskovskikh87] V. A. Iskovskikh: On the rationality problem for conic bundles. Duke Math. J. **54** (1987) 271–294

[Iskovskikh-Manin71] V. A. Iskovskikh, Yu. I. Manin: Three-dimensional quartics and counterexamples to the Lüroth problem. Math. USSR Sbornik **15** (1971) 141–166

[Johnston89] B. Johnston: The uniform bound problem for local birational nonsingular morphisms. Trans. A.M.S. **312** (1989) 421–431

[KaMaMa87] Y. Kawamata, K. Matsuda, K. Matsuki: Introduction to the Minimal Model Problem. In: T. Oda (ed.), Algebraic Geometry. Sendai Adv. Stud. Pure Math., vol. 10., Kinokuniya - North-Holland 1987, pp. 283–360

[Kawamata84] Y. Kawamata: The cone of curves of algebraic varieties. Ann. of Math. **119** (1984) 603–633

[Kawamata91] Y. Kawamata: On the length of an extremal rational curve. Inv. Math. **105** (1991) 609–611

[Kawamata92] Y. Kawamata: Boundedness of $\mathbb{Q}$-Fano threefolds. Proc. Int. Conf. Algebra. Contemp. Math. **131** (1992) 439–445

[Keel-McKernan99] S. Keel, J. McKernan: Rational curves on quasi-projective surfaces. Mem. Amer. Math. Soc. (to appear) 1999

[Kleiman66] S. Kleiman: Toward a numerical theory of ampleness. Ann. of Math. **84** (1966) 293–344

[Kleiman74] S. Kleiman: An intersection theory for divisors. 1974 (Manuscript)

[Kobayashi-Ochiai70] S. Kobayashi, T. Ochiai: On complex manifolds with positive tangent bundles. J. Math. Soc. Japan **22** (1970) 499–525

[Kodaira86] K. Kodaira: Complex manifolds and deformations of complex structures. Springer, Berlin Heidelberg New York 1986, pp. 43–76

[Kodaira-Spencer58] K. Kodaira, D. C. Spencer: On deformations of complex analytic structures, I–II. Ann. Math. **67** (1958) 328–466; III. Ann. Math. **71** (1960) 43–76

[Kollár81] J. Kollár: Fano varieties of large index. Vestnik M.G.U. (1981) 31–34 (in Russian)

[Kollár86a] J. Kollár: Higher direct images of dualizing sheaves I. Ann. of Math **123** (1986) 11–42

[Kollár86b] J. Kollár: Higher direct images of dualizing sheaves, II. Ann. of Math **124** (1986) 171–202

[Kollár87a] J. Kollár: Subadditivity of the Kodaira dimension: Fibers of general type. Advanced Studies in Pure Math. **10**, Algebraic Geometry Sendai **85**. Kinokuniya-North Holland 1987, pp. 361–398

[Kollár87b] J. Kollár: The structure of algebraic threefolds – an introduction to Mori's program. Bull. AMS **17** (1987) 211–273

[Kollár89] J. Kollár: Flops. Nagoya Math. J. **113** (1989) 14–36

[Kollár90] J. Kollár: Minimal Models of Algebraic Threefolds: Mori's Program. Astérisque **177-178** (1990) 303–326

[Kollár91a] J. Kollár: Extremal Rays on Smooth Threefolds. Ann. Sci. ENS **24** (1991) 339–361

[Kollár91b] J. Kollár: Flips, Flops, Minimal Models, etc. Surv. in Diff. Geom. **1** (1991) 113–199

[Kollár91c] J. Kollár: Cone Theorems and Cyclic Covers. In: ICM-90 Satellite Conference Proceedings. Springer, Berlin Heidelberg New York 1991, pp. 101–110

[Kollár91d] J. Kollár: Flip and Flop. Proc. Int. Congr. Math. Kyoto 1990. Springer, Berlin Heidelberg New York 1991, pp. 709–714

[Kollár92] J. Kollár: Cone Theorems and Bug-eyed Covers. J. Alg. Geom. **1** (1992) 293–323

[Kollár93] J. Kollár: Effective Base Point Freeness. Math. Ann. **296** (1993) 595–605

[Kollár95a] J. Kollár: Nonrational hypersurfaces. J. AMS **8** (1995) 241–249

[Kollár95b] J. Kollár: Flatness criteria. J. Algebra **175** (1995) 715–727

[Kollár97] J. Kollár: Singularities of pairs. In: Algebraic geometry – Santa Cruz 1995. Proc. Sympos. Pure Math., vol. 62, Amer. Math. Soc. 1997, pp. 221–287

[Kollár98] J. Kollár: Low degree polynomial equations. In: European Congress of Math. Birkhäuser 1998, pp. 255–288

[Kollár et al. 92] J. Kollár (with 14 coauthors): Flips and Abundance for Algebraic Threefolds, Vol. 211. Astérisque, 1993

[Kollár-Matsusaka83] J. Kollár, T. Matsusaka: Riemann-Roch type inequalities. Amer. J. Math. **105** (1983) 229–252

[KoMiMo92a] J. Kollár, Y. Miyaoka, S. Mori: Rational Curves on Fano Varieties. In: Proc. Alg. Geom. Conf. Trento, Lecture Notes in Mathematics, vol. 1515. Springer, Berlin Heidelberg New York 1992, pp. 100–105

[KoMiMo92b] J. Kollár, Y. Miyaoka, S. Mori: Rationally Connected Varieties. J. Alg. Geom. **1** (1992) 429–448

[KoMiMo92c] J. Kollár, Y. Miyaoka, S. Mori: Rational Connectedness and Boundedness of Fano Manifolds. J. Diff. Geom. **36** (1992) 765–769

[Kollár-Mori92] J. Kollár, S. Mori: Classification of three dimensional flips. J. AMS **5** (1992) 533–703

[Kollár-Mori98] J. Kollár, S. Mori: Birational geometry of algebraic varieties. Cambridge Univ. Press 1998

[Kontsevich-Manin94] M. Kontsevich, Yu. I. Manin: Gromov-Witten classes, quantum cohomology and enumerative geometry. Comm. Math. Phys. **164** (1994) 525–562

[Kovács94] S. Kovács: The cone of curves of a $K3$ surface. Math. Ann. **300** (1994) 681–691

[Lang52] S. Lang: On quasi algebraic closure. Ann. Math. **55** (1952) 373–390

[Lang86] S. Lang: Hyperbolic and Diophantine analysis. Bull. AMS **14** (1986) 159–205

[LangW79] W. Lang: Quasi-elliptic surfaces in characteristic 3. Ann. Sci. ENS **12** (1979) 473–500

[Lauritzen96] N. Lauritzen: Embeddings of homogeneous spaces in prime characteristics. Amer. J. Math. **118** (1996) 377–387

[Lazarsfeld84] R. Lazarsfeld: Some applications of the theory of positive vector bundles. Complete Intersections (ed. S. Greco, R. Strano). Lecture Notes in Mathematics, vol. 1092. Springer, Berlin Heidelberg New York 1984, pp. 29–61

[Levine81a] M. Levine: Deformations of uni-ruled varieties. Duke Math. J. **48** (1981) 467–473

[Levine81b] M. Levine: Some examples from the deformation theory of ruled varieties. Amer. J. Math. **103** (1981) 997–1020

[Luo89] T. Luo: Riemann–Roch type inequalities for nef and big divisors. Amer. J. Math. **111** (1989) 457–487

[Manin66] Yu. I. Manin: Rational surfaces over perfect fields. Publ. Math. IHES **30** (1966) 55–114

[Manin72] Yu. I. Manin: Cubic forms. Nauka (1972) (in Russian)

[Manin-Tsfasman86] Yu. I. Manin, M. A. Tsfasman: Rational varieties: algebra, geometry and arithmetic. Russian Math. Surveys **41** (1986) 51–116

[Matsuki87] K. Matsuki: A criterion for the canonical bundle of a 3-fold to be ample. Math. Ann. **276** (1987) 557–564

[Matsumura80] H. Matsumura: Commutative Algebra. Benjamin – Cummings, 1980 (second edition)

[Matsumura86] H. Matsumura: Commutative Ring Theory. Cambridge University Press, Cambridge 1986

[Matsusaka68] T. Matsusaka: Algebraic deformations of polarized varieties. Nagoya Math. J. **31** (1968) 185–245

[Matsusaka70] T. Matsusaka: On canonically polarised varieties (II). Amer. J. Math. **92** (1970) 283–292

[Matsusaka72] T. Matsusaka: Polarised varieties with a given Hilbert polynomial. Amer. J. Math. **94** (1972) 1027–1077

[Matsusaka86] T. Matsusaka: On polarized normal varieties, I. Nagoya Math. J. **104** (1986) 175–211

[Matsusaka91] T. Matsusaka: On numerically effective divisors with positive self-intersection number. J. Fac. Sci. Univ. Tokyo **38** (1991) 165–183

[Matsusaka-Mumford64] T. Matsusaka, D. Mumford: Two fundamental theorems on deformations of polarised varieties. Amer. J. Math. **86** (1964) 668–684
[Mazur86] B. Mazur: Arithmetic on curves. Bull. AMS **14** (1986) 207–259
[Megyesi98] G. Megyesi: Fano threefolds in positive characteristic. J. Algebraic Geom. **7** (1998) 207–218
[Mella99] M. Mella: Existence of good divisors on Mukai varieties. J. Alg. Geom. (to appear) 1999
[Miyaoka87] Y. Miyaoka: The Chern Classes and Kodaira Dimension of a Minimal Variety. Advanced Studies in Pure Math. 10, Algebraic Geometry Sendai 85. Kinokuniya-North Holland, 1987, pp. 449–476
[Miyaoka88a] Y. Miyaoka: On the Kodaira dimension of minimal threefolds. Math. Ann. **281** (1988) 325–332
[Miyaoka88b] Y. Miyaoka: Abundance conjecture for threefolds: $\nu = 1$ case. Comp. Math. **68** (1988) 203–220
[Miyaoka92] Y. Miyaoka: (Lecture at the University of Utah) 1992
[Miyaoka-Mori86] Y. Miyaoka, S. Mori: A Numerical Criterion for Uniruledness. Ann. of Math. **124** (1986) 65–69
[Moishezon67] B. Moishezon: A criterion for projectivity of complete abstract algebraic varieties. Amer. Math. Soc. Transl. **63** (1967) 1–50
[Mok88] N. Mok: The uniformisation theorem for compact Kähler manifolds of nonnegative holomorphic bisectional curvature. J. Diff. Geom. **27** (1988) 179–214
[Mori75] S. Mori: On a generalization of complete intersections. J. Math. Kyoto Univ. **15** (1975) 619–646
[Mori79] S. Mori: Projective Manifolds with Ample Tangent Bundles. Ann. of Math. **110** (1979) 593–606
[Mori82] S. Mori: Threefolds whose Canonical Bundles are not Numerically Effective. Ann. of Math. **116** (1982) 133–176
[Mori83] S. Mori: Cone of Curves and Fano 3-folds. Proc. ICM, Warszawa 1983, pp. 747–752
[Mori87] S. Mori: Classification of higher-dimensional varieties. Algebraic Geometry Bowdoin 1985, Proc. Symp. Pure Math., vol. 46, 1987, pp. 269–332
[Mori88] S. Mori: Flip theorem and the existence of minimal models for 3-folds. J. AMS **1** (1988) 117–253
[Mori89] S. Mori: Birational Classification of Algebraic 3-folds. Algebraic Analysis, Geometry, and Number Theory, Proc. of JAMI Inaugural Conf. Johns Hopkins University Press, 1989, pp. 307–311
[Mori91] S. Mori: Birational Classification of Algebraic Threefolds. Proc. ICM Kyoto, Springer, Berlin Heidelberg New York 1991, pp. 234–248
[Mori-Mukai81] S. Mori, S. Mukai: Classification of Fano threefolds with $B_2 \geq 2$. Manuscr. Math. **36** (1981) 147–162
[Mori-Mukai83] S. Mori, S. Mukai: On Fano threefolds with $B_2 \geq 2$. In: S. Iitaka (ed.), Algebraic Varieties and Analytic Varieties, Adv. Stud. Pure Math., vol. 1. Kinokuniya and North-Holland, 1983, pp. 101–129
[Mukai88] S. Mukai: Problems on characterization of the complex projective space. Birational geometry of algebraic varieties, Open problems (Katata), 1988, pp. 57–60
[Mukai89] S. Mukai: Biregular classification of Fano 3-folds and Fano manifolds of coindex 3. Proc. Nat. Acad. Sci. U.S.A. **86** (1989) 3000–3002
[Mumford66] D. Mumford: Lectures on curves on an algebraic surface. Princeton University Press, 1966
[Mumford68] D. Mumford: Abelian Varieties. Tata Inst. Lecture Notes, 1968
[Mumford76] D. Mumford: Algebraic Geometry I. Springer, Berlin Heidelberg New York 1976

[Mumford-Fogarty82] D. Mumford, J. Fogarthy: Geometric invariant theory, (2nd edition). Springer, Berlin Heidelberg New York 1982

[Nadel90] A. M. Nadel: A finiteness theorem for Fano 4-folds. (Unpublished)

[Nadel91] A. M. Nadel: The boundedness of degree of Fano varieties with Picard number one. J. AMS **4** (1991) 681–692

[Nagata55] M. Nagata: On the normality of the Chow variety of positive 0-cycles of degree m in an algebraic variety. Mem. Coll. Sci. Univ. Kyoto **29** (1955) 165–176

[Nakai63] Y. Nakai: A criterion of ample sheaf on a projective scheme. Amer. J. Math. **85** (1963) 14–26

[Nishimura55] H. Nishimura: Some remarks on rational points. Mem. Coll. Sci. Univ. Kyoto **29** (1955) 189–192

[Paranjape94] K. Paranjape: Cohomological and cycle theoretic connectivity. Ann. of Math. **139** (1994) 641–660

[Paranjape-Srinivas92] K. Paranjape, V. Srinivas: Unirationality of complete intersections. Flips and Abundance for Algebraic Threefolds. Astérisque **211** (1992) 241–247

[Pourcin69] G. Pourcin: Théorèmes de Douady audessus de S. Ann. Sci. E.N.S. Pisa **23** (1969) 451–459

[Pukhlikov87] A. V. Pukhlikov: Birational isomorphisms of four dimensional quintics. Inv. Math. **87** (1987) 303–329

[Pukhlikov89] A. V. Pukhlikov: Birational automorphisms of a double space and a double quadric. Math. USSR. Izv. **32** (1989) 233–243

[Queen71] C. Queen: Non-conservative function fields of genus one, I. Arch. Math. Basel **22** (1971) 612–623; II. Arch. Math. Basel **23** (1972) 30–37

[Ramanan73] S. Ramanan: The moduli spaces of vector bundles over an algebraic curve. Math. Ann. **200** (1973) 69–84

[Ramanathan85] A. Ramanathan: Schubert varieties are arithmetically Cohen–Macaulay. Inv. Math. **80** (1985) 283–294

[Ramero90] L. Ramero: Effective bounds for Unirationality of Complete Intersections. Manu. Math. **68** (1990) 435–445

[Ran93] Z. Ran: Hodge theory and the Hilbert scheme. J. Diff. Geom. **37** (1993) 191–198

[Raynaud78] M. Raynaud: Contre-examples au "Vanishing theorem" en caracteristique $p > 0$. C. P. Ramanujam – a tribute. Springer - Tata Inst., 1978, pp. 273–278

[Raynaud-Gruson71] M. Raynaud, L. Gruson: Critères de platitudes et de projectivité. Inv. Math. **13** (1971) 1–89

[Reid80] M. Reid: Canonical Threefolds. In: A. Beauville (ed.), Géométrie Algébrique Angers. Sijthoff & Noordhoff, 1980, pp. 273–310

[Reid87] M. Reid: Young person's guide to canonical singularities. Algebraic Geometry Bowdoin 1985, Proc. Symp. Pure Math., vol. 46, 1987, pp. 345–416

[Reid94] M. Reid: Nonnormal Del Pezzo surfaces. Publ. RIMS Kyoto Univ. (1994) 695–728

[Remmert-V.d.Ven61] R. Remmert, A. Van de Ven: Über holomorphe Abbildung projektiv-algebraischer Mannigfaltigkeiten auf komplexe Räume. Math. Ann. **142** (1961) 453–486

[Rojtman80] A. A. Rojtman: The torsion of the group of 0-cycles modulo rational equivalence. Ann. Math. **111** (1980) 553–569

[Rosenlicht56] M. Rosenlicht: Some basic theorems on algebraic groups. Amer. J. Math. **78** (1956) 401–443

[Roth55] L. Roth: Algebraic Threefolds. Springer, Berlin Göttingen Heidelberg, 1955

[Sacks-Uhlenbeck81] J. Sacks, K. Uhlenbeck: The existence of minimal immersions of 2-spheres. Ann. Math. **113** (1981) 1–24

[Samuel55] P. Samuel: Méthodes d'Algèbre Abstraite en Géométrie Algébrique. Springer, Berlin Heidelberg New York 1955

[Sarkisov81] V. G. Sarkisov: Birational automorphisms of conic bundles. Math. USSR Izv. **17** (1981) 177–202

[Sarkisov82] V. G. Sarkisov: On the structure of conic bundles. Math. USSR Izv. **120** (1982) 355–390

[Sato93] E. Sato: A criterion of uniruledness in positive characteristic. Tohoku Math. J. **45** (1993) 447–460

[Schoen93] C. Schoen: On Hodge structures and non-representability of Chow groups. Comp. Math. (1993) 285–316

[Segre50] B. Segre: Questions arithmétiques sur les variétés algèbriques. Algèbre et Théorie des Nombres, CNRS, 1950, pp. 83–91

[Serrano95] F. Serrano: Strictly nef divisors and Fano threefolds. J. f.r.u.a. Math. **464** (1995) 187–206

[Serre64] J.-P. Serre: Cohomologie Galoisienne. Lecture Notes in Mathematics, vol. 5. Springer, Berlin Heidelberg New York 1964

[Serre79] J.-P. Serre: Local fields. Springer, Berlin Heidelberg New York 1979

[SGA1] A. Grothendieck: Revêtements Etales et Groupes Fondamental. Lecture Notes in Mathematics, vol. 224. Springer, Berlin Heidelberg New York 1971

[Shafarevich72] R. I. Shafarevich: Basic Algebraic Geometry. Nauka, 1972 (in Russian)

[Shepherd-Barron92] N. I. Shepherd-Barron: The rationality of quintic Del Pezzo surfaces – a short proof. Bull. London. Math. Soc. **24** (1992) 249–250

[Shepherd-Barron97] N. I. Shepherd-Barron: Fano threefolds in char. p. Compositio Math. **105** (1997) 237–265

[Shioda74] T. Shioda: An example of uniruled surfaces in characteristic p. Math. Ann. **211** (1974) 233–236

[Shioda-Katsura79] T. Shioda, T. Katsura: On Fermat varieties. Tohoku Math. J. **31** (1979) 97–115

[Siu-Yau80] Y.-T. Siu, S. T. Yau: Compact Kähler manifolds of positive bisectional curvature. Inv. Math. **59** (1980) 189–204

[Snapper60] E. Snapper: Polynomials associated with divisors. J. Math. and Mech. **9** (1960) 123–129

[Spottiswoode1879] W. Spottiswoode: On the twentyone coordinates of a conic in space. Proc. London Math. Soc. **10** (1879) 138 ff.

[Swinnerton-Dyer72] H. P. F. Swinnerton-Dyer: Rational points on del Pezzo surfaces of degree 5. Algebraic geometry, Oslo 1970. Wolters–Noordhoff, 1972, pp. 287–290

[Szpiro81] L. Szpiro: Propriétés numerique du faisceau dualisant relatif. Astérisque **86** (1981) 44–78

[Szurek-Wiśniewski90] M. Szurek, J. Wiśniewski: Fano bundles on P^3 and Q_3. Pacific J. Math. **141** (1990) 197–208

[Thaddeus94] M. Thaddeus: Stable pairs, linear systems and the Verlinde formula. Invent. Math. **117** (1994) 317–353

[Timmerscheidt82] K. Timmerscheidt: On deformations of threedimensional rational manifolds. Math. Ann. **258** (1982) 267–275

[Traverso70] C. Traverso: Seminormality and Picard Group. Ann. Sci. E.N.S. Pisa **24** (1970) 585–595

[Tsai93] I-H. Tsai: Rigidity of holomorphic maps from compact Hermitian symmetric spaces to smooth projective varieties. J. Alg. Geom. **2** (1993) 603–634

[Tsen36] C. Tsen: Quasi-algebraisch-abgeschlossene Funktionenkörper. J. Chinese Math. **1** (1936) 81–92

[Tsuji96] H. Tsuji: Global generation of adjoint bundles. Nagoya Math. J. **142** (1996) 5–16

[Voisin96] C. Voisin: On a conjecture of Clemens on rational curves on hypersurfaces. J. Differential Geom. **44** (1996) 200–213

[Weil62] A. Weil: Foundations of algebraic geometry. Amer. Math. Soc., 1962

[Wentzel93] C. Wentzel: Rationality of G/P for a nonreduced parbolic group scheme P. Proc. AMS **117** (1993) 899–904

[Wiśniewski90] J. Wiśniewski: On Fano manifolds of large index. Manuscr. Math. **70** (1991) 145–152

[Wiśniewski91a] J. Wiśniewski: On a conjecture of Mukai. Manuscr. Math. **68** (1990) 135–142

[Wiśniewski91b] J. Wiśniewski: On contractions of extremal rays on Fano manifolds. J. f.r.u.a. Math. **417** (1991) 141–157

[Zariski39] O. Zariski: The reduction of singularities of an algebraic surface. Ann. Math. **40** (1939) 639–689

[Zariski58a] O. Zariski: On Castelnuovo's criterion of rationality $p_a = P_2 = 0$ of an algebraic surface, Ill. J. Math. **2** (1958) 303–315

[Zariski58b] O. Zariski: Introduction to the problem of minimal models in the theory of algebraic surfaces. Math. Soc. Japan, 1958

Index

A blank space or a $*$ stands for a variable. T.i refers to the Terminology section at the end of the Introduction.

Printing and Binding: WB-Druck, Rieden/Allgäu

www.ingramcontent.com/pod-product-compliance
Lightning Source LLC
Chambersburg PA
CBHW051921171225
36953CB00003B/177
* 9 7 8 3 5 4 0 6 0 1 6 8 5 *